Taking The Waters

Soil and Water Conservation among Settling Beja Nomads in Eastern Sudan

Johan A. van Dijk

Taking The Waters

Soil and Water Conservation among Settling Beja Nomads in Eastern Sudan

ACADEMISCH PROEFSCHRIFT

ter verkrijging van de graad van doctor
aan de Universiteit van Amsterdam,
op gezag van de Rector Magnificus
prof. dr P.W.M. de Meijer
ten overstaan van een door het college van dekanen ingestelde
commissie in het openbaar te verdedigen in de Aula der Universiteit
op woensdag 3 mei 1995 te 13.30 uur
door

Johan Aad van Dijk

geboren te Schiedam

Promotores:

prof. dr G.A. de Bruijne
dr A.J. Dietz

ISBN 1 85972 162 1

Contents

Abbreviations

Ar	Arabic
BAPP	Border Area Pilot Project
CARE	Co-operative for American assistance and Relief Everywhere (NGO)
c: ca	catchment to cultivated area ratio
CDCS	Centre for Development and Cooperation Services (Amsterdam Free University)
COR	Commissioner of Refugees
DAC	Development Assistance Committee (of the OECD)
DGIS	Directorate-General for International Cooperation (of The Netherlands government)
DC	District Commissioner
DSA	Durham Sudan Archive (University of Durham)
EEC	European Economic Community
EPLF	Eritrean People's Liberation Front
fd	*feddan* (0.42 ha)
FAO	Food and Agricultural Organization (of the UN)
FDES	Fuelwood Development for Energy in Sudan
FNC	Forestry National Corporation
GDAC	Gash Delta Agricultural Corporation
GIS	Geographic Information System
GON	Government of The Netherlands
GOS	Government of The Sudan
GRASS	Geographic Resources Analysis Support System
GTZ	*Gesellschaft für Technische Zusammenarbeit*
HADA	Hameskoreib Area Development Agency (NGO)
HEIA	High External Input Agriculture
IARA	Islamic African Relief Agency (NGO)
IFAD	International Fund for Agricultural Development
ILO	International Labour Organization
IMF	International Monetary Fund
iSWC	indigenous soil and water conservation
KADA	Kassala Area Development Activities (programme)
KARS	Kassala Agricultural Research Station
LMMC	Livestock and Meat Marketing Corporation
LEIA	Low External Input Agriculture
L&E	Livelihood and Environment
MANR	Ministry of Agriculture and Natural Resources
MFEP	Ministry of Finance and Economic Planning

MSF	*Médecins Sans Frontières* (NGO)
msl	mean sea level (at Alexandria)
MT	metric tons (1,000 kg)
NCDRWR	National Corporation for the Development of Rural Water Resources
NCR	National Council for Research
NDVI	Normalized Difference Vegetation Index
NGO	Non-Governmental Organization
noSWC	no soil and water conservation
nopSWC	no project introduced soil and water conservation
NRO	National Records Office (Khartoum)
PHC	Primary Health Care
pSWC	project soil and water conservation
rc	run-off coefficient
RPM	(Department of) Range and Pasture Management
RRC	Relief and Rehabilitation Commission
SALPA	Semi-Arid Lands Project Amsterdam
SPSS	Statistical Package for Social Sciences
SRC	Sudanese Red Crescent (NGO)
SwRC	Swiss Red Cross (NGO)
SSA	Sub-Saharan Africa
SWC	Soil and water conservation
SCLUWP	(Department of) Soil Conservation, Land Use and Water Programming
Tb	*To Bedawee* (language of Hadendowa)
TLU	Tropical Livestock Unit
TNO	Institute of Applied Geoscience
TOT	Transfer-of-Technology
UN	United Nations
UNDP	UN Development Programme
UNEP	UN Environmental Programme
UNHCR	UN High Commission for Refugees
UNICEF	UN International Children's Education Fund
UNESCO	UN Educational, Scientific and Cultural Organization
UvA	University of Amsterdam
WADS	Water resources Assessment and Development project in the Sudan
WARK	Waterspreading Research Kassala
WFP	World Food Programme
WID	Women in Development

£	Sterling pound
£s	Sudanese pound
$	US dollar
Dfl.	Dutch guilder

Official foreign exchange rate (£s/US$) for selected years

1983	1.27	1986	2.50	1989	4.50
1984	1.35	1987	2.50	1990	4.50
1985	1.52	1988	3.75	1991	12.20

Local measures

Length

1 *feddan* length = 60-65 m
1 *kolei* = length covered by normal throw with a stick; variable
1 *jedah* (Tb) = same as *kolei*
1 *durah* = 1 elbow length, or approx. 0.5 m
1 *goleen* (Tb) = same as *durah*
1 *gassaba* = 7 *durah*, or approx. 3.5 m

Area

1 *feddan* area = 4,200 m^2

Volume

1 *ruba* = 8.25 litre; 1 *ruba* grain equals 1/16 sack, or approx. 5.6 kg
1 *tumna* = 1/8 *ruba*
1 *ardeb* = 2 sacks (see below), or 198 litres
1 (liquid) *rottle* = 0.45 litre

Weight

1 *rottle* = 0.449 kilogram
1 sack sorghum = 88.7-94.4 kilogram (variety dependent)
1 sack millet = 91.4 kilogram

Source:
Livelihood & Environment field research, Tothill (1952,941-955), Ausenda (1987,569), MANR (1991,i).

Colloquial Sudanese Arabic and To Bedawee

Ads Sudani	pigeon pea, *Cajanus cajan*
Bamia	okra, *Hibiscus esculentus*
Berseem	lucerne, *Medicago sativa*
Berish	mat, or tent made of dried leaves of the dom palm
Betig	watermelon, *Citrullus vulgaris*
Boka	small rills used to manipulate run-off flows
Buda	witchweed, root parasite, *Striga hermontica*
Doum	branch palm, *Hyphaene thebaica*
Dukhn	millet, *Pennisetum typhoideum*
Dura	sorghum, *Sorghum bicolor and S. vulgare*
Eilab	annual grass, *Aristida adscensionis*
Esh	bread, also used to refer to sorghum
Girgir	rocket cress, garden rocket, *Eruca sativa*
Guar	cluster bean, *Cyamopsis tetragonobola*
Haboob	sand-laden dust storm
Hafir	excavated earth tank for domestic water supply
Jamam	shallow well dug in riverbed
Jebel	mountain
Karkadeh	rosella, *Hibiscus sabdariffa*
Khalwa	Koran school
Khodra	jew's mallow, *Corchorus olitorius*
Khor	water course, stream-bed (Tb origin)
Komfood	mixture of sorghum, wheat bran and groundnut shells (livestock forage)
Libish	brushwood
Lubia	fodder bean, *Dolichos lablab*
Mazareef	standard field allowance
Muttar	rain, also used to refer to rainfed cultivation
Nafir	voluntary work group
Nazir	leader of tribe
Ombaz	sesame or cotton seed cake (livestock forage)
Omda	leader of tribal section
Regla	purslane, *Boerhavia repens*
Ruda	mixture of sorghum and wheat bran (livestock forage)
Saaf	dried leaves of the branch palm
Seed	wildflooding
Sedtorabi	earth dam
Senna mekka	pods of *Cassia acutifolia*, *C. oblovata*
Shayote	irrigation canal

Sheikh	leader of tribal section
Shok	brushwood
Sim sim	sesame, *Sesamum indicum*, *S. orientale*
Teras, pl. terus	u-shaped earth bund for water harvesting
Ushar	Dead Sea apple, *Calatropis procera*
Ushur	tithe traditional taxation
Wadi	water course, stream-bed
Zaka	alms giving

Note on orthography

The text does not follow any "official" or "authorative" transcription rule Arabic-English, other than a notation which is applied consistently. When a colloquial name is given, this is Arabic. When a colloquial name is followed by the suffix *Tb*, this word derives from the To Bedawee language. Certain Arab words have been admitted to the English language and are usually given English plurals, such as khors instead of the Arabic *kheiran*. These anglicized plurals are also used in this text.

Preface and acknowledgements

This book is an account of field research into soil and water conservation techniques in eastern Sudan. The research was part of the "Livelihood and Environment" programme of the faculty of Environmental Sciences of the University of Amsterdam. It was carried out between 1991-1993 in collaboration with the Institute of Environmental Studies of the University of Khartoum, The Sudan. This research was funded by the Netherlands Foundation for the Advancement of Tropical Research WOTRO of the Netherlands Organization of Scientific Research NWO. However, this book at the same time draws from two earlier professional engagements in Sudan. The first was between 1987-1989 in the Sudanese-Netherlands rural development programme "Kassala Area Development Activities" MFEP/KADA. This was co-financed by the Netherlands Directorate-General for International Cooperation DGIS. Technical assistance was provided by DHV Consulting Engineers. The second was between 1989-1991 in the "Waterspreading Research Kassala" project WARK of the Sudanese National Council for Research NCR. This project was part of the "Field Studies in Sudan's Rainlands Programme" financed by the Ford Foundation. I would like to thank Dr. Sien Thio of DHV and Dr. Richard Longhurst of the Ford Foundation, firstly, for their support and, secondly, for their permission to use the data also for the purpose of this book. In 1991, research in the framework of the Livelihood and Environment programme began. I express my gratitude to my promotores Prof. Dr. Ad de Bruijne and Dr. Ton Dietz of the Department of Human Geography who, first of all, made it possible to continue working in The Netherlands. They furthermore facilitated the organization of additional research, both in the field and at the department. Finally, their stimulating discussions have greatly contributed to the production of this book. I thank Dr. Yagoub Abdalla Mohamed of the Institute of Environmental Studies for his constructive suggestions made in Khartoum.

Field research for the Livelihood and Environment programme could never have been carried out without the help of many people. Foremost among these are the people of Um Safaree, Hafarat, Ilat Ayot and Telkook. I remember with great joy living with them during parts of the year. My sincere thanks for their hospitality go to all sheikhs and village leaders. Particularly, I would like to mention Omer Osman Mohamed and Tahir Aderab Mohamed in Ilat Ayot, Mustapha Mohamed Faki and Hameed Monei in Telkook, Idrees Hameed Ali and Ali Karar Tahir in Um Safaree, and finally Ibrahim Mohamed AbdelRahman and Saeed Ali Mohamed in Hafarat. The Ministry of Agriculture and Natural Resources in Kassala provided all necessary support to carry out research in the

Border Area. I owe much to the Director General AbdelHaleem El Hassan and the Director of the Department of Soil Conservation Land Use and Water Programming Hashim Mohamed El Hassan. AbdelRahman Khalid AbdelRahman and Mohamed Hassan Ahmed Habib of the same department, and Ali Eisa Hussein of the Department of Range and Pasture Management were always with me in the field. At one occasion, we literally had to dive to hide from Ethiopian MIG's patrolling so low over Sudanese territory that we could spot the pilot's drooping moustache. Anyhow, I cherish the time we worked together as something precious. This was not only in the Livelihood and Environment research programme, but also earlier in MFEP/KADA and WARK. Dr. Mustapha Baasher shared his great knowledge of the Beja people and area with me in long evening sessions in Kassala. Suliman Ali Betai let me into the secrets of the Hadendowa way-of-life and accomplishments of the religious movement established by his father. Anita van Dam, John Houtkamp and David Niemeijer, students of the University of Amsterdam and Wageningen Agricultural University, travelled to Sudan in October 1990. They supported me in field research under difficult conditions. With rising political tensions at the eve of the Gulf War, the return journey came earlier than foreseen. The repatriation of all of us in January 1991 caused much brain-racking. However, you all managed to complete your thesis. My complements for that. David even liked it so much he decided to continue with soil and water conservation research in Wageningen. I myself had the opportunity to take up research again for a last three months' round in October 1991. Esther Moonen and Herald Voorneveld provided invaluable assistance in the field on this occasion.

Several colleagues at the University of Amsterdam are acknowledged for their suggestions and critical remarks during the phase of writing the manuscript. In particular, I thank Johan Berkhout of the Department of Physical Geography and Soil Science. With never- ending enthousiasm, he introduced me to new literature and research fields and taught me GIS. Sjoerd de Vos of the Department of Human Geography helped me to judge several methodological issues. Andries Klaasse Bos of ISMOG made valuable comments on several of the economic aspects addressed in this book. Paolo de Mas of IMES shared with me his knowledge of the Islamic world. Martijn van Leusen and Willy Simons provided assistance in working with GIS. However, if errors in fact or interpretation in any of these fields have crept into this book, these are entirely on my account. I thank Dr. Jan Mansvelt Beck for having roused in an early stage my special liking for water affairs. The Livelihood and Environment home team *"Jong Oranje"* occasionally had to pull me through. Thank you, Anja, Annemieke, Annette, Antje, Coen, inmate Fred, Gonda, Hans, Henk, Isa, Jeanine, Johan Post, Josanne, Leo, Maarten, Marcel, Marijk, Micky, Nina, Paul, Pyt, and Wim (we are all young, aren't we?).

I am grateful to Mr. C.D. Smit of the DGIS section DAF/NF for allowing me to consult several background reports on the Sudan projects. Mrs. Jane Hogan of the University of Durham helped me to search in the Sudan Archive. Dr. Mike Hulme of the University of East Anglia is acknowledged for granting his permission to reproduce figure 3.2. Chris Reij at CDCS of the Amsterdam Free University pointed out literature of interest, and allowed me the use of their library with an enormous wealth of reports on soil and water conservation. Jaap van der Velden of HVA International is acknowledged for allowing me the use of data collected during my consultancy assignments in 1993 and 1994 in the framework of their Gash Delta Project. Dr. M.A. Mohamed Salih of the Institute of Social Studies wrote the abstract in Arabic. My wife and colleague Annick Alexander wrote the abstract in French. Nancy Bishop corrected the English of the manuscript. Christian Smid and Hans de Visser of BRON drew the maps and figures. Frank Perk (Stichting RAAP) prepared the lay-out of the book.

The early Sudan years have provided me with an invaluable basis for this research. For this reason, I return for a moment to these years. I thank Johan Post of the Department of Planning and Demography of the University of Amsterdam for bringing me to the country in the first place. This was in 1987 as a research assistant in the Economic Activity and Town Planning Project. In the years that would follow, many other persons provided assistance in the context of the WARK and MFEP/KADA programmes. I would like to thank the staff of NCR supervising WARK in Khartoum, Prof. Dr. Saeed Abbadi, Dr. Ekhlas Abdel Bari and Dr. Mahdi Amin El Tom for their supervision and guidance. I also thank all people related to MFEP/KADA: Yousif Yagoub, Abdellah Mohamed El Mosbah, Abdellah El Haj, Dr. Mohamed El Naseeh Osman, El Fatih Abu Reid, Jabir Hussein, Salah El Tayed Mohamed, and Abdelmagheed Abdel Salaam for creating a stimulating working environment. I thank Babiker El Amin of the Soil Conservation Administration and Dr. Ali Mohamed Osman of the Department of Archaeology for their regular support and advice in Khartoum. I am grateful to Ingrid and Guus Cosijn who set me on the track of soil and water conservation in Kassala. Guus taught me the tricks of the drylands. Still, he had to send food on camel-back when we got stuck for days, because of too much water in the desert. Woutje and Hans de Vries greatly stimulated me to continue research. Without their fierce support, this book probably would never have been begun. Dick Nauta, at the time MFEP/KADA co-Director in Kassala, had as part of his additional job as "civil defence warden" the difficult task of getting the students and myself out of the country in January 1991. He was left with our shambles in Kassala by way of thanks. This is to make up for it. I thank Margot and Joost Geijer for providing accommodation in Kassala in 1991-1992.

The greatest gratitude of all goes to my wife Annick. Occasionally, the writing of this book left little time to spend together. However, this also provided you with an excuse for a most welcome travel south. So-called, because the grass needed mowing. You were quite right. I dedicate this book in memory of my parents.

Introduction

The relevance of directing attention to crop production in the marginal drylands of the world is strikingly expressed by Barrow (1993,197-198) who writes:

> "[..] At the start of the next century roughly four-fifths of the world's arable farmland will be non-irrigated. By that time much of the world's easily developed irrigated land will have been opened up. The problem will be how to intensify, diversify, extend and sustain the production of numerous, remote, small cultivators who will probably have little or no money, no access to groundwater or supplies from canals and streams. The answer must largely be the development and promotion of improved rainfed cultivation, run-off collection and concentration and the use of ephemeral flows and floods, together with the cultivation of new moisture-efficient crops and improved varieties of existing rainfed crops. Innovations must help farmers to make optimum use of available moisture, and the improvement of security and sustainability of harvest must be given equal, if not higher, priority than obtaining increased and higher-value crops [..] the routes to improved rainfed cultivation, the use of streamflow, run-off, or floodwaters will have to be cheap and accessible, yet today's research and investment is often directed at developing high-technology, costly cultivation systems, frequently unsuited to the environmental conditions in which most developing country farmers operate. Much agricultural development, especially of irrigation, has sought to modify the environment to suit particular crops [..] A more sensible approach would be to select crops and cultivation practices to suit particular developing country environments and human needs. [..] Two encouraging developments are that researchers have begun to study ancient farming systems in the hope that this may help provide strategies suitable for poor, remote smallholders, and that a number of research institutions are seeking strategies and crops for small farmers in marginal lands [..]".

The population of 50 million in one particular marginal dryland region namely the Horn of Africa, is among the poorest in the world. The national economies of Sudan, Ethiopia, Djibouti and Somalia are doing badly, even by African standards. These governments spend significantly more on arms than on public health and education together. On top of that, two great droughts have occurred in this region in the last decade. War and drought have caused movements of millions of refugees. After the first drought in 1984-1985, Lloyd Timberlake extensively reported in his book *Africa in Crisis* about a continent on the brink. Seven years later after the second drought of 1990-1991, the TIME International magazine of September 1992 featured *Africa: The Scramble For Existence*. The continent is portrayed as an immense illustration of chaos-theory. It makes one wonder to precisely what extremes the people who are living here can go in their capacity to cope with normal stress, and to survive in times of crises. However, there is also hope forming on the margins. Africa's genius for extremes, now in a positive sense, was illustrated in Paul Harrison's book of 1987 *The Greening Of Africa*. There are ways to "[..] break through in the battle for land and food [..]". This is only possible when the specific problems of the continent are addressed by what are sometimes called "African solutions" (*cf.* UN-NGLS 1990). Such specific African solutions with respect to soil and water conservation (SWC) for crop production are the focus of this study. Two research questions address respectively the achievements of government SWC intervention measures, and the local importance of indigenous SWC in the livelihoods of recently settled nomadic peoples. The area we discuss in this study is located in eastern Sudan in a region called the Border Area.

Stage for SWC research

The Border Area which is described at greater length in chapter 3 is characterized in the Environmental Profile Study of the Netherlands Government DGIS (DHV/IES 1989) as a low potential area. Its natural resources are being overused in pastoralism and forestry, but are still being underused in crop production. The local population of the Beja group is under considerable stress because these marginal environmental conditions have a direct and negative effect on their economic positions. It is typically in these situations that the niches and combinations in subsistence means become important to sustain local household livelihoods (Chambers 1990). The Livelihood and Environment (L&E) programme of the Department of Human Geography of the University of Amsterdam carried out research in the Border Area in 1991-1992. In this research, particular attention is paid to indigenous SWC as one among several livelihood activities. Geographical research in Amsterdam traditionally places livelihoods in local environments and regional contexts (De Bruijne 1989), and due attention is given to the influence of the physical environment (Van Schaik & Reitsma 1992). The study in Sudan followed earlier research, monitoring and evaluation work of SWC techniques in the area. Firstly in the framework of the Sudanese-Netherlands MFEP/KADA programme (1987-1989), this mainly focussed on SWC techniques introduced in projects. Secondly in the Sudanese programme Waterspreading Research Kassala (WARK) of the National Council for Research (1989-1991), attention was also given to indigenous SWC. Already in those years, it had become apparent that on the one hand existing government interventions fell short in their performances. The local indigenous techniques, on the other hand, had never been seriously studied. The main rationale of all subsequent research was therefore to collect more detailed information about local indigenous SWC use. This was expected to contribute to the improvement of future interventions in this field.

Livelihood in this context means "[..] a combination of people, resources and environment in which the stocks and flows of food and cash are used to meet the basic needs of people [..]" (Reijntjes *et al.* 1992,214). Indigenous means "[..] originating from and naturally produced in an area [..]" (Chambers 1991,83). Technique means "[..] a combination of knowledge, inputs and management practices which are deployed together with productive resources to produce a desired output [..]" (Reijntjes *et al.* 1992,217). SWC consists of a range of techniques. Water harvesting is the most important in this study. It is defined as "[..] the collection and concentration of run-off before it reaches seasonal or perennial streams [..]" (Critchley *et al.* 1992,6-7). At the onset of the research in Sudan in 1991, two major fields of constraining factors were expected to be important with respect to crop production and SWC interventions. The first derives from the physical setting of the Border Area. The expected constraints are water availability and soil fertility. These are limitations largely inherent to semi-arid and arid environments. The second derives from the socio-economic setting of the Border Area. The expected constraints are labour availability and instutional-organizational capacity. These are limitations associated with, respectively, the need of households to combine livelihood activities, and the decline of public management capacities in the region.

Organization of this book

This book contains descriptive and analytical sections. The first five chapters are descriptive. In chapter 1, we discuss the main lines in the literature about irrigation and SWC. We will illustrate the changing outlooks on dryland development over the years. These can be typified as initially preoccupied with a mainly Transfer-of-Technology approach. Currently, these outlooks involve more participatory methods of working. The role of the constraining factors water, soil nutrients, labour, and institutions is also discussed in more detail here. The framework for study is given in chapter 2. The two research questions are presented in detail. These focus on the effects of government SWC interventions and the importance of indigenous SWC in household livelihoods in four selected villages in the Border Area. The techniques used in data collection and processing are also discussed in this chapter. The regional setting of the Border Area in physical and socio-economic terms is discussed in chapter 3. The SWC techniques applied in this area are presented in detail here. The historical evidence on early crop production is given in chapter 4. We cover the period which starts at the end of the 19th century, and report on emerging indigenous and introduced SWC techniques. Also the changes in the local Beja livelihoods are considered in more detail here. Chapter 5 is an introduction to the natural resources and economic sectors in the Border Area of the present-day. Chapter 6 is the first chapter of the analytical section. The research data are discussed at the level of households, household members, and landholdings. A first concise presentation of the findings on the role of indigenous SWC is given here. The answering of our research questions is the central theme in chapter 7. We discuss the effects of government SWC interventions in the Border Area in terms of common project evaluation measures of effectiveness, efficiency, impact, and project-sustainability. The importance of indigenous SWC in household livelihoods is assessed in terms of income, coverage of cultivated lands, the allocation of labour time, and the perception of land users. The conclusions and policy considerations are presented in chapter 8. We also turn in this chapter to the role of the formulated constraints, and the potential contribution of indigenous SWC to dryland development in other world regions.

1

Dryland rural development

Most countries in Sub-Saharan Africa (SSA) gained their independence in the 1960s. The main rural development outlooks discussed in this chapter will be considered for the period starting from this date. This is not to deny the critical role which Colonial Rule has played in the region, but is led by considerations to emphasize the role of new actors in the development of independent states. Two important events demarcate this new era. Firstly, the International Bank for Reconstruction and Development (World Bank) started to include developing countries in its lending policy in 1956. Secondly, the United Nations General Assemblee announced the start of the "First Development Decade" in 1961. These events have announced the beginning of the international development aid era. The group of new actors has come to include a motley array of large and small agencies and non-governmental organizations (NGOs). Over this same period, many SSA administrations have weakened, especially since the 1980s, while the volume of the international resources received continued to grow. This has also been the case in the Sudan where the Official Development Assistance (ODA) received amounted to $ 639 million in the early 1980s, and to over $ 1 billion in the mid 1980s. International assistance declined over the 1990s to $ 850 million in 1990/1991 (OECD 1992) mainly as a result of international disapproval of the country's domestic policy. External factors, by consequence, have increasingly come to determine the course of internal economic and political developments in SSA (Gulhati 1990). This means that what is sometimes represented as an African policy is frequently to an important extent shaped and steered by the ideas of the international donor community. This also applies to policy formulations in the field of land and water resources development in general, and government SWC interventions in particular.

1.1 Main viewpoints

Rural development is defined by Sterkenburg (1987,20) as: "[..] a process of change in rural areas leading to better living conditions and a greater security of existence for the population, which process comprises: (i) the growth of production and productivity and the diversification of the production activities within the agricultural sector; (ii) the increase in complexity and the linkages in the rural economy as a result of the expansion of non-agricultural production activities, rural industries in particular; (iii) the improvement of the employment situation and the rise of incomes for broad segments of the rural population; (iv) the expansion and amelioration of agro-support and community services; (v) the improvement of environmental conservation as a form of preserving natural resources, which is essential to sustain the process over a longer period of time [..]". Rural development, furthermore, can be viewed by governments in the developing countries from different convictions. Sterkenburg mentions five[1], of which "(agricultural) production increase" and "an integrated approach to reach higher living standards" are the most relevant to the discussion in this study (*cf. Ibid.* 1987,13-16). The production increase view will be referred to here as "National Growth Strategy". We add to these another recent viewpoint which has emerged in the late 1980s called "Sustainable Development". Given these three main rural development perspectives, their significance to this study of introduced and indigenous SWC can be appreciated best by discussing their (i) prevailing approaches; (ii) working areas, and their outlooks on (iii) water resources development; (iv) application of indigenous knowledge; and (v) the role of external inputs (table 2.1). The general ideas that exist on the latter fields will be turned to first before we discuss their more specific contents in each of the main rural development viewpoints discerned.

Approaches, working areas, water resources development, indigenous knowledge, external inputs

Chambers (1991) divided the rural development approaches into Transfer-of-Technology (TOT) and Participatory Approaches. The latter had earlier also been called "Farmer First" approaches. The main objective of TOT is the introduction of new technologies. Its main characteristic is that the analysis of local needs and priorities of the people is made by outsiders. The research location is usually the experimental station, laboratory, or greenhouse. The techniques developed here are transferred as precepts, messages, and packages of practices of a "fixed menu". However, the main objective in Participatory Approaches is the empowerment of land users. Local needs and priorities are assessed by the land users themselves. The outsiders ideally only provide assistance to achieve this objective. The research and development location is the field. The techniques that have been developed here are dissiminated as guiding principles, methods, and as a "basket of

---

[1] The rural development viewpoints presented in Sterkenburg (1987, 13-17) interpret the process respectively as (i) more or less synonymous to (agricultural) production increase, (ii) welfare improvement, (iii) basic-needs satisfaction, (iv) structural transformation, and (v) an integrated approach to reach higher living standards.

choices". Land users would, in this approach, work *à la carte* (*Ibid.* 1991,182). The idea of Participatory Approaches is young and constantly changing. Even a new "Beyond Farmer First" approach has recently been proposed which introduces further refinements and places the concept in a wider context (*cf.* Scoones & Thompson 1993).

The working areas can be divided into high potential and low potential regions, for example along lines discussed in DHV & IES (1989). Such qualitative judgements are usually based on combined characteristics of the physical setting including the status of its natural resources and climate, and the socio-economic setting including factors like the status of its infrastructure and institutions (*cf. Ibid.* 1989).

The development of water resources for crop production can be in the form of large-scale and small-scale irrigation. SWC techniques which contain a dominant water harvesting component also serve this purpose. The references to the scale of implementation in the first two cases of irrigation development have never been consistent. However, small usually refers to irrigation schemes of less than 50-100 ha in size (FAO 1986b).

The importance given to the incorporation of indigenous knowledge in rural development can range from low to high. The land users are considered at either end of this continuum respectively as adopters or rejectors of the innovation introduced, and as the originators of their own innovations (Scoones & Thompson 1993).

Table 1.1 **Dryland rural development views, approaches, working areas, water resources development techniques, role of indigenous knowledge and external inputs**

	NATIONAL GROWTH STRATEGIES	INTEGRATED RURAL DEVELOPMENT	TWO EMERGING VERSIONS OF SUSTAINABLE DEVELOPMENT	
			I	II
APPROACH	TOT	TOT, Participatory	TOT	Participatory
WORKING AREA	hp	hp, lp	hp	lp
WATER RESOURCES DEVELOPMENT	large-scale irrigation	small-scale irrigation and pSWC	irrigation rehabilitation	small-scale irrigation, iSWC
INDIGENOUS KNOWLEDGE	absent-low	low-medium	low	high
EXTERNAL INPUTS	HEIA	HEIA, LEIA	HEIA	LEIA, (HEIA)
PERIOD	1960s-1970s	1970s-late 1980s	late 1980s	late 1980s-present

Note: TOT means Transfer-of-Technology, hp/lp means high/low potential area, HEIA/LEIA means high/low-external-input-agriculture, iSWC/pSWC means indigenous/project-introduced soil and water conservation. Source: various sources mentioned in the text.

The level of external inputs in rural development may range from low to high. The two ends of this continuum scale are usually called low-external-input-agriculture (LEIA) and high-external-input-agriculture (HEIA) when the discussion is about agricultural development in general, and crop production in particular.[2] Kieft (1993) typified the HEIA conditions as generally involving homogeneous locations which have a good infrastructure, use large-scale farming systems that are market-oriented and capital intensive, which systems generally have a low degree of diversity. Strategies of yield and income maximalization prevail under HEIA. The level of output is medium to high. In order to achieve these objectives, the level of external inputs defined as inputs originating from outside the farm such as fertilizers, pesticides, machines and pumped irrigation water, must be high. The LEIA conditions, on the contrary, usually involve heterogeneous locations which have a poor infrastructure. The farming systems are usually small-scale, labour-intensive, food production-oriented, and characterized by a high degree of diversity. The dominant strategy under LEIA conditions is risk minimalization. The level of output is medium to low. The level of the external inputs required is also low. Mineral fertilizer is usually a common external input and can be used to classify the different systems. An application of 100 kg per ha is commonly considered to distinguish LEIA and HEIA systems in this respect (*Ibid.* 1993,128).

Rural development viewpoints

The National Growth Strategies which have placed a great emphasis on the TOT approaches prevailed in the 1960s-1970s. The rural development interventions at this time were mainly implemented in the high-potential regions. Large-scale irrigation was the most common means of water resources development. Indigenous knowledge was usually not included in its design. The level of external inputs in this period was high. These National Growth Strategies formulated for SSA were largely based on the ideas of development which had emerged in Europe after World War II. These regarded development as a linear process of transition through a series of fixed growth phases. After completion of the first stage of national growth, regional development was supposed to follow suit through a "trickling-down" of benefits over the entire society in the long run. Typical examples of National Growth Strategies in central Sudan in the field of rural development are the large-scale gravity irrigation schemes in the Nile, Rahad and Atbara rivers (*cf.* section 3.2 *economy and development*).

Integrated Rural Development gained importance in the early 1970s largely in response to the disappointing results of the National Growth Strategies (Hunter 1982). This viewpoint was supposed to advocate Participatory Approaches. However, its factual working methods must now be considered to have been largely ruled by TOT approaches (Chambers 1991). The working areas were in both low and high potential regions. Small-scale irrigation was favoured with respect to water resources development, but also the first

---

[2] The LEIA concept is sometimes also called LEISA where "S" stands for Sustainable. It is remarkable that the term HEISA is never used, wrongly implying that these systems can not be sustainable (*cf. Rural development viewpoints* in this section).

SWC techniques were introduced under international programmes at this time. The importance of indigenous knowledge was usually recognized, and LEIA and HEIA systems were propagated together. International development aid significantly influenced the contents of the programmes. The main objective of Integrated Rural Development had become to interlock different activities, sectors, areas, and institutions in the rural zones of the developing countries (Sterkenburg 1987). The working field was usually at regional scale. The vocabulary which caught on was one of "target groups" and "target regions". Other (claimed) characteristics of Integrated Rural Development include flexibility, periodic adjustment, and open-endedness of its programmes. Several critics have argued that the approach was frequently inappropriate, and sometimes even negatively affected the rural sectors involved (Horowitz & Salem-Murdock 1987). A typical example of Integrated Rural Development in eastern Sudan is the MFEP/KADA programme (*cf.* section 4.2.3).

Sustainable Development became an instantly accepted concept after the publication of the report of the World Commission on Environment and Development. It is defined as "[..] a process of change in which the exploitation of resources, the direction of investments, the orientation of technological development and institutional change are all in harmony and enhance both current and future potential to meet human needs and aspirations [..]", WCED (1987,46). The ideas behind Sustainable Development date back to formulations of the World Conservation Strategy in 1980 (IUCN 1980). They allegedly even date back to some African-based conferences held in the 1960s (Brookfield 1991). The new element is that development and environment are considered as strongly interlinked. Despite this clear position adopted, still no generally accepted definition of Sustainable Development exists (Holmberg & Sandbrook 1992). The above definition given by WCED is global in outlook, but remains little more than a statement of good intent. We discuss three attempts that have been made to operationalize the concept. The first is of the Netherlands DGIS. This definition is general and covers different scale levels. The second is used by economists like Pearcey *et al.* (1990). This definition equally covers different scale levels, but mainly the higher levels. Finally, the third is used by agronomists and soil scientists. The definition is ecological in outlook and covers different scale levels, but mainly the lower levels.

The DGIS formulations largely adhere to the WCED definition. A framework is formulated in which especially the linkages are stressed between the different fields of environment and development, different levels of scale, and the developments over time (IOV 1992,9). Arntzen *et al.* (1993) provide a set of directives to evaluate whether or not development should be considered as sustainable. These include: (i) the use of renewable resources may not exceed their regeneration and may not affect the regenerative-capacity of the resources; (ii) pollution and waste production may not exceed the degradability-capacity of the environment; (iii) the potential for substitution of non-renewable resources by renewable ones must be enhanced; and (iv) bio-diversity must be maintained. These are still general formulations. Their main purpose is policy direction and formulation.

Pearce *et al.* (1990,3-4) propose to use a general definition of the concept and a set of more precise minimum conditions to be fulfilled. Sustainable Development is defined as "[..] the general requirement that a vector of development characteristics may not be decreasing over time [..] the elements to be included in this vector are open to ethical debate [..] and the relevant time horizon for practical decision-making is similarly indeterminate outside of agreement on inter-generational objectives [..]". The necessary condition is the premiss of "constancy of the natural capital stock". This stock is operationalized in terms of either specific physical stock, the economic value of this stock, its price, or the total economic value of all stocks which allows mutual substitutions to maintain the constancy of stock.

Van Reuler & Prins (1993) present a collection of papers where Sustainable Development is mainly formulated in terms of the soil nutrient balance (*cf.* section 1.2). The same general WCED principle of inter-generational equity also underlies this definition. However, now only soil fertility, being usually the most limiting resource to crop production, is considered. Sustainable (crop production) Development accordingly is defined as the situation where the rate of nutrient inflow into a given system at the level of a field, farm, or country does not exceed its rate of outflow. This formulation in fact is a specialization of the premiss of the constancy of natural capital stock. The efforts to maintain the nutrient balance can be priced so that sustainability defined in this way can also be expressed in economic terms (*cf.* Van der Pol 1992).

Confusion sometimes arises where sustainable development focussed on rural development is presented as almost exclusively based on LEIA systems (*cf.* Reijntjes *et al.* 1992, footnote 2). This raises the suggestion that HEIA can not make any contribution in this respect at all. The common counter-argument of the HEIA supporters, accordingly, is that LEIA approaches alone can not meet the challenges that are now being imposed by the growing populations in the developing countries (*cf.* Van Reuler & Prins 1993, *passim*). This misunderstanding can be solved when LEIA and HEIA are considered at their appropriate levels of implementation scale and context. The emerging outlooks on Sustainable Development currently distinguish between high and low potential regions. These have their own characteristics and demand their own approaches. This may include a combined LEIA and HEIA systems use (Blokland & Van der Pol 1992). The main objective of rural development in a low potential area under LEIA systems of land use is not its contribution to the national granary, but to sustain and improve local livelihoods.

Sustainable Development usually advocates Participatory Approaches in the low potential areas. Small-scale irrigation and indigenous and improved SWC are favoured for the purpose of water resources development. The importance of indigenous knowledge and LEIA is high. However, also HEIA systems are propagated which generally are to a lesser extent based on the use of this type of knowledge (DGIS 1992). In the high potential areas, on the contrary, TOT approaches and HEIA systems prevail. Water resources development usually takes the form of large-scale irrigation. In the recent decade, however, the emphasis has changed from building new schemes to rehabilitating existing ones. The main reason is that large-scale irrigation is notorious for its high costs, low returns and vast range of negative effects on the environment and society, particularly in SSA (Moris 1987). The contribution of indigenous knowledge in high potential areas is increasingly being considered, especially in issues of irrigation management (Adams 1990). Its overall importance remains modest, however. A typical example of a Sustainable Development initiative in the low potential area of eastern Sudan is the current SWC programme of the regional Ministry of Agriculture and Natural Resources (MANR, or the Ministry of Agriculture) and its Department of Soil Conservation Land Use and Water Programming (SCLUWP, or the Department of Soil Conservation) in the Border Area (*cf.* section 4.2.2). A similar example for the high potential region in eastern Sudan is the rehabilitation of the Gash Delta irrigation scheme (*cf.* section 4.1.3).

The sustainability of government SWC interventions is evaluated in this study in terms of project-sustainability. This concept is defined in section 7.1.3. The sustainability of indigenous SWC is also briefly turned to in section 6.2.4 and chapter 8 in terms of the evidence available on its soil nutrient balance.

1.2 Technical requirements

Main technical constraints to crop production under SWC in the Border Area are expected to relate to the availability of water and soil nutrients.

Crop water requirements

Water is a constraining element to plant growth because it is essential in the building of tissue and sustenance of the life functions. Water is available in the form of rainfall, dew, mist and run-off. It can also be supplied in a processed form as irrigation water. Although parts of the water-soil-crop system can be influenced by land users, the resource itself has no substitute. The water requirements of plants are a function of various factors, including the characteristics of the crop, soil, and climate. The crop water requirements of evapotranspiration can be approximated by several formulae. Calculations based on the Blaney-Criddle method (*cf.* Doorenbos & Pruitt 1977) indicate that the demand of sorghum is about 600 mm per annum under the conditions found in the Border Area. Since annual rainfall in this same area is only half this amount, crop production can only be practiced when extra water is made available. Water has frequently been considered to constitute the first factor limiting crop production in SSA (Falloux & Mukendi 1988). However, it is currently argued that the supply of soil nutrients is more limiting in most situations.

Soil nutrient status

Nutrients may be constraining to plant growth because a total of 13 mineral elements is essential. There are a few more elements that have a beneficial effect on the growth of some plants. The major macro-nutrients are nitrogen, phosphorus, and potassium. It is the first two which are usually the main limiting nutrients in the soils found in SSA (Van Reuler & Prins 1993). The nutrients are provided either by the soil itself, or by organic and mineral fertilizers. The organic fertilizers include besides naturally occuring organic matter and living micro-organisms (i) farm wastes (crop residues, animal manures, compost and green manures which are nitrogen-fixing plants); (ii) residues from the processing of plant products (fibers, woody materials, molasses); (iii) residues from the processing of animal products (blood, horn, bone meal, leather dust); and (iv) urban wastes (composted household refuse, sewage sludge). The mineral fertilizers can be classified according to their method of production, number of nutrients and types of combinations, and their physical condition and mode of action (*cf.* appendices 4 and 5 in Van Reuler & Prins 1993,157-160). Soil nutrients should be supplied in the balanced composition required. There are no substitutes for the essential mineral elements. The status of the flow of nutrients can be expressed in a soil nutrient balance. A large body of literature has been directed to its quantification under different land use conditions (Pieri 1989, Wolf *et al.* 1991). Most soils in SSA are inherently poor in chemical fertility. Negative nutrient balances have now been found to prevail and to commonly constitute the first limiting factor to crop production in this region (Smaling 1993, Van Reuler & Prins 1993). Soil nutrient depletion, in addition, is also an important factor in dryland degradation in SSA (Oldeman *et al.* 1990).

1.3 Organizational consequences

The main organizational constraints to crop production under SWC in the Border Area are expected to relate to the availability of labour and suitable institutions to implement SWC interventions.

Labour

Labour may be constraining with respect to physical availability and timeliness. This latter situation refers to the availability of labour at the right time. Labour can be provided by members of the own household, hired labour paid in cash or kind, and through formal and informal types of local organization. The latter may range from workgroups which are established for the occasion, to the cultural mores of a community which call upon the virtue of mutual assistance. The availability of labour in the own household depends on its size, composition and life-cycle stage. The ability to hire additional labour is mainly determined by household wealth. The opportunities of households to participate in organizations depend on a range of various social, cultural and economic factors. Timeliness was interpreted by Carlstein (1982) in terms of the constraints imposed on the coupling, capacity, and capability of activities and organizations over time. These constraints are particularly important in water resources development and crop production because of their highly seasonal character (*Ibid.* 1982, 257-301). Timeliness of a particular activity of households and organizations depends on the range of other activities they are engaged in, and the priorities given to these other activities. These priorities tend to be evaluated in terms of opportunity costs (*cf.* section 2.1.2). Labour, including labour time, is an economic production factor. Timeliness can be "purchased" by hiring casual labour. Labour can to some extent also be substituted by mechanization. Labour demands for cultivation arise from land preparation, sowing, weeding, harvesting, processing, transport, harvest-storage activities, and from travel to the cultivated lands. Additional demands on labour arise from the use of SWC techniques because special artefacts to manipulate run-off such as bunds, rills, dams, stone lines, and catchments must be constructed and regularly maintained. The role of labour constraints in crop production has been discussed by Ruthenberg (1980) and in SWC techniques by Stocking & Abel (1992).

Institutions

The constraining effects of water, soil nutrients, and labour all have their own organizational and institutional dimension. Firstly, where water is a limiting factor but more can be made available locally, its volume usually demands a communal type of action. This, in turn, implies higher levels of organization. Secondly, where nutrients are limiting, this situation can be improved by the external input of fertilizers. This demands elaborate regional and national organization of infrastructure and marketing channels. Thirdly, where labour is limiting, the constraints can usually be alleviated to some extent by local organization and the formation of work groups. However, when we speak of institutional constraints in this study, we confine ourselves to those associated with the organization of government SWC intervention measures.

Two levels on which these constraints may be found are discerned in this respect. The first is the level of relationships between the international organizations and SSA governments. The second is the level of relationships between the SSA government and the local organizations or target groups in the development process. These institutional constraints will be discussed against the background of Sustainable Development for low potential regions, where Participatory Approaches are envisaged, and indigenous knowledge makes important contributions in either LEIA or HEIA systems (*cf.* table 1.2).

Institutional constraints derive from the absence of room for Participatory Approaches and interactive learning environments, as defined by Pretty & Chambers 1993. In conventional settings of international organizations and SSA governments, the mode of decision-making is usually highly centralised and standardized. Their planning is static in design. The delivery of technologies and services comes in fixed packages, which are pushed by supply instead of led by demand. Field-learning is by so-called "rural development tourism". This means that visits to projects are usually of extremely short duration. By consequence, these institutions tend to receive misleading feedbacks from the field, which sustain falsely favourable impressions of the impact of their intervention measures. Finally, the linkages and alliances are usually poor: it is common that all institutions work in relative isolation (Pretty & Chambers 1993,54). These constraints can not easily be changed because they are controlled by political interests. An important activity in this field is therefore lobbying. A good example of the latter way of removing institutional constraints is the work of the International Institute for Environment and Development (IIED), and its Sustainable Agriculture and Drylands programmes. The institutional constraints with respect to small-scale SWC also tighten as a result of opinions of SSA administrations. These may regard simple techniques as not important enough a symbol of progress and modernity. Moris (1987) referred in this context to the favoured position which the SWC pendant of conventional irrigation takes, being usually considered as a "privileged solution".

Table 1.2 **Labour demands in indigenous and project SWC techniques (demands in terracing refer to initial construction, demands in other techniques refer to annual requirements), in man-days per ha, various countries**

TECHNIQUE	LABOUR REQUIREMENT	REMARKS
Semi-circular earth bunds (hoops)	10	density 40-50 per ha
Individual hand-dug basins	12	density 200 per ha
Earth contour ridges	32-50	15-20 cm high
Semi-circular earth bunds, (demi-lunes type)	39-78	design density 313 per ha
Live barriers	40-43	depending on slope
Contour dikes	100	
Infiltration ditches	57-205	
Hillside ditches with drains	84-143	depending on slope
Reinforced contour bund	60-190	in workdays per ha
Terrace (Fanya juu type)	136-281	depending on slope
Terrace (Bench type) with drains	238-283	depending on slope
Contour furrows	110-328	
Rockwall barriers with drains	253-310	depending on slope
Terrace (Bench type)	496-500	
Organic contour farming terrace	741	30-45 % slopes
Terrace (Bench type)	750-1,800	slopes up to 50 %
Terrace	1,181	with stone side slopes

Source: IFAD (1986,18), various sources in Stocking & Abel (1992,214), Critchley et al. (1992,passim), Remonde et al. (1992). Note: the number of working hours per man-day is usually not given. Additional rule-of-thumb data on SWC labour demands include 1 man-day per m^3 for earth work and 1.5-2.5 man-day per m^3 for stone (Critchley et al. 1992).

The general institutional constraints in water resources development have been discussed by Adams (1992), and those in SWC development by Cullis & Pacey (1992). The higher level of relationships on which institutional constraints may occur in SWC development in the Border Area concern the regional Ministry of Agriculture and Natural Resources (MANR) and its implementing Department of Soil Conservation. The lower level of relationships concern the households in the Border Area (who are usually not defined as a target group for SWC intervention beforehand, however) (*cf.* section 6.3.3, 6.5.3).

1.4 SWC in dryland development

The technical and organizational constraints discussed above are interrelated. The constellation of these relationships is now usually considered as a political-economic phenomenon (Blaikie 1985, Blaikie & Brookfield 1991). Poor land users, being either cultivators, herdsmen or collectioners with little power and restricted access to resources, are frequently left with no other choice than to mine on their environment. By consequence, the fragile drylands and vulnerable livelihoods enter a process of vicious circles. The main links are formed by land degradation, demographic pressure with production decline and poverty, in the way shown in figure 1.1 (Horowitz & Salem-Murdock 1987, Kates & Haarmann 1992). Sustainable Development which distinguishes high and low potential regions might be interpreted as providing defendable arguments why people should be told to avoid these hazardous lands. Such a simple recipe, however, is no longer possible. The drylands cover 43 % of the African continent, and provide a habitat to some 67 % of its population (Toulmin *et al.* 1992).

Figure 1.1 **Main links in the vicious circle of dryland managment**

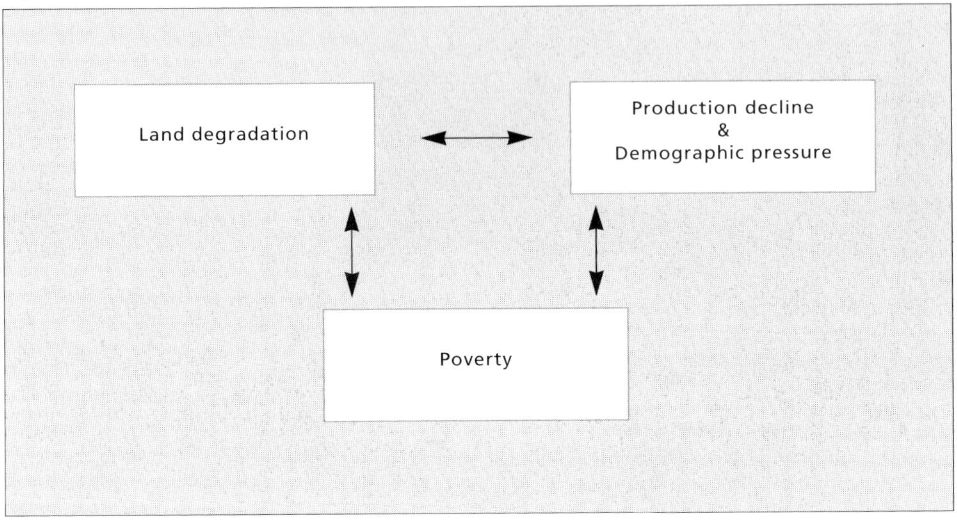

Source: after Horowitz & Salem-Murdock (1987), Kates & Haarman (1992).

The SWC techniques propagated for low potential areas are small-scale and the development locations in these areas are usually geographically remote. This means that a significant intervention impact can only be achieved when the adoption of these techniques is rapid and widespread. This is the chief reason why indigenous techniques have been selected as a starting point for SWC intervention programmes in such areas (IFAD 1992). Another compelling argument is that indigenous techniques are usually well-adjusted to the local environmental and socio-economic niches. They are therefore likely to be less influenced by the common constraining factors in these fields than techniques introduced from other settings. There are also two important limitations to building on indigenous techniques in marginal areas. Firstly, the improvement of indigenous SWC can not be regarded as an end in itself. Reijntjes *et al.* (1992) call it therefore a "starter technique", and IFAD (1992) considers it in the function of a "delaying mechanism". Both these conceptions refer to a transition process that must be set in motion to divert dryland livelihoods away from the inherent insecurities. A similar long term view has been expressed by Blokland & Van der Pol (1992). Secondly, it is expected that a substantial share of indigenous SWC techniques in SSA can not just be taken up and expanded. Many would technically still be functional, but have run out of use because the external conditions have changed. Several regional examples are given by Reij (1992,7-10) who refers to the effects of changes in political stability and population densities, and the development of market forces.

In this study, the organizational constraints will generally be given more attention than the technical constraints. This is in the first place a professional preoccupation. Secondly, this is also influenced by another matter. It is increasingly being acknowledged that the sciences concerned with dryland rehabilitation have now arrived at a point where more technical data does not add new information essential for the design of successful intervention measures (Belshaw *et al.* 1991). The present challenges lie not in technical solutions, but in the field of organization (*cf.* Stroosnijder 1993).

1.4.1 Soil and water conservation

Land users in SSA apply a range of techniques to improve the supply of water for the production of crops. Generally, this would be called irrigation. However, when this water supply depends on local rainfall alone, such techniques are called water harvesting (Reij *et al.* 1988). Although the collection of water is usually the main objective, frequently also sediment is harvested during these activities. It is for this reason that we refer in this study collectively to soil and water conservation techniques. We furthermore limit our scope to SWC use for crop production, leaving out other applications for domestic water use, range land rehabilitation, and forestry. Other techniques sometimes considered as SWC, which fields also lie beyond the scope of this book, include specific tillage practices (Unger 1984), biological measures (Hudson 1987), and agro-forestry (Huxley 1983).

1.4.2 Principles and practices

We confine ourselves in this section to a discussion of the principles and practices of the water harvesting component in SWC. Relatively little was known about water harvesting until the mid 1980s. The available experiences were based on pilot projects in countries such as Israel (Evenari 1971) and the United States (Frasier & Myers 1983). There were also successful commercial applications in dryland Australia (UNEP 1979). Pacey and Cullis wrote in 1986 one of the first books on water harvesting in the developing countries. They still commented in the second edition published in 1989 that in Africa "[..] no serious investigation of the potential of run-off farming in many of these areas [..]" is made, and that "[..] information about existing traditions is inadequate nearly everywhere [..]", (*Ibid.* 1989,127). This situation has changed since the first inventories of the distribution of these techniques were made, like in IFAD's "Soil and Water Conservation in SSA Study" of 1986. These studies resulted in many definitions which are usually area-specific, and show great overlaps for a more general application. Definitions of water harvesting usually classify the techniques on the grounds of (i) water source (snow, mist, dew, precipitation, seasonal river); (ii) origin and/or type of run-off (roof tops, overland flow, flood spates); (iii) water use purpose (crop production, grazing, forestry, natural resources conservation); (iv) catchment size (small, large); (v) catchment type (roof, ground, streams, mountain ranges) which catchments can be either internal or external to the landholding; (vi) slope length (long, short); (vii) means of water conveyance (ground, ephemeral streams, man-made channels); (viii) characteristics, medium and duration of water storage (storage in reservoir, tank, basin, soil; long and short term storage; storage in soils at field capacity or at saturation point); (ix) geomorphological setting of the entire SWC system (water courses, drainage divides, hill sides); and (x) SWC artefacts used (dam, channel, bunds, ridges), (*cf.* Bruins *et al.* 1986, Pacey & Cullis 1989, Critchley & Siegert 1991 and various sources mentioned in Reij *et al.* 1988). A comparison of these definitions shows that many terms are being used for the same practices. Alternatively, also similar practices are being given different names. The following consensus seems to have been reached on what are characteristic main elements of the water harvesting component in SWC (Reij *et al.* 1988,8):

1. *Water harvesting is applied in arid and semi-arid regions where run-off often has an intermittent character. Because of the ephemerality of flow, storage is an integral part of water harvesting systems.*

2. *Water harvesting is based on the utilization of surface run-off and requires a run-off producing and a run-off receiving area.*

3. *Most water harvesting systems use water only near to where it falls. They therefore do not include the storage of river water in large reservoirs or the mining of groundwater.*

4. *Water harvesting systems are relatively small-scale operations in terms of catchment area, volume of storage and capital investment.*

Water harvesting accordingly has been defined in two World Bank Technical Papers (Reij *et al.* 1988, Critchley *et al.* 1992) as: "[..] the collection and concentration of surface run-off for plant production before it reaches seasonal or perennial streams [..]". The same definition is used in this study. We explicitly add, however, that it is the land users who collect and concentrate this water. This places the use of water collected *in situ*, like in natural depressions, outside the definition of water harvesting used here (see below). Surface run-off, in addition, can be defined as "[..] water flowing over the ground surface to reach the sea [..]" (Strahler 1975,201). In the context of SWC application in the Border Area, two additional entries are important. The first relates to the type of water source, and the second to the mode of SWC implementation. We distinguish between, firstly, techniques of rainwater harvesting and floodwater harvesting. This is a common dividing principle also in the earlier classifications. It is of particular relevance in the Border Area because both are applied, but place entirely different demands on local organization. Rainwater and floodwater harvesting techniques are respectively mainly private and communal types of enterprise (*cf.* section 3.3). Secondly, we distinguish between indigenous SWC (abbreviated as iSWC), and techniques introduced by the government and international programmes. These will be called project SWC (abbreviated as pSWC).

Water harvesting is based on the use of a so-called "rainfall-multiplier effect". When areas of collection (catchment) and concentration (cultivated area) are purposefully combined, more irrigation water can be made available for crop production than would locally be provided from on-the-spot rainfall alone. However, the proper ratio of the catchment to cultivated area (*c: ca* ratio) must be chosen for this multiplier to become effective. Land users engaged in iSWC will normally assess this by trial and error. Critchley & Siegert (1991) elaborate on the basic principles involved with concepts derived from western irrigation science (*Ibid.* 1991,36):

(i) Water harvested = Extra water required.

where:

Water harvested =
catchment area x design rainfall x run-off coefficient x efficiency factor.

and

Extra water required =
cultivated area x (crop water requirements – design rainfall).

By substitution in (i) it holds that:

(ii) Catchment area x design rainfall x run-off coefficient x efficiency factor = cultivated area x (crop water requirements – design rainfall).

If (ii) is rearranged, it holds that:

(iii) Crop water requirements – design rainfall divided by design rainfall x run-off coefficient x efficiency factor = catchment area divided by cultivated area (or c: ca).

Design rainfall is the amount of rainfall at which the technique provides sufficient run-off to meet the crop water requirements. The design rainfall is calculated at an arbitrary probability of occurrence. Usually a probability of 80 % is considered, which indicates the amount of rainfall that can be expected in 8 years out of 10. The run-off coefficient (*rc*) indicates the share in total rainfall which arrives in a given area as run-off. The value of *rc* can be either measured or estimated from existing tabulations (*cf.* Schwab *et al.* 1981). The efficiency factor indicates the share of run-off which is lost through evaporation, seepage, and deep percolation. This factor is usually estimated on the basis of local experience. The crop water requirements indicate the moisture needs (usually on a daily or 10-day period basis) of the plants. These depend on the type of crop, its growing stage, climate, and locality. Crop water requirements also can either be calculated or estimated (*cf.* Doorenbos & Pruitt 1977).

Figure 1.2 **Soil and water conservation techniques in Africa**

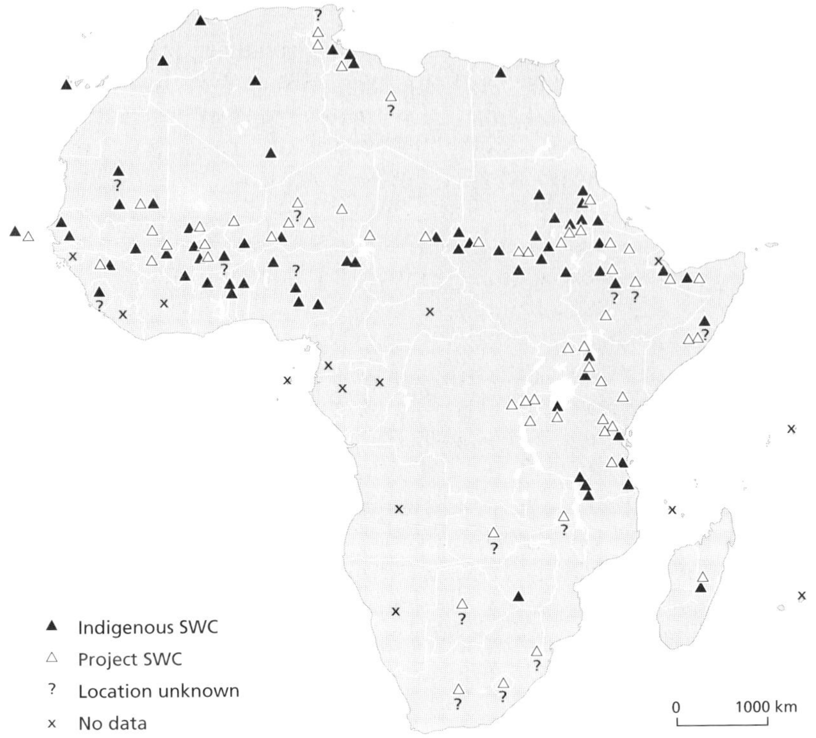

Algeria: FAO (1986b), Klemm & Prinz (1989). **Benin**: Angladette & Deschamps (1974), Reij (1992). **Botswana**: Willcocks (1979), Hudson (1987), Carter & Miller (1991). **Burkina Faso**: Pacey & Cullis (1986), Piquemal (1991), Van Driel & Vlaar (1991). **Burundi**: Jones & Egli (1984). **Cameroon**: Beauvilain (1992), Witsenburg (1994). **Cape Verde**: Reijntjes (1986), Haagsma & Reij (1993). **Chad**: FAO (1985), Sommerhalter (1987). **Egypt**: Barrow [1987](1993), **Ethiopia (Eritrea)**: Hurni (1986), Hudson (1987), Kejela (1992). **Gambia:** IFAD (1990). **Ghana**: Bonsu (1981). **Guinea Conakry**: IFAD (1988b), Martin (1990), Landeck (1991). **Kenya**: Morgan (1974), UNEP (1987), Critchley (1991b), Cullis & Pacey (1992). **Lesotho**: FAO (1986b), IFAD (1987). **Libya**: FAO (1986b), Barrow [1987](1993). **Madagascar**: CTFT (1979). **Malawi**: Hudson (1987). **Mali**: Hallam et al. (1985), Klemm & Prinz (1989), Critchley (1991b), Rands (1992). **Mauritania**: FAO (1986b), Rochette (1989). **Morocco**: Kutsch (1983). **Niger**: Reijntjes (1986), Barrow [1987](1993). **Nigeria**: Bennet (1939). **Rwanda**: Jones & Egli (1984). **Senegal**: Diemer (1990). **Sierra Leone**: Richards (1985). **Somalia**: McCarthy et al. (1985), Pacey & Cullis (1986). **South Africa**: Ross (1947). **Sudan**: see separate detailed entry. **Swaziland**: Reij (1984), Nsibandze (1987), Hudson (1987). **Tanzania**: Burra & Van Klinken (1987). **Togo**: De Haan (1988). **Tunisia**: Evenari (1975), Pacey & Cullis (1986), Hudson (1987). **Uganda**: Reij (1992). **Zaire**: Jones & Egli (1984). **Zambia**: Norén & Nordén (1983), **Zimbabwe:** Reij (1990).

1.4.3 Distribution in Africa and Sudan

The Centre for Development and Cooperation Services (CDCS) of the Amsterdam Free University has been engaged in several studies of SWC. These include the World Bank "Sub-Saharan Water Harvesting Study", and IFAD's "Special Programme for Sub-Saharan African Countries Affected by Drought and Desertification". The reports which relate to these programmes (IFAD 1986, Reij *et al.* 1988, Van der Wal & Zaal 1990, Critchley *et al.* 1992, IFAD 1992), and literature in the CDCS library have been used to make an overview of the distribution of SWC application in Africa. The list includes techniques of indigenous SWC, and projects which contain a dominant component of water harvesting for crop production (figure 1.2). Several existing lists were consulted in Hale (1966), Ludwig (1968), Rochette (1989), and Reij (1990, 1992). The presented overview is not intended to be complete. Firstly, my own last review of these sources dates from September 1992. Secondly, several reports provide insufficient descriptions either to classify the technique, or put it on the map. Thirdly, no reports have been encountered for Angola, Central African Republic, Republic of Congo, Djibouti, Equatorial Guinea, Gabon, Guinea Bissau, Ivory Coast, Liberia, Namibia, and for the island groups off the continent of Comoros, Mauritius, São Tomé and Principé, and Seychelles. Several of these countries have no significant stretches of drylands (Republic of Congo). Others largely consist of arid and hyper-arid lands (Djibouti), or have for a long time been in a state of conflict which makes reporting difficult (Angola). The main references per country are illustrated in figure 1.2 on page 18 (cf. References p.293).

Indigenous and project SWC techniques, accordingly, are shown to appear in nearly all dryland regions of Africa. Their distribution displays a striking parallel with the lattitudinal characteristics of climate, especially in the Sahel and part of the Horn of Africa. A similar overview for the Sudan alone is given in figure 1.3. The main iSWC techniques in Sudan which contain a dominant water harvesting component include: (i) stone terracing. This is exclusively found in the mountainous western regions of the country; (ii) the construction of bunds of earth (in the entire country), of stone (exclusively in western Sudan), and of brushwood (exclusively in eastern Sudan) to capture run-off from adjacent catchments; (iii) earth dams to divert floodwaters; and (iv) tanks excavated in the ground for domestic water use. Only in western Sudan, the bottom of these tanks is also being cultivated using the residual soil moisture (Ibrahim, pers. comm., March 1992). The tanks themselves can be found on clay soils in the entire country. The main references on the distribution of SWC techniques in Sudan are presented by region (West, central and East Sudan), by SWC artefacts used (terraces, bunds, dams), and by mode of implementation (iSWC, pSWC). Not included in this list are the country-wide references given in Tothill (1952), Barbour (1961), Lebon (1965), and Ibrahim (1988).

Figure 1.3 **Soil and water conservation techniques in Sudan**

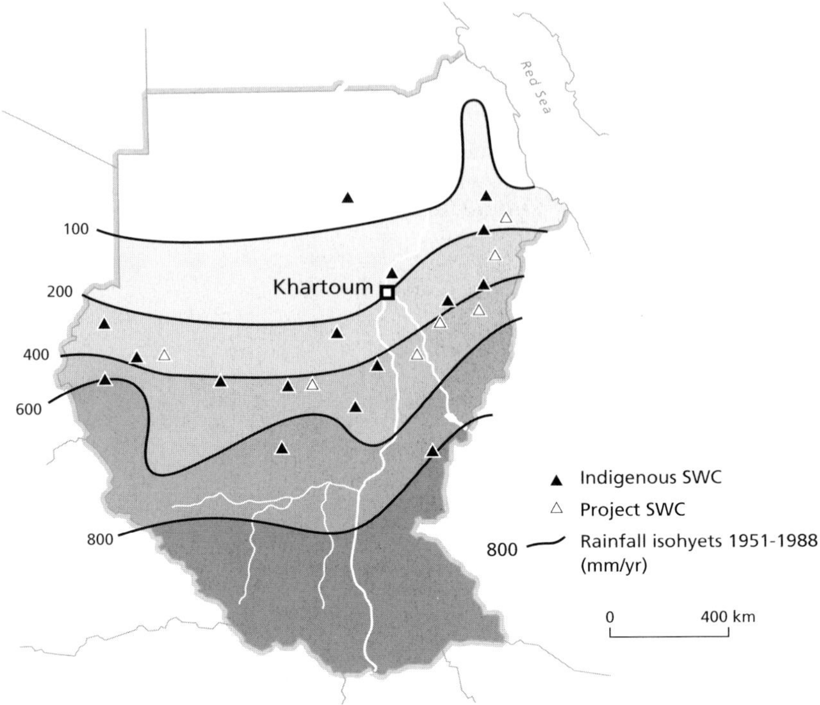

West Sudan
Indigenous SWC: UNDP/FAO (1968), Miehe (1988) [terraces in the Jebel Marra area]; Colvin (1939), Nadel (1949) [terraces in the Nuba mountains]; Barbour (1961) [ridging in the Nuba mountains]; El Arifi (1971), Tubiana & Tubiana (1977) [terraces in other western regions]; Martin (1985) [stone bunds in the El Fasher area]. El Sammani (1988) [earth bunds and dams in the El Fasher area]. Harmond Associates (1991) [earth bunds in the El Obeid area]; Al Jahli (1973), Wiadanapathirana (1986) [dams in the El Fasher area]. **Project SWC**: El Jahli (1973), Kanno (1985), Widanapathirana (1986), El Sammani (1988) [government introduced dams in *i.a.* the El Fasher area]; Attiya & Nivose (1966), DECARP (1976) [international projects using various introduced techniques]. Omer & Madibo (1990), Omer & El Amin (1991) [introduced earth bunds in North Kordofan area].

Central Sudan
Indigenous SWC: Davie (1924), MoA (1963, 1965), De Vadja (1966), El Dishouni (1989) [earth bunds in the Nile region] Abu Sin (1970), Born et al. (1971), Sörbö (1985) [earth bunds in the Butana area]. **Project SWC**: UNDP (1990) [improved earth bunds and introduced dams in the Butana area].

East Sudan (excluding the Border Area which is discussed separately in section 1.2.3)
Indigenous SWC: Born et al. (1971), SOGREAH (1982), Reij et al. (1988), Critchley (1990), Critchley et al. (1992), Van Dijk & Ahmed (1993) [earth bunds in the Kassala area]. Stern (1985), Cole (1989) [dams in the Red Sea province]. **Project SWC**: Osman & Dabalob (1990), Van Dijk (1991), Mohamed (1992) [government dams in the Red Sea province].

When the regional SWC distribution in Sudan is compared with thematical maps in Barbour (1961), Lebon (1965) and Craig (1991), the pattern seems mainly related to climate, physiographic and drainage conditions, soil types, and population distribution. SWC techniques are predominantly practiced in the zone of 250-500 mm annual rainfall. For obvious reasons, terraces are associated with relief. The use of bunds prevails in areas of poor surface drainage, and the use of dams in areas of well-developed drainage ways. These drainage characteristics, in turn, are correlated with soil types and physiography. Poor drainage conditions occur on sandy soils, and in areas without rock outcrops that may act as larger natural catchments. Better drainage conditions occur on clayey soils, especially near rock outcrops. SWC techniques, finally, are predominantly applied in the more populated dryland areas. Their distribution on a smaller geographical scale is importantly influenced by the local availability of drinking water (*cf.* Lebon 1965, Born 1965, Born *et al.* 1971 for a discussion of factors influencing the distribution of indigenous SWC in the Sudan).

Other Sudanese practices which also conserve soil and moisture for crop production include *in situ* moisture conservation. Use is made of the natural depressions in the landscape called *maya*, or *nagaa*, *rahad*, *turda*, *fula*, *birka*, and *shagg* (Ibrahim 1988). Another type is the construction of man-made closed basins, called *hod*. The techniques are mainly applied in the Nile region of central Sudan (Randell 1963, Born *et al.* 1971), and in western Sudan (Dumont & Stevens 1987). Land users in these two situations do not collect and concentrate run-off, but merely use the ponded water and precipitation where it has fallen. These techniques therefore do not qualify as water harvesting. The dividing-line between *in situ* moisture conservation and water harvesting clearly is fluid.

2

Framework for research

The research framework to assess the effects of government SWC interventions in the Border Area and the importance of local techniques of indigenous SWC is presented in this chapter.

2.1 Research questions and design

Before Independence of the Sudan in 1956, the Anglo-Egyptian Administration had made only a few isolated SWC interventions in the Border Area of eastern Sudan (*cf.* chapter 3). However, from the early 1980s onwards, the Sudanese government embarked on a special programme for this purpose. The implementation of this programme was under supervision of the Ministry of Agriculture based in the regional capital Kassala, and was executed in the field by its Department of Soil Conservation. These interventions greatly increased in their scope after the 1984-1985 drought, when support was received from the government of The Netherlands through the rural development programme "Kassala Area Development Activities" (KADA). The programme will be referred to next as MFEP/KADA, where the abbreviation stands for Ministry of Finance and Economic Planning which was the main Sudanese counterpart. The achievements of the early interventions of the department, based on the construction of earth dams, were evaluated by an external expert as poor (Mulder 1990). The achievements of the later interventions, based on the construction of earth embankments for which the department had received international support, had also remained below expectation (*Ibid.* 1990). In addition, no impact assessment of these interventions had ever been made. This was judged by a Sudanese-Netherlands evaluation mission as a serious drawback (GOS/GON 1990).

At this same time, indigenous SWC, applied in the Border Area by settling nomadic communities of the Beja group, had been given more attention as a possible alternative for the techniques introduced by the government and international programmes (Van Dijk 1991). Crop production was already practised by the Beja in the 19th century. The adoption of SWC techniques for this purpose probably emerged in the 1950s, when their nomadic lifestyles were giving way to more sedentary forms of living (*cf.* chapter 4). In the course of the following decades, their household economies changed accordingly. Although the number of livestock in the area still increased between the mid 1950s and mid 1970s, crop production gradually also gained in importance (Green Mission 1985). When in the drought of 1984-1985 not only considerable numbers of these livestock were lost, but also crop production failed, these households were forced to turn to other economic activities. The Beja livelihoods of the 1980s and early 1990s, therefore, were no longer pastoral, and frequently not even agro-pastoral, but could best be typified as "multi-resources economies" (*cf.* chapter 4).

Against this background of emerging government SWC interventions in the Border Area and changing Beja livelihoods, the following two main questions for research as part of the Livelihood and Environment programme of the University of Amsterdam have been formulated. Firstly, the impact of the government interventions in SWC is addressed as follows:

1. *"What have been the achievements of different government SWC intervention measures implemented in the Border Area from the early 1980s onwards?"*

Secondly, the potential contribution of indigenous SWC to future rural development programmes is addressed by considering its position in the newly emerging household economies. The rationale is that either a greater or smaller importance of these techniques has implications for the design of government programmes for SWC intervention based on them. The second research question is formulated as follows:

2. *"What has been the relative importance of indigenous SWC in crop production and in the household livelihood in the Border Area from the early 1980s onwards?"*

The formulation in relative figures, firstly, allows comparisons to be made in the Border Area between households in different villages, and between their different types of multi-resources economies. Indigenous SWC is usually one among several crop production methods applied, and crop production is usually also one among several means to sustain the household livelihood. Secondly, we expect that within these different types of livelihood profiles, also these relative shares of indigenous SWC differ in function of the distance to regional urban centres. We argue in this study that the outcomes concerning precisely these relative shares have important implications for the design of SWC interventions.

2.1.1 Operationalizations

Government SWC interventions
An accepted framework for monitoring and evaluation of international programmes and interventions is given in the "Principles for effective aid" of the Development Assistance Committee (DAC). Its elements of evaluation include relevance and fulfilment of objectives, developmental efficiency, effectiveness, impact, and sustainability (*cit.* in IOV 1993, 84-87). This DAC framework was recently used in two comparative studies of programmes financed by the Netherlands DGIS (IOV 1993) and co-financing agencies (SIM 1990, 1991). The same elements are also used in this study to assess the impact of government SWC interventions in the Border Area. In terms of SIM (1990) and IOV (1993), these are named (i) accordance with policy; (ii) effectiveness; (iii) efficiency; and (iv) impact and sustainability. The first three will be used on the basis of reports of evaluation and technical backstopping missions, and publications and files of the Soil Conservation Department. The impact and sustainability of interventions is mainly assessed by means of data collected in the L&E programme, referred to next as L&E Survey data.

The relative importance of indigenous SWC
The significance of indigenous SWC in crop production and in the household livelihood is considered in terms of (i) its contribution to the household income. The incomes are gross figures because calculating net incomes on a yearly basis for nomadic and recently settled communities is complex (*cf.* Abdullahi 1990,98-126); (ii) the amount of allocated cultivated land; (iii) the amount of allocated labour time; and (iv) perception. For this purpose, land users have been asked to rank all livelihood activities according to the importance they attach to them. These four fields will be referred to concisely in the text as the "Income", "Land", "Time" and "Perception" dimensions. They have been operationalized by a set of 9 variables for 2 datum years providing a total of 18 research variables (appendix 2.1). These 18 represent the set of dependent variables used in this study to assess the importance of indigenous SWC in the Border Area.

2.1.2 Selection of villages

The selection of villages is made in such a way that the scores of dependent variables can also be evaluated in function of two main independent factors. These are (i) the occurrence of government SWC intervention; and (ii) the village distance to the regional urban centre. The first factor is introduced to address our research question on the achievements of the government SWC intervention measures. The second to address the expected differences in the relative importance of indigenous SWC.[1] Geographic distance represents economic costs. More in particular, these may include benefits forgone in the form of the opportunity

[1] We acknowledge that also the achievements of government SWC interventions are influenced by distance to the urban centre (*cf.* Chambers 1991, *passim*). We will occasionally comment on this, but the relationship itself is not a subject for research.

costs of labour (Cox 1972). The basic premise of our research framework is that the relationship between distance to urban areas and the level of labour-opportunity costs is a negative one (*Ibid.* 1972,265-266). This means that the latter decrease with an increasing distance to the location where these opportunities are being provided. This negative relationship has also been described for the case of the Border Area (Kuhlman 1990,199-121). The second premise is that the relationship between the level of labour-opportunity costs and the relative importance of indigenous SWC is equally of a negative type. This means that with an increasing level of the labour-opportunity costs, the importance of indigenous SWC is likely to decrease. This second assumption has no general validity like the first. This type of relationship is commonly being referred to for SWC use in general (Kerr & Shangi 1991), and for its application in dryland multi-resources economies in SSA in particular (Little 1992, Cullis & Pacey 1992, Tiffin *et al.* 1994).

From these first two premises, it now follows that the relationship between distance to urban areas and relative importance of indigenous SWC is likely to be positive. This means that the importance of indigenous SWC in crop production and household livelihood is likely to increase with distance to urban areas. The main regional centre in the Border Area is the town of Kassala. The selection of our research villages was made with respect to the distance to this town. A matrix (table 2.1) can now be constructed with intervention (present versus absent) on one axis, and distance (small versus great distance to Kassala) on the other. The selection of the village of Hafarat was entirely led by our choice to include the well-documented international SWC programme of MFEP/KADA in this study (*cf.* chapter 4). The selection of Telkook as a village where the Soil Conservation Department made SWC interventions was mainly led by considerations of logistics. Telkook offered the opportunity of using the department's guesthouse during fieldwork. The non-intervention villages Um Safaree and Ilat Ayot were selected on grounds of pronounced indigenous SWC application. This was assessed from the most recent aerial photographs of the Border Area taken in December 1986.

Table 2.1 **Research villages in the Border Area**

	GOVERNMENT SWC INTERVENTION	
	PRESENT	ABSENT
DISTANCE TO KASSALA TOWN		
GREAT	Telkook	Ilat Ayot
SMALL	Hafarat	Um Safaree

The relationship between on the one hand, geographical location and distance to urban areas, and on the other hand economic performance has been subject of several theoretical debates. We mention two. Firstly, Von Thünen demonstrated in his classic work of 1826 *Der isolierte Staat in Beziehung auf Landwirtschaft und Nationalökonomie* that the spatial distribution of land use systems is a function of what he called Economic Rent.[2] The level of Economic Rent decreases with increasing distance to the market place (town). Land users adjust to this level of Economic Rent through conscious changes in the level of their inputs (for example of labour and fertilizer), and their choice of the products cultivated (for example in function of their labour demands). Generally, the result will be that the land use intensity decreases with an increasing distance to the town. This relationship will be stronger when a region can be considered as being more "isolated". Although in the present-day open economies also other factors than distance are important, the basic ruling principles are argued to have remained valid. These, accordingly, continue to determine land use patterns around urban areas in the long term (Chisholm 1973,24). Secondly, a number of scholars has studied the relationships between urban areas and their rural hinterlands. Hoselitz (1960, *cit.* in Harvey 1973,233) introduced the pair of terms of "generative" and "parasitic" towns based on characteristics of the prevailing urban-rural relations. In the first situation, the net transfer of wealth is towards rural areas and is at the expense of urban areas. In the second situation, this net transfer is towards urban areas at the expense of the rural hinterland. Generative towns can therefore be important driving forces behind rural development, technological change, the expansion of land use, and introduction of new techniques. Parasitic towns, on the contrary, can frustrate and eventually even undermine these same developments. This is because, on balance, fewer investments are being made in the surrounding rural lands. There is proof of the existence of both types of town in SSA[3], although parasitic traits seem to prevail in Sudan (Yassein 1967, Abu Sin 1991).

These theories add yet another dimension to this discussion. The postulated positive relationship between distance to urban centres and the importance of indigenous SWC (based on labour-opportunity costs) produces the same pattern as generated by parasitic towns (based on urban-rural exhange relations). In both situations, indigenous SWC is of smaller importance near the urban centre. When the generative traits prevail, however, the opposite situation results, at least near urban centres. These relationships have also been commented on from the specific viewpoint of labour-opportunity costs in SWC. Kerr & Shanghi (1991) state that urban incomes usually impede land users to invest in SWC because of relatively unfavourable returns to labour in the latter activities. Similar arguments have been raised for zones adjacent to the urban agglomerations in northern Africa (IFAD 1986). In other settings, however, it was found that indigenous SWC

---

[2] Economic Rent was defined by Von Thünen as the return that can be obtained per unit of land above that which can be obtained from the land which is at the margin of economic cultivation (*cf.* Chisholm 1973, 23-24).

[3] Gugler & Flanagan (1978) and Evans & Ngau (1991) for example emphasize the role of prevailing generative traits of towns in SSA. Lipton (1977) and Harriss & Harriss (1984) emphasize their parasitic traits.

particularly flourishes near urban areas. This was reported for example by Piquemal (1991) for Burkina Faso, and Beauvilain (1992) for Cameroon. It is for this reason that the outcome of the relationship for the Border Area can not be meaningfully hypothesized beforehand.[4]

2.2 Collection and processing of geographical data

Maps indicating recent topographical and thematical information are lacking in the Sudan[5]. This information must be collected for smaller areas by field work, and for larger areas by remote-sensing techniques. The latter include the interpretation of aerial photographs and processing of satellite imageries in a Geographic Information System (GIS). Remote-sensing techniques are discussed by Richards (1986) and GIS techniques by Dana Tomlin (1990). The applications in dryland environments are discussed by Prince *et al.* (1990), and for SWC studies in particular by Tauer & Humborg (1992). Remote-sensing techniques are employed in this study to collect base-line data on indigenous SWC use in the Border Area. The outcomes do not directly contribute to the answering of our two main research questions.

2.2.1 Remote-sensing materials

Aerial photographs

Aerial photographs are used for detailed delimitations of farming zones and to assess land use changes over time. The material available at the Khartoum Survey Department includes vertical aerial photographs at varying scales from 1950 onwards. The department also supplies photo-compilations photographed and printed on single sheets at an approximate scale of 1:250,000 (airphoto-mosaics). MFEP/KADA implemented an aerial survey in 1986. This provides the most recent and detailed 1:20,000 coverage of the Border Area available at present (*cf.* appendix 2.2). Airphoto-interpretation was alternatively done at the Survey Department in Khartoum and the MFEP/KADA office in Kassala. Area-measurements were made using transparant millimeter paper and a planimeter at these two locations respectively.

Satellite imageries

Digital information of satellite imageries is used in a multi-temporal land use analysis of the Border Area. 1983 and 1988 are the main datum years used in this study (*cf.* section 2.3.1). Since there were no suitable imageries for 1983, a 1987 scene was selected instead as best alternative given the prevailing agro-climatological conditions. A combination of

---

4 In methodological terms, this is expressed by non-directional testing of the hypothesis based on the research questions in this study (*cf.* section 2.3.3).

5 The last revision of 1:250,000 topographical maps which were surveyed in the first decade of this century date from the mid 1970s. A new series at a scale of 1:100,000 is now under preparation. The published maps of the Border Area are still based on 1960s and 1970s aerial photographs, however.

Landsat TM and SPOT XS imageries on magnetic tape and CD-ROM was used. The two sensor types complement each other which yields extra information. SPOT displays more geometric detail, while Landsat has a more favourable spectral differentiation[6] (Prince *et al.* 1990).

2.2.2 GIS methods

The Geographic Resources Analysis Support System (GRASS) (USACERL 1993) is used for the processing of satellite imagery data in a GIS.[7] The first objective of analysis is to determine a set of priority features of main relevant land units for crop production and indigenous SWC use in the Border Area. The units selected include Rock Outcrop, Piedmond Plain and Khor Alluvium after Van der Kevie & Buraymah (1976). The second objective is to make a multi-temporal analysis of land use under indigenous SWC for this area. Land use classification by means of satellite imageries is constrained by two factors. The first is geometric resolution. This may reduce the scope of classification when features of interest cannot be detected. The second is spatial and temporal variability in the main determining factors of spectral reflectance (greenness and brightness). This is caused by variations in biomass, soil type, and soil moisture contents (Prince *et al.* 1990). The effects of these constraints can be reduced to a minimum, however, when the sensor types of SPOT and Landsat are combined (Simons 1994). The Border Area is covered by two SPOT scenes referred to as northern and southern and one underlying Landsat scene, as shown in figure 2.1. The working approach in GIS is discussed next.[8]

Selection and acquisition of satellite imageries
The Landsat TM full-scene of 12 October 1987 represents typical Border Area dry year conditions. This imagery was taken after nearly all rainfall in this year (271 mm) had been received. The last rainday (defined as daily fall of over 10 mm) before scene recording was 24 August. The scene predominantly displays dry soils. The cropping details are known for selected Border Area locations from progress reports of MFEP/KADA. The two SPOT XS full-scenes of 24 August 1988 were taken after 66 % of all rainfall in this year (396 mm) had been received. The last rainday was 10 August. The scenes predominantly display moist soils. Detailed cropping characteristics are known from ground observation at selected locations cultivated under indigenous SWC and other techniques.

---

[6] The *Système Probatoire d'Observation de la Terre* SPOT XS has a ground resolution of 20 m x 20 m against 30 m x 30 m (120 m x 120 m for thermal IR) in the Landsat Thematic Mapper (TM). Landsat TM however provides 7 channels against 3 available in the SPOT XS system.

[7] GRASS is a public domain GIS developed at the Environmental Division of the US Army Construction Engineering Research Laboratories (USACERL) in Illanois, USA. Versions 4.0 and 4.1 are installed at the Faculty of Environmental Sciences of the University of Amsterdam and are run on a UNIX workstation with a X windows graphic interface.

[8] Johan Berkhout of the Department of Physical Geography and Soil Science is acknowledged for indicating the main working approach in GIS followed in this study.

Figure 2.1 **Border Area coverage by satellite imageries**

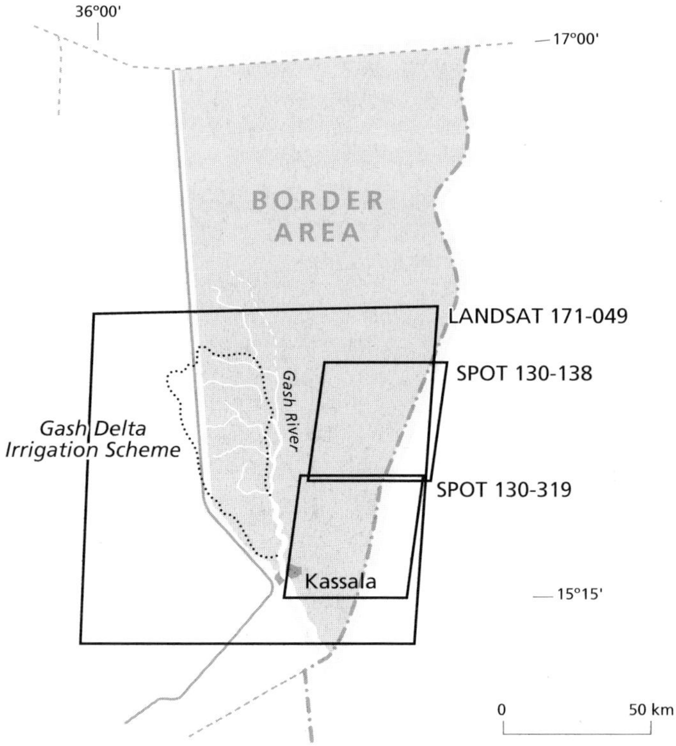

Initial image processing

Before satellite imagery raster type of data can be used in a GIS, geo-referencing and geometric correction must be applied. The dual purpose is to match the different scenes with true ground control points and to match the pixel resolutions of different sensor types. Prominent landmarks of *inselberg* rock outcrop and khor bifurcations on 1:100,000 topographical maps were used for geo-referencing. Landsat TM data in UTM geographical coordinates were resampled to higher resolution SPOT XS data. The approximate study region in the Border Area was selected by excluding the Gash delta irrigation scheme, Gash basin horticultural areas, and part of Eritrea by means of GRASS masking procedures. No further image optimalization procedures were applied. A computer-assisted or "unsupervised" classification was followed in nearly all subsequent data processing phases. The map layers of the original unprocessed data are called SPOT_1, SPOT_2, SPOT_3 and Landsat_1, Landsat_2, Landsat_3, Landsat_4, Landsat_5 and Landsat_7 (table 2.2).

Classification

The total of 10 channels available from Landsat TM and SPOT XS together share common information which is redundant. Firstly, the thermal-infrared channel 6 in Landsat is not used in this application and was excluded from analysis. Secondly, a principle component analysis (pca) was performed on the remaining 9 channels of the two sensors to further reduce the amount of correlated information. These channels contain the combined data of the normal-to-dry year 1987 and wet year 1988. The outcome is a set of 9 uncorrelated principal components (*cf.* Ingebritsen & Lyon 1985 and Goossens 1992 on the application of pca in satellite imagery classification). The library of (compiled) GRASS functions and procedures available at our work station provides no tools to determine the total variance contained in each component. Commonly, the first four which account for some 90 % of the total variance are used when two different images are processed in a GIS (Ingebritsen & Lyon, 1985). GRASS clustering and Maximum Likelihood procedures were used next to search for an initial arbitrary number of 10 spectral classes from these first 4 principal components, and to classify the cell spectral reflectances accordingly. The classes were interpreted visually from their brightness and distribution in the picture. These were given tentative semantic definitions. The map layer thus created is called Principle_Components.

Table 2.2 **Map layers and map overlays produced in a GIS-based land use classification of the Border Area**

MAP NAME	REMARKS
SPOT_1,2,3	3 map layers of unprocessed data
Landsat_1,2,3,4,5,7	6 map layers of unprocessed data
Principle_Components	Map layer created by unsupervised classification
Land_Units	Map layer created by reclassification of Principle_Components and using Rock_Outcrop as an overlay
Rock_Outcrop	Overlay created by supervised procedures
Best-fit_Teras	Overlay created by supervised and unsupervised procedures
NDVI_Biomass	Overlay created by unsupervised procedures

Ground checks, verification and improved classification

The 10 initial classes were merged based on field knowledge into 3 main and easily recognizable classes of (i) low soil moisture content areas, which mainly consist of the physiographic unit Piedmont Plain (elevated land); (ii) high soil moisture content areas, which mainly consist of the physiographic unit Khor Alluvium (depressions); and (iii) Rock Outcrop based on the physiographic unit of the same name.

Final reclassifications and production of image overlays

The next procedures also involve "supervised" classifications where interpretations are not computer-assisted, but made by the GIS operator. Firstly, reclassifications were made so as to allow the use of 2 of the 3 new classes in a separate map layer called Land_Units. Secondly, two new single-factor maps, or "overlays", were made of (i) the existing unit Rock Outcrop, and (ii) the new unit *Teras* SWC. The first was made because the spectral reflectance of rock outcrop was not consistently shown in any of the classes created in the pca. This overlay was made by digitizing raster maps using 1: 100,000 topographical maps as base reference. This new overlay map is called Rock_Outcrop. The second overlay indicating *teras* SWC occurence in the Border Area was made according to the following procedures. Firstly, known *teras* locations under known cropping patterns (training sites) were identified in an easily interpretable map layer of good resolution (we used the near-IR map layer SPOT_3). Secondly, the corresponding spectral characteristics (signature) of these locations were assessed in one of the ten initial classes of the map layer Principle_Components. Only the most important brightness values of SPOT_3 were considered and were defined for this purpose as values with an arbitrary minimum occurrence of 5 % for *teras* in the southern part of the Border Area, and 10 % for *teras* in the northern part. These lower limits are based on the frequency distributions of brightness values of all training sites considered (3 in each southern and northern part of the Border Area). Thirdly, these signatures were used to search for similar locations in the wider region. The resulting map-layer is called Best-fit_*Teras* because *teras* occurrence is only approximated and not fool-proof. A multi-temporal land use analysis was made by assessing the biomass occurrence in the overlay Best-fit_*Teras*. The standard dimensionless indicator of biomass Normalized Difference Vegetation Index (NDVI) was used to compare normal-to-dry year and wet year situations. NDVI is calculated as the brightness value in the infrared channel minus the brightness value in the red channel divided by the brightness value in the infrared channel plus the brightness value in the red channel. The red and infrared channels in Landsat TM are 3 and 5, in SPOT XS these are channels 2 and 3 (Prince *et al.* 1990). The median of these NDVI scores of *teras* locations was used to divide them in two classes high and low. Map calculation procedures in GRASS were used to determine the combined scores of high, medium and low over the two datum years. The new overlay map is called NDVI_Biomass.

Digital linkage of data sets
The geographical occurrence and acreage of *teras* SWC in the Border Area were assessed from the overlay map Best-fit_*Teras*. The multi-temporal *teras* land use characteristics were assessed from the overlay map NDVI_Biomass.

2.3 Collection and processing of socio-economic data

A recurrent procedure in data processing and analysis of the set of 18 socio-economic variables is the determination of a household total score, and the computation of a sectoral share in this total. An example of this is the calculation of the share which indigenous SWC contributes to the total household crop production income in 1983 (*cf.* variable SWCCROP3 in appendix 2.1). This implies that a relatively rigid data structure must be maintained. Missing values for household total scores not only reduce the number of valid cases for these variables, but also for the selected variables to be calculated from them. A first examination of the raw field data collected in the L&E Survey indicated that such shortcomings were likely to occur. This especially applied to data on typically sensitive subjects such as income, and difficult to measure subjects such as the allocation of household labour time. At this point two decisions could be made. The first would be to accept the mounting number of missing values in the research. The second would be to model these data so that all information remains functional to the analysis. The second approach has been followed in this study.

A model is defined by Breman & De Ridder as "[..] the mathematical description of a process which allows an evaluation of outcomes without engaging in experiments or real-life observation [..]" (*Ibid.* 1991,453). Models exist in different types and varying degrees of complexity. The applications in natural resources sciences are fairly common and are used in land evaluation (Van Diepen *et al.* 1991), assessments of agricultural production (Van Keulen & Wolf 1986), of rangeland productivity (Breman & De Ridder 1991), and of soil nutrient status (Smaling 1993). Applications in the socio-economic sciences occur in decision-making under uncertain dryland conditions (Bakker 1993), and general economic and financial analysis of so-called modal or pattern farms (Price Gittinger 1984). Models need not always be as complex as in the applications referred to above. Van Diepen *et al.* (1991) discuss different methods to determine levels of crop production in the field suggesting that relatively simple procedures already contain model-like elements.[9] For our purposes, it suffices to use only a prescript by which one series of scores is replaced by another. This is an equation-type of model, and sometimes more specifically called a lineair transformation (Davis 1986). We also use one slightly more elaborate application. The L&E Survey data are only modelled when real scores of variables are missing.

2.3.1 Survey design

Three levels of analysis are used in this study (i) the household; (ii) the household member; and (iii) the landholding. The household is our conventional unit for the aggregation of data. These data must also be available for individual household members because allocation of labour time is member-specific, and for individual landholdings because the use of cultivation techniques is location and landholding-specific. This study uses two main research years 1983-1984 and 1988-1989. The first was a relatively normal-to-dry year. The second a relatively wet year. This allows a comparison of land use under different agro-climatological conditions. The selection of these years has been guided by the relative ease with which they can be remembered by respondents. The first year is recalled in the Border Area as "the year before the last great drought" [which started in 1984]. The second year is recalled as "the year of abundant water", or alternatively, as "the year of malaria". Besides these two research years, additional data of lesser detail have also been collected for the relatively dry years 1989-1990, and very dry years 1990-1991. The terms season and year are used interchangeably in this study, sometimes with reference to the first year only. 1983 accordingly refers to the crop production season which starts in 1983 and ends in 1984.

---

[9] These methods range from intuitive estimates (informal procedures based on local experience) and reasoned intuitive estimates (hypothetical rating schemes and "systematized farmer knowledge"), to data collection (through farmer interviews and other sources), and finally to empirical models (for example based on regression analysis), and deterministic models (for example based on dynamic simulations of factors that control crop growth) (*Ibid.* 1991:178-179).

Research methods
A questionnaire-type of survey with closed answer categories has been conducted for a random sample of households in Um Safaree, Hafarat, Ilat Ayot, and Telkook (table 2.3). The research population, or universe *N*, consists of all village-based households. The research units, or sample *n*, consist of all households drawn in this sample. The L&E Survey respondent is the member in the household responsible for land use decision-making.

Table 2.3 **Characteristics of the research population, universe N and sample size n (in number of households)**

	1ST-ROUND			2ND-ROUND	
	UNIVERSE N	SAMPLE n	% n/N	REVISIT r	% r/N
Ilat Ayot	307	58	19	32	10
Um Safaree	328	64	20	39	12
Hafarat	230	61	27	61	27
–of which nopSWC		35		35	
–of which pSWC		26		26	
Telkook	916	61	7	61	7
–of which nopSWC		37		37	
–of which pSWC		24		24	
Total	1,781	244		193	

Source: L&E Survey. Note: pSWC means project SWC, nopSWC means no project SWC (*cf.* section 6.1 for definitions).

A sample frame had to be made before these samples could be drawn. The households of the Border Area are not registered officially anywhere. Rough village maps were drawn for this purpose from 1986 aerial photographs (1:20,000) indicating tracks, compounds, and other relevant features for local orientation. The maps were checked and corrected in the field. The inhabited compounds were numbered resulting in a list which provided the sample frames for Um Safaree and Ilat Ayot. Another approach had to be followed in Hafarat and Telkook. Sampling had to be stratified in these villages with strata built according to household access to land with and without government SWC intervention. The same initial procedures were followed in Hafarat. A purged register of participants in the MFEP/KADA project in this village was added as a separate list containing all households with access to project SWC land (Van Dam & Houtkamp 1992).

Different procedures were followed in Telkook. More strict social and religious rules did not allow outsiders to enter any part of the village other than its public market. Since no map of the residential area could be produced, its sample frame has been based on masterlists of ration book holders. Ration books are issued by the government to all households for the purchasing of certain foods. Local shopkeepers keep a masterlist of names of their clientele. All these masterlists together provide a list of resident households in the village. For Telkook, it was decided to leave the stratification of households with and without access to project SWC land entirely to the distribution that would result from random sampling. These strata sizes finally appeared to take approximately equal shares in Hafarat (26 households in SWC project out of 61, or 43 %) and Telkook (24 out of 61, or 39 %) (table 2.3 and section 6.1).[10] The different groups with and without access to project SWC land proved relatively homogeneous per village on key-indicators of average landholding size and livestock wealth.[11] The sample size was set for all research villages at some 20 % in accordance with available research time. An exception was made for the more populous Telkook where the sample fraction is 7 %.

A first survey round was made between March-September 1990 in Um Safaree and Ilat Ayot, and between October-December 1990 in Hafarat. This survey was part of the WARK project (*cf.* section 4.2.3). The second survey round was part of the L&E programme and was carried out between October 1991-January 1992 in Telkook. In the latter period, the first-round group was also revisited for an update. The number of households present in the villages and available for interviews was, for various reasons, lower than in the first round, however (*cf.* table 2.3).

The surveys were made in collaboration with the staff of the Soil Conservation Department in Kassala, trained local and expatriate research assistants, M.A.-students of the University of Amsterdam, and Arabic and To Bedawee interpretors employed locally. All surveys were conducted in private. Various consistency checks on the data were made in the field. A village-specific calendar of events was used as an aid during the surveys for references in time. The collected data are considered reliable. This is supported by the observation that typically the most sensitive information that can be collected from former nomadic groups, namely on livestock (Dahl & Hjort 1976), shows congruent outcomes when households, villages, and trends over years are compared.

Definitions

The following definitions of research units, household livelihood categories, geographical areas, and units of measurement will be used in this study. A household and its members are defined as "[..] a person or a group of persons generally bound by ties of kinship who live together under a single roof or within a single compound and who share a community of life in that they are answerable to the same head and share a common source of food [..]" (Casley & Kumar 1988,60). In addition, we consider spatially separate dwellings of one or more wives of polygamous households as one single compound. Temporarily absent members like labour migrants are only considered as a household member when their livelihood complies with this definition for a period of at least one month of the year. A landholding is defined as a contiguous unit of crop production comprising all land used completely or partly for such purpose, without regard to legal ownership. The land should

[10] The only difference in sampling is that in Hafarat the universe was first sub-divided into two strata before random samples were drawn from each of them.

[11] The average livestock wealth is in Hafarat slightly higher among households with access to project SWC land than among those without. However, none of the differences in Telkook and Hafarat were found statistically significant in T-tests with $\alpha=0.05$.

be currently in use for crop production, or having once been in use, or be lying fallow with the intention of being used again at a later date (*Ibid.* 1988,61).

With respect to household livelihood categories, (i) crop production and the techniques applied in the Border Area are discussed in sections 3.4 and 3.5. We furthermore discuss the following other categories: (ii) livestock husbandry; (iii) labour migration; (iv) local off-farm employment; and (v) networking incomes. Livestock husbandry concerns the household herd only. It includes all income-generating activities related to the processing of its produce such as fat, hides, milk, and meat. Household livestock wealth is recorded in head of stock for the mid-season situation as per January. Labour migration is defined according to the length of stay at the place of destination of employment. We distinguish between daily, temporal, and long-term migration types. In the first situation, the migrant returns home daily and spends the night in the village. Temporary migration covers trips of a duration of more than 1 day up to 1 year. Finally, absence for a consecutive period of more than 1 year is called long-term labour migration. The destinations must be outside the village domain (see below), otherwise these activities are called local off-farm employment. Herdsmen on trek are considered as labour migrants if they work with herds other than their own, otherwise these activities are called livestock husbandry. Local off-farm employment is defined as all income-generating activities carried out inside the village domain. The latter reference is included to distinguish these activities from labour migration. Paid herdsmen or "contract-herdsmen" who operate inside the village domain are counted as local off-farm workers. These activities are called labour migration when they exceed these boundaries. Networking income is defined as all informal transfer in cash or kind between households and between households and village leaders.

With respect to geographical areas, a village domain is defined as all land where access for its use is controlled by the traditional leaders of the village. The cultivated lands are called local, when these are located inside this village domain. They are called non-local, when these are located outside this domain. The cultivated lands are usually located in geographically separate farming zones. A farming zone is defined as an area of contiguous cultivated lands. It is frequently under one single technique for cultivation, but this is no hard and fast rule. The households in the Border Area are usually entitled to different landholdings located in one or more local and non-local farming zones.

With respect to the units of measurement, the official Sudanese standards are adhered to as much as possible. A variety of local measures was used during the surveys in the field (*cf.* Local measures p.vi). The *feddan* (fd) is the standard unit of area (4,200 m²). The same name is also given to a unit of length of about 60-65 m. Other local measures of length are also used to demarcate land. These include the *gassaba* (1 elbow-length, approximately 0.5 m), and the *jedah* (Tb) (the distance covered by a normal throw with a stick, approximately 60-70 m). The Tropical Livestock Unit (TLU) of 250 kg live weight-equivalent equals for camels 1.0, cattle 0.7, goats and sheep 0.2. These are standards of the Sudanese Livestock and Meat Marketing Corporation (LMMC). The same values were also used in the National Livestock Survey of 1975-1976 and ERGO (1990). All prices are presented in current Sudanese pounds (£s) unless otherwise indicated. Inflation corrections are made to allow the comparison of different research years. The official consumer price-index figures for Khartoum "low-income groups" have been used for this purpose (MANR 1985, MFEP 1990). The 1988 consumer prices are deflated by 160 % to obtain real prices in constant 1983 Sudanese pounds.[12] One single deflator is used here for reasons of data availability. There are indications that the degree to which inflation is being translated into local prices is not uniform over villages, livelihood activities, and products in the rural Border Area. Village grain prices, for example, quickly adjust to mainly urban-based price developments. Price developments in other activities of low interference with the urban economy, like the local selling of animal produce and networking transfers, usually lag behind in this respect.

2.3.2 Modeling approach

The information on the household livelihoods in the Border Area was organized in a database built on the set of 18 variables which represent the Income, Land, Time, and Perception dimensions (*cf.* appendix 2.1). The research units are households. The scores of variables at the household level can be broken down to lower scale levels of (i) household members; and (ii) landholdings. This condition is essential to the answering of our research questions, and is maintained in all phases of data processing.[13] With respect to individual variables, we consider the scores to be built up of one or more "components". For example, the variable total household income may include one or more of the following components of crop production income, livestock husbandry income, labour migration income, local off-farm income, and networking income. We consider these components in turn to be built up of one or more "elements". For example, the first crop production component may include the elements of production frequency, number of units of produce, and unit price of produce. Under a set of strict rules to be discussed below, a missing value for an element in the component may be transformed into a valid score that can be used for further analysis. This modeling is only applied to variables of the Income and Time dimension. Variables in the Land and Perception dimensions have almost no scores missing.

---

[12] Early consumer price-index (cpi) figures are all Khartoum-based and derive from series which are regularly callibrated with data collected by means of household surveys. The last two surveys of 1978/80 and 1988/90 can not be directly linked because their base-years were changed. The 160 %-deflator was calculated for the July 1983 to June 1989 period with adjusted monthly data taken from the two unlinked cpi-series. Since 1989, cpi data is also being collected in the regional capitals, including Kassala, and shows a similar trend when compared with the Khartoum "low-income" groups data of the same years.

[13] Three different system files have been created for this purpose in the statistical package for the personal computer SPSS version 4.01. The cases respectively represent households, household members, and landholdings. The system files all share one common variable which points to the household to which individual landholdings and household members belong. Procedures for the aggregation of data over individual cases and joining different system files also make use of this common variable.

The dimension Income

Income variables are used to assess the share of indigenous SWC income and crop production income in the total household income. The total household income is calculated by the summation of income gained from all livelihood categories (income components) engaged in by all members in the household over a given period of time. This calculation of household income can also be expressed by equation 1: for households k; 1 to l, household members j; 1 to m, and income components i; 1 to n:

$$\sum_{i=1;\ j=1,\ m;\ k=1,\ l}^{n} \textit{Income component}_i \quad (1)$$

where the calculation of each of the individual income components i can be expressed by its elements in the following equation 2:

$$\sum_{i=1}^{n} \textit{Income component}_i = \textit{produce frequency over time}_i \times \textit{number of units}_i \times \textit{unit price}_i \quad (2)$$

There are two situations where L&E Survey data do provide information, but where this is not in a suitable format to calculate income components and household total incomes. The first situation arises when respondents do not remember the income component totals. This was anticipated in our research design. The individual elements in the component (use frequency, number of units, unit price) are normally remembered much better. The questionnaire list for this reason goes stepwise through these elements so that all can be recorded individually. Still, usually some data remain missing for one or more elements. For example, respondents frequently do not recall the precise unit price for a given year. The second situtation arises when the components contain elements which have not been expressed in monetary terms, but in kind. Certain fields in the household economies still remain largely outside the sphere of market relations. Accordingly, unit prices either do not exist, or can not be given with reasonable accuracy by the respondents. Information on the use and produce frequency, and the number of units involved in this, are commonly the only data available. For example, paid herdsmen are allowed the free consumption and use of milk from the herd they look after. This provides them an income which can not easily be priced. It is for this reason that also village normals of prices of certain goods, services, and produce have been collected in separate surveys. The monetary equivalents can be used in these situations instead.

The following rules must be adhered to before any missing values for elements in the components of income variables are transformed into valid scores. The scores remain coded as missing in all other situations. The gifts-in-kind include, among other gifts, the standard package of field allowance called *mazareef* (*cf.* 5.2.2).

1. *A maximum of one element is transformed per component and per household. Exceptions are gifts-in-kind, and milk consumption incomes for which composite indices are used.*

2. *The village normals and monetary equivalents are only used in data transformations for the villages where these data have been collected. Typical price differentials observed between the villages impede the use of regional average values for this purpose.*

3. *Kassala normal scores are used in two situations only. Firstly, when the prices have no relation to the villages, as holds for example for livestock sales by Border Area households at the Kassala market; for the prices of goods that are part of gifts-in-kind transactions and which goods are purchased in the town; and for free milk consumption by paid herdsmen. Secondly, when the village normal scores are strongly correlated with the Kassala normals, as holds for example for the prices of certain grains that can only be purchased in Kassala.*

The main procedure for the data transformations can briefly be described as:

If
the produce frequency over time, or the number of units, or the unit price is unknown

then
replace these respectively by the village/Kassala normal produce frequency over time, or the village/Kassala normal number of units, or the village/Kassala normal monetary equivalents.

The resulting procedures are in fact simple models of the equation type and have the following general form:

$$\boldsymbol{X'_i = aX_i} \quad \textbf{(3)}$$

where:

a = village/Kassala normal (v/K.n.) produce frequency, or v/K.n. number of units produced, or v/K.n. monetary equivalent respectively
X_i = component element Income
X'_i = transformed component element X_i

The use of data transformations of this type is not new. When applied elsewhere, the outcomes are usually called "ascribed", "imputed", "attributed", "inferred", or "transferred". Only rarely, however, is the list of underlying assumptions made explicit when the transformations are not being presented as simple models in the first place. Table 2.4 lists all the village and Kassala normals that will be used in this study. The greater part of these are unit prices of grain, stalks, and livestock. A use-frequency element is introduced in one case on the number of months that purchased fodder is provided to livestock in the village. Gifts-in-kind and milk consumption are calculated as composite indices. The partially modeled data can not provide more than an approximation of household incomes. Common weaknesses associated with the questionnaire survey method remain (*cf.* Gill 1993). Expectedly, not all information has been disclosed by the respondents, while other information is biased. Illegal trade is a regular practice in the region (Eisa 1978, Cole 1989). At the time of the research, the official sanctions were extremely severe including executed sentences of capital punishment. We expect that not all illicit activities have been reported to us during the field surveys.

The dimension Time

Labour-time variables are used in two ways. Firstly, to assess the importance of indigenous SWC techniques in terms of the share of labour time allocated to them when compared to allocations to other cultivation techniques. Secondly, to assess the importance of crop production in livelihood in terms of the share of labour time which remains available when the allocations to all other livelihood engagements of the household have been made. The first variable is called "SWC-time *S*" and the second "Potential crop production time *C*". The allocations of SWC-time *S* and Potential crop production time *C* can not be compared directly, because their units of measurement differ substantially. SWC-time *S* is expressed in man-hours and Potential crop production time *C* in months. Two different frameworks are therefore used to model these variables.

SWC-time S

Allocations of labour time to SWC are assessed for all members in the household. Only allocations to a selected set of land preparation and cultivation activities are considered for computation. These include general maintenance, gap-filling, weeding, and thinning. The reason is that we are only interested in the relative figures of labour-time allocations, and more in particular in their pattern over the research villages as influenced by other factors of distance and government intervention. The selected activities have been chosen on grounds of (i) being influenced by explicit land user decision-making; and (ii) being discriminative between the different cultivation techniques used locally. The collected data on SWC labour time require to some extent similar transformations as are used for the income variables. The respondents hardly ever recall the precise number of man-hours spent on a given landholding, in a given year. The L&E Survey data however do indicate, when we take the weeding activities as an example, the number of weeding rounds applied, and the amount of cultivated land involved, by landholding and year. The missing values on the actual amount of labour time allocated can therefore under certain rules be transformed into valid scores. The components in this situation are labour-time allocations over different cultivation techniques. The elements in the component are the frequency of application over time, cultivated area, and labour time per unit cultivated area.

The calculation of the total amount of labour time allocated to selected activities in SWC per household can be expressed by equation 4. For households *k*; 1 to *l*, landholdings *j*; 1 to *m* and selected land preparation and cultivation practices (maintenance, weeding, gap-filling and thinning) *i*; 1 to *n*:

$$\sum_{i=1;\ j=1,\ m;\ k=1,\ l}^{n} \textit{Labour-time allocation}_i \qquad \textbf{(4)}$$

where the calculation of each of the individual labour-time allocations *i* can be expressed by its elements as in equation 5:

$$\sum_{i=1}^{n} \textit{Labour-time allocation}_i = \textit{application frequency over time}_i \times \textit{area}_i \times \textit{labour-time per unit area}_i \qquad (5)$$

Sub-sets of landholdings are determined by means of standard selection operations in the computer programme SPSS. These sub-sets are used to assess labour-time allocations, per household, by technique for the purpose of comparing indigenous SWC with other cultivation techniques applied. There is one rule that must be adhered to before the missing values for elements in the components of the labour-time variables are transformed into valid scores. The score remains coded as missing in the situation where this rule can not be met.

A maximum of one element is transformed per component and per household.

Village normal scores of labour-time demand for selected land preparation and cultivation activities have been collected in separate surveys. However, these data proved incomplete for thinning and gap-filling. For these two activities, Border Area average figures must be substituted.[14] We use in both situations the labour demand per unit area for a second weeding round (table 2.5). The individual techniques presented in this table will be discussed in section 3.4. Gap-filling, in addition, is assumed to be carried out over 30 % of the cultivated area. The transformation can be represented as in equation 3:

$$X'_s = aX_s \qquad (6)$$

where:

- a = village/Border Area normal of labour-time demand per unit area for maintenance, weeding, gap-filling, and thinning respectively
- X_s = component element Time S
- X'_s = transformed component element X_s

---

[14] We have no reason to believe that these labour demands per unit area substantially differ between the Border Area villages. The greatest variation in labour time is related to what we called application frequency over time (Bokkers & Dabloub 1986, De Leeuw 1987). These data are not missing in the L&E Survey.

Table 2.4 **Values used in data transformations for variables of the dimension Income (US = Um Safaree, HA = Hafarat, IL = Ilat Ayot, TE = Telkook), (1983, 1988, 1989, 1990)**

1983 PRODUCE	UNIT	US	HA	IL	TE	KASSALA
Sorghum	£s/sack	56	45	53	71	
Sorghum	£s/ruba	5	5	8	9	
Millet	£s/sack					66
Straw	£s/bundle	1.03	1.03	1.02	1.04	
Fodder	no. of months	6	6	6	8	
Camel	£s/head					1,840
Cattle	£s/head					590
Sheep	£s/head					240
Goat	£s/head					120
Gift-in-kind	£s/month				31	
Milk cons.	£s/¼rottle/day					0.56

1988 PRODUCE	UNIT	US	HA	IL	TE	KASSALA
Sorghum	£s/sack	147	120	91	214	
Sorghum	£s/ruba	7	9	12	36	
Millet	£s/sack					226
Straw	£s/bundle	1.18	1.18	1.18	1.15	
Fodder	no. of months	4	4	2	6	
Camel	£s/head					3,460
Cattle	£s/head					2,370
Sheep	£s/head					440
Goat	£s/head					240
Gift-in-kind	£s/month					78
Milk cons.	£s/¼rottle/day					1.25

1989 PRODUCE	UNIT	US	HA	IL	TE	KASSALA
Sorghum	£s/sack	215	168	288	275	
Millet	£s/sack					315
Straw	£s/bundle	2.1	2.1	2.2	2.8	

1990 PRODUCE	UNIT	US	HA	IL	TE	KASSALA
Sorghum	£s/sack	906	834	1,200	1,000	
Millet	£s/sack					1,300
Straw	£s/bundle	5.0	5.0	4.0	3.0	

Source: L&E Survey. Note: 1989 and 1990 income calculations are less detailed and hence fewer transformations are required. Millet prices are based on adjusted sorghum prices according to Gedaref price ratios as: millet price=sorghum price + 19 % in 1983, and sorghum price + 65 % in 1988. These data are taken from MANR (1985), but actual Kassala millet prices are used for 1989 and 1990. Livestock prices are based on market prices in Kassala. Gifts-in-kind are based on a package of 1 rottle of coffee, sugar and oil, 1 ruba of sorghum and 7.5 rottle of free milk consumption per month. Milk consumption is based on a daily 0.25 rottle consumption and Kassala prices. Grain is more expensive when purchased by ruba than by sack.

Table 2.5 **Values used in data transformations for variables of the dimension Time (SWC-time S) for selected land preparation and cultivation practices in man-hour per feddan, Border Area averages**

TYPE OF ACTIVITY	VALUE (man-hour/fd)	UNIT AREA REFERENCE
MAINTENANCE		
– cleaning teras catchment	10	catchment size of teras
– earth bund repair	25	bunded area of teras
– repair, raising of brushwood panels	25	area under brushwood command
WEEDING		
– 1st-round	32	cultivated area
– 2nd-round	25	cultivated area
– 3rd-round	18	cultivated area
– 4th-round	11	cultivated area
THINNING [A]		
– plant thinning	25	cultivated area
GAP-FILLING [B]		
– gap-filling	7.5	cultivated area

Source: L&E Survey. Note: [A] Based on the labour-time demand of 2nd-round weeding. [B] Based on the labour-time demand of 2nd-round weeding and thinning carried out over 30 % of the cultivated area of the landholding.

Potential crop production time C

During initial field surveys in the Border Area, it was frequently observed that household members were absent for at least a part of the year, including the months of the growing season. A detailed assessment of labour-time allocations to crop production would for this reason not have been useful. The approach was changed to one through which we determine how much time eventually would remain for crop production, given the pattern of livelihood activities of the individual household members reported. Crop production time *C*, accordingly, is a measure of potential availability of household labour for cultivation. We know from the L&E Survey data, per household, for all its members whether or not they are participating in a certain activity by month. This means that the outcomes must be treated as indicative because, obviously, livelihoods are seldom organized according to calendar months. The framework for data processing for this particular application comes closest to that of a simple empirical model (*cf.* Huijsman 1986,84). The following steps are taken in the computer programme SPSS to calculate *C*:

1. *Create a matrix of household members (rows) by months of the research year (columns).*
2. *Assign an arbitrary value to all cells in the starting matrix (we use a score of 1).*
3. *Recode values in the matrix cells for relevant months according to the type of livelihood category household members are engaged in for that particular month.*

4. *Aggregate the individual household member scores by households and by months (and continue with the following computations).*

The initialization to 1 is arbitrary, but the value does relate to other transformations used. A score of 1 indicates after completion of the data-processing procedures a situation of full-time household presence in the village, where all its members are potentially available for local cultivation. A score of zero, alternatively, indicates a situation of full-time absence of the household in the village, where no single member is available for local cultivation. The other weights for the transformation of data (*viz.* for the recoding of score 1) are given in table 2.6. The same type of transformation as in equations 3 and 6 is again used. In this case, the transformations are executed per cell of the data matrix:

$$X'_c = aX_c \quad (7)$$

where:

a = coefficient of physical presence in the village
X_c = member/month matrix cell Time C
X'_c = transformed data of member/month cell X_c

The weights of coefficients a are also arbitrary. However, these do reflect to some extent the degree to which individual members of the household are physically present in the village area. For example, engagement in local livestock husbandry is recoded from 1 to 0.75. This activity is normally performed early in the morning and late in the afternoon. There potentially remains sufficient time for crop production, which in this example is considered to be some three-quarters of the working day. Since only livelihood activities which may compete with crop production are of interest in this study, the computations of C are only made for the months of the growing season June to and including October. A score of 1 remains default when no recoding is made in the matrix. The total household score of Potential crop production time C is divided by the number of economically active members in the household. The activities of women have been included in these where applicable. The value of this fraction, finally, is used as the research variable. This variable therefore indicates the household labour availability of economically active members during the months of the growing season. This measure provides no real value of potential labour availability, but is a tool to make comparisons between research villages.

The dimensions Land and Perception

Land and Perception variables indicate the importance of indigenous SWC and crop production in terms of cultivated land, and the priority given to different livelihood activities by the respondents. These data are directly available from the L&E Survey.

Table 2.7 finally provides an overview of the available data by research dimensions, years, and application of data transformations.

Table 2.6 **Values used in data transformations for variables of the dimension Time (potential crop production time C), weighing coefficients for different livelihood categories, Border Area averages, growing season months June-October**

TYPE OF LIVEHOOD ENGAGEMENT	COEFFICIENT
PROLONGED VILLAGE ABSENCE	
– long term camel herding	0.00
– long term cattle herding	0.00
– permanent labour migration	0.00
– temporary labour migration	0.00
PART-TIME VILLAGE ABSENCE	
– school-attendance	0.25
– daily labour migration	0.25
– local off-farm activities	0.50
– local camel herding	0.75
– local cattle herding	0.75
– local sheep herding	0.75
– local goat herding	0.75
NO LIVELIHOOD ENGAGEMENT	
– village presence assumed	1.00

Source: based on local village observations.

Table 2.7 **Research data availability by dimensions, years and data transformation occurrence, u means untransformed data, t means partially transformed data**

DIMENSION	1983-1984	1988-1989	1989-1990	1990-1991
Income	u,t	u,t		
Land	u	u	u	u
Time				
– SWC-time S	u,t	u,t	u,t	u,t
– Potential crop production time C	u,t	u,t		
Perception	u	u		

Source: L&E Survey.

2.3.3 Data analysis

The impact of these data transformations on the final outcomes can be evaluated in a sensitivity analysis. This tests the behaviour of outcomes under changing data-processing conditions (*cf.* De Graaf 1993,124). In this section, it will be used to determine the sensitivity of outcomes to changes in the value of elements in the transformed components. Usually, an increase of 10 % is introduced as an arbitrary change to evaluate the impact (*cf.* Smaling 1993,112). The analysis is only made for the income variables because the procedures affect the scores of labour-time variables all in exactly the same way. Table 2.8

first lists the elements in the data transformation and shows the relative weight data-processing procedures take in terms of the percentages of cases transformed (which equal the percentages of cases which initially had a missing value for one element in the income component). The table shows that the transformations affect only a relatively small group of elements. In two cases of gifts-in-kind and milk consumption, this is a group of elements contained in a composite index. Gifts-in-kind, alternatively, consist of either livestock, or grain, or packages of food allowances. The unit prices used to give these a monetary value were presented in table 2.4. The gift-in-kind package of *mazareef* is set to include 1 *rottle* of coffee, sugar and cooking-oil, 1 *ruba* of sorghum, and 7.5 *rottle* of free milk consumption per month. Milk consumption as part of this package and as separate income component of contract-herdsmen is based on the Border Area normal of a daily consumption rate of 0.25 *rottle*.

Table 2.8 furthermore shows that these operations most frequently transform the price of sorghum and stalks, and of goods in the gifts-in-kind index. The first two influence the outcomes at about 20-35 % of the landholdings (N=496) for the situation in 1983 and 1988 respectively. The transformations of the gifts-in-kind index influence the outcomes for about 17-21 % of the households (N=244) in these same years. The transformations of millet prices are important in 1988, affecting the outcomes at 16 % of the landholdings. Although these individual percentages of households and landholdings influenced are substantial, the overall impact on the calculated total household income can be considered negligible. Table 2.9 shows that if the value of all transformed elements is increased by 10 %, the resulting variation in the total household income is in all cases less than 2 %.

Table 2.8 **Transformed elements in variables of the dimension Income (in % of transformed cases), 1983 and 1988**

VARIABLE NAME	ELEMENTS			UNIT IN TRANSFORMATION	% CASES 1983			% CASES 1988		
	f	n	p		hh	hm	lh	hh	hm	lh
Gift-in-kind	x	x	x	composite index	16.8			21.3		
Milk cons.	x	x	x	composite index		2.4			0.3	
Sorghum yield			x	sack price			23.0			35.1
Millet yield			x	sack price			9.3			15.9
Stalk yield			x	bundle price			21.6			31.7

Note: f= Produce frequency element, n= number of units element, p= unit price element; hh= % cases over total number of households (N=244), hm= % cases over total number of household members (N=623), lh= % cases over total number of landholdings (N=496). The composition of the gift-in-kind and milk-consumption indices is given in table 2.4.

The same analysis is made for the variables used in the research: SWCCROP3, SWCCROP8, CROPLIV3, CROPLIV8 (*cf.* appendix 2.1). Table 2.10 shows that also in this situation, the effect of a 10 %-increase produces only modest changes in the range of – 4 % to + 1.5 % in the final outcomes of the variables. The greatest impact results from the use of price normals for stalks in the SWCCROP3 and SWCCROP8 variables.[15] We conclude that the proposed transformations are acceptable. The order of magnitude of change in the survey outcomes is importantly smaller than of the change introduced in the sensitivity analysis.

Table 2.9 **Sensitivity analysis. Variation in total household income as a result of a 10 %-increase in the value of transformed elements of variables in the dimension Income, 1983 and 1988**

VARIABLE NAME	UNIT IN TRANSFORMATION	% VARIATION 1983	% VARIATION 1988
Gift-in-kind	composite index	0.2	0.0
Milk consumption	composite index	0.2	0.0
Sorghum yield	sack price	0.9	1.3
Millet yield	sack price	0.2	0.5
Stalk yield	bundle price	1.3	0.0

Note: The composition of the gift-in-kind and milk-consumption indices is given in table 2.4.

Table 2.10 **Sensitivity analysis. Variation in the research variable % SWC income in total crop production income (SWCCROP3, SWCCROP8) and % crop production income in total household income (CROPLIV3, CROPLIV8) as a result of a 10 %-increase in the value of transformed elements of variables in the dimension Income, 1983 and 1988**

ELEMENT IN	% VARIATION		% VARIATION	
TRANSFORMATION	Swccrop3	Swccrop8	Cropliv3	Cropliv8
Gift-in-kind composite index	n.a.	n.a.	– 0.2	– 0.2
Milk consumption composite index	n.a.	n.a.	– 0.1	– 0.1
Sorghum price per sack	+ 1.1	+ 0.3	+ 1.3	+ 0.7
Millet price per sack	+ 0.5	+ 0.6	+ 0.4	+ 0.2
Stalk price per bundle	– 2.1	– 3.7	– 0.8	+ 0.9

Note: The composition of the gift-in-kind and milk-consumption indices is given in table 2.4, n.a. means not applicable.

The scores of the set of 18 variables (*cf.* appendix 2.1) can now be evaluated on the effects of the independent factors of distance and government SWC intervention. The four research villages and their households are arranged for this purpose in groups of two representing (i) areas at a short distance to the town of Kassala (Um Safaree and Hafarat) versus areas at a long distance to Kassala (Ilat Ayot and Telkook); and (ii) areas with government SWC intervention (Hafarat and Telkook) versus areas without such intervention (Ilat Ayot and Um Safaree). Two statistical tests will be used to analyse the pattern of household scores in these different village configurations. The first is the Mann-Whitney U-test which is used to compare the score distributions of households, household

[15] The negative sign moreover indicates that stalk production contributes less to the crop production income under SWC techniques than under other techniques. A higher bundle price results in smaller % SWC shares (and, alternatively, in higher % non-SWC shares which are not listed in the table).

members and landholdings in identical research years. The second is the Wilcoxon Matched-pairs Signed-ranks test which is used to compare the same score distributions, but this time of identical research units of households, household members and landholdings over different research years. These tests are among the more powerful for this purpose (Hammond & McCullagh 1982). Their use in SPSS is discussed in Norušis (1988,b176-b180).[16] We formulate the H_0 hypotheses according to our research questions described in section 2.1 as:

1. *Scores of variables have equal distributions for households who have access and households who do not have access to cultivated land under government SWC intervention.*

2. *Scores of variables have equal distributions for households in villages located at a greater and households in villages located at a smaller distance to the urban centre of Kassala town.*

The H_1 hypotheses are formulated as:

1. *Scores of variables have different distributions for households who have access and households who do not have access to cultivated land under government SWC intervention.*

2. *Scores of variables have different distributions for households in villages located at a greater and households in villages located at a smaller distance to the urban centre of Kassala town.*

These hypotheses will be tested with a 95 %-confidence interval (α=0.05). The outcomes will be discussed in chapter 7.

[16] The Mann-Whitney U-test considers the hypothesis that two independent samples come from populations with the same distribution. The Wilcoxon Matched-pairs Signed-ranks test considers the direction and magnitude of the relation of two related samples, and considers the hypothesis that these samples have the same distribution. The two tests are distribution free, require aselect sampling, and at least ordinal scale levels.

3

The Border Area and its regional setting

The Border Area is delimited by us for the purpose of this study as the region between the Kassala-Port Sudan tarmac road in the west, the Kassala-Red Sea provincial boundary line in the north, the Gash river in the south, and finally the international boundary line with Eritrea in the east (figure 3.1). It covers a geographical area between latitudes 15°-17° North and longitudes 36°-37° East. It measures a total area of some 8,600 km². The administrative units that have constituted the Border Area over various periods of different years are shown in figure 3.5. The Border Area can also be described as roughly that part of the Gash drainage basin in Sudan located east of this seasonal river. However, a geographical demarcation of the study region on either strict administrative or physiographic grounds was considered inappropriate. The first type of boundaries regularly change in Sudan. The second do not adhere to important socio-economic dimensions. The wider physical and socio-economic setting of this Border Area is discussed next with a topical focus on land and water resources.

3.1 Physical setting[1]

We refer in this section to the central Sudan, tentatively defined as located between lattitudes 12°-20° North as wider physical setting.

---

[1] The regional data used in this section were taken from the various contributors in Tothill (1952) and Craig (1991), and from Barbour (1961), Lebon (1965), TNO (1982), Blokhuis (1993).

Figure 3.1 **Border Area in eastern Sudan**

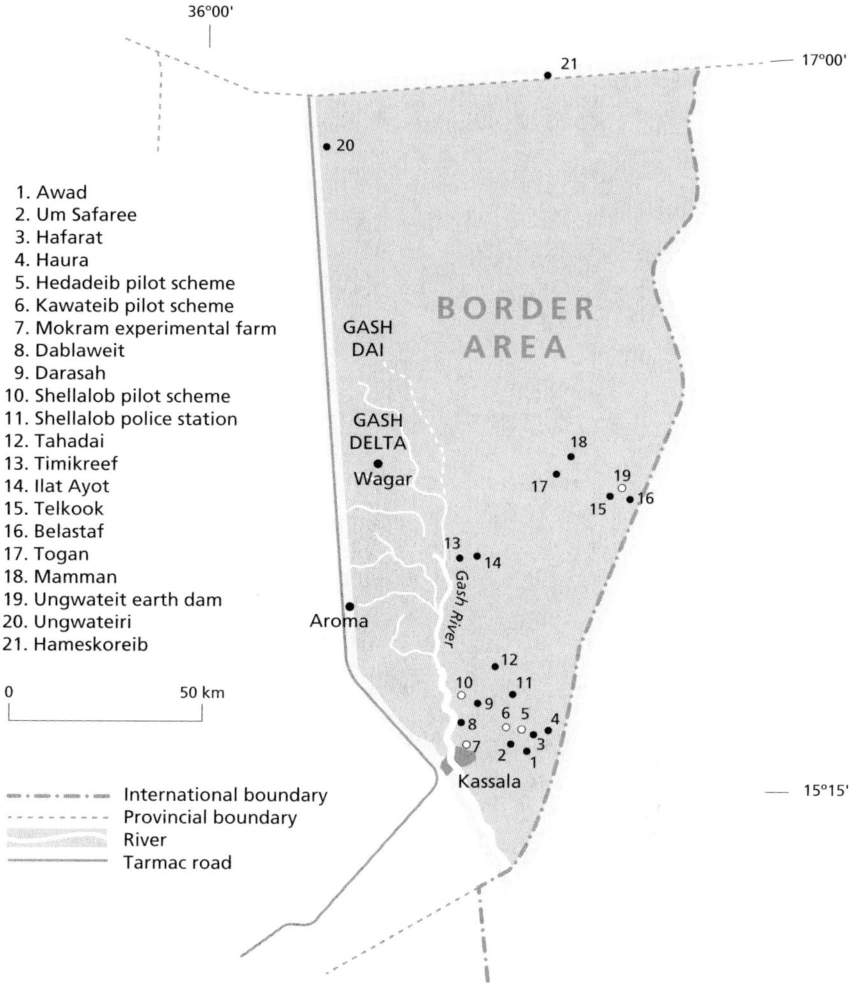

Climate

The climatic region of which the Border Area is part is the tropical semi-arid climate of low precipitation and high temperatures. This climatic region is characterized by Walsh (1991) as having an annual rainfall of 300-500 mm (Kassala 1901-1991 mean: 299 mm), a limited number of 0.5-2 wet months (Kassala 1951-1980 mean: 1.0) where a wet month is defined as receiving more than 100 mm rainfall per annum, and a high index of seasonality between 1.10-1.25 (Kassala 1946-1985 mean: 1.14) where a score of zero indicates regimes of equal rainfall in all months, and the maximum score 1.83 indicates the concentration of rainfall in one single month. The monthly average temperatures are generally over 18 °C (Kassala 1941-1970 mean: 29 °C). Another climatic region which equally exerts some influence in the eastern part of central Sudan and the northern Border Area is referred to by Walsh as the Red Sea Desert climate. The total annual rainfall in this climate is under 200 mm. The number of wet months is lower than in the tropical semi-arid climate (under 0.5), but the seasonality index reaches comparable levels (1.10-1.25). The most important trait of the Red Sea Desert climate is its rainfall maximum in the winter. Other classifications of climate refer to central Sudan as Semi-Desert zone (Lebon 1965) and as Hot Steppe BSh (400-800 mm annual rainfall, winter-dry and a mean annual temperature over 18 °C) and Hot Desert BWh (less than 400 mm annual rainfall, winter-dry and a mean annual temperature of over 18 °C) according to the Köppen-system (Barbour 1961). A list of selected climatic parameters for the Border Area is given in appendix 3.1.

The main driving force behind climate on the African continent is the Inter Tropical Convergence Zone (ITCZ) which movement follows the migration of the overhead sun. Air masses of different circulation systems and different moisture and temperature characteristics meet in this zone, and rise. They produce, in this process, heavy seasonal rainfall. The ITCZ location is near the equator in the winter, and it moves north during the summer months. It arrives at the southern and northern borders of the Sudan in late March and in August-September respectively. It heads south again on its retreat over the months that follow (El Tom 1975). This ITCZ movement produces distinct North-South gradients in features like rainfall and vegetation belts over much of the continent. Clear deviations in this pattern occur in the Horn of Africa. These disturbances are caused by the Ethiopian Highlands and all gradients take a more northeast-southwest heading in this region (Walsh 1991).

The semi-arid climate is not only characterized by a low moisture availability in absolute terms, but also by a high degree of variability in supply over months and years (Heathcote 1983). The monthly rainfall is mainly a function of the ITCZ movement. In central Sudan, this results in summer rainfall between April-October with August generally being the wettest month. In addition, the northeastern tradewinds generate winter rainfall in the coastal region of central Sudan between November and January. The greatest fall is received here in November or December. The monthly variabilities in rainfall are indicated by the seasonality index. The short-term variability in annual rainfall can be expressed by a coefficient of variation. This is defined as the standard deviation divided by the mean annual rainfall and reaches levels of 30-60 % in the central Sudan (Kassala 1951-1980 mean: 27 %). The long-term trend in rainfall in Sub-Saharan Africa was studied by Rasmusson (1987). He found series of relatively dry and wet years to alternate in distinct cycles. Long-term dry regimes of about 10 years and shorter 1-3 year episodes of intense drought have been discerned. Again, deviant patterns were found for the Horn of Africa. The shorter drought cycles in this region are longer with maxima up to 6 years. They also have a more patchy geographical occurrence. The possible causes of these processes were discussed by Farmer & Wigley (1985), Mattsson & Rapp (1991) and Hulme *et al.* (1992) to include land degradation as a result of human action. There is still no real proof for this as yet, however. General agreement among these scholars does exist on the occurrence of a downward trend in annual rainfall since the 1960s. This trend is superimposed on the inter-annual variabilities. This process of rainfall decline intensified in the 1980s. It is clearly shown in figure 3.2 which depicts standardized rainfall (Z scores) for stations in central Sudan between lattitudes 12°-16° North (Hulme 1991).

Figure 3.2 **Annual rainfall index for central Sudan (12° to 16° N), 1900-1990. The deviations are from the 1921-1980 mean and the smooth line is a 10-year filter. (Hulme 1991,73; reproduced with permission of the author)**

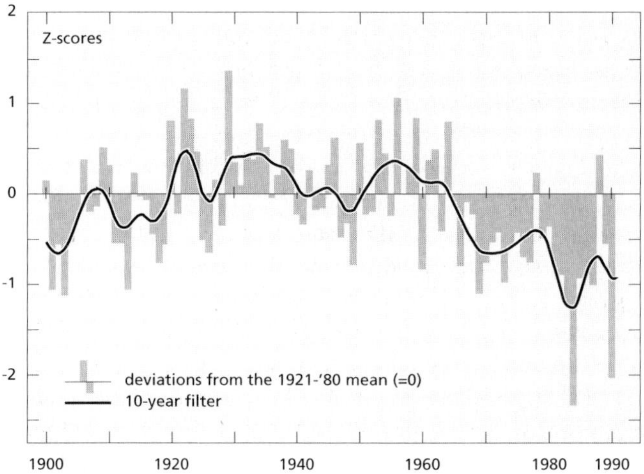

Three distinct epochs can be identified in this figure. These were described by Walsh 1991,39) as (i) a relatively dry first two decades of the 20th century; (ii) a wet epoch from the 1920s to the 1960s; and (iii) a very dry period from 1965 to date. When these are compared with rainfall data collected at the Kassala meteorological station, it is shown that the 1910s and 1920s on the contrary were relatively wet in the Border Area, as was the period up to the 1960s. Only the very dry period which started in 1965 is also shown in the Kassala data. Meteorological drought years based on Kassala data and defined as years with rainfall 1 standard deviation below the mean, will be discussed in more detail in section 4.3.3. When dry periods are considered over an even longer time-span starting at 4,000 B.P., it appears that the 1920s-1960s wet period is anomalous and that the dry periods shown in figure 3.2 represent the typical situation for the Sudan over the past thousand years (Walsh *et al.* 1988, Walsh 1991).

In the course of the dry 1965-to-present period, the position of the 200 mm and 300 mm isohyets which bound critical areas for SWC use (*cf.* section 1.4.3) have retreated southward over a distance of some 120 km. Also certain characteristics of the wet season which are critical to crop production have changed unfavourably. The pattern for the central Sudan indicates (i) a shortening of the length of the rainy season; (ii) an increase in number of years for which no rainy season has materialized at all; and (iii) a lower reliability of rainfall as a result of an increasing number of rainfall interrupts in the mid and late growing season periods (Hulme 1989, Walsh 1991).

Geology and physiography

The Sudan forms an integral part of the African Shield of the Precambrian formation of the Basement Complex (Whiteman 1971). This formation was elevated by tectonism, and folded in various later periods producing the mountain ranges characteristic of the eastern central Sudan. Sedimentation phases dominated in the late Mesozoic. These produced the Nubian sandstone formation in the northwestern central Sudan. Phases of alternating continental deposition and renewed tectonism have occurred since. These formed the Red Sea hills in the late Tertiary Pliocene and Miocene. The Clays of the Plain formation, also called Central Clay Plain, was formed by the Blue Nile depositions in the Quaternary. The material to build this plain was mainly removed from the Ethiopian Highlands. The recent geological history is dominated by processes of alluvial deposition. The geological formations in the Border Area include the Basement Complex, the Clays of the Plain, and recent sediments. The first mainly consists of granitic and hornblendic gneisses. The formation regularly outcrops in the form of *inselbergs*. More generally, it is covered by its recent weathering products and deposits of local alluvial and colluvial origin. The latter produce a distinct feature in the Border Area called "outwash aureole" at the lower pediment slopes surrounding these outcrops. The Clays of the Plain formation overlies the Basement Complex. It is partially covered by the same alluvial and colluvial depositions. The recent alluvial sediments of the seasonal river Gash, in turn, overlie the Clays of the Plain and Basement Complex formations. Sand and dust-laden storms, or *haboob*, are a common phenomenon and form the most recent deposits. These are of particular importance in the northern part of the Border Area (Saeed 1972).

The general elevation of the central Sudan is between 400-1,200 m above mean sea level (msl). The Jebel Marra Plateau in the west rises to some 3,000 m, the Red Sea Hills in the northeast reach some 2,000 m. The foothills of the Ethiopian Highlands in the southeast along the international border generally reach some 1,500 m above msl (figure 3.3). Between these mountain ranges, vast stretches of almost level pediplain dominate the landscape. These are dissected by the Nile and its tributaries, and by seasonal water courses called *wadis* or *khors* (Tb).[2] The pediplain can be subdivided into "degradational" and "aggradational" units, according to the local prevalence of processes of either erosion or sedimentation (Blokhuis 1993). The Border Area is a pediplain gently sloping to the northwest at an average 3 % gradient. It would be called aggradational in the terms used by Blokhuis. The plain has a "terraced-shape" appearance with gradients between subsequent level sections regularly steepening in the direction of the slope. Were a first transect to be made in this direction, the following main land units would be encountered (i) colluvial slope; (ii) channel slope; and (iii) alluvial flat. This sequence repeats again starting with the colluvial slope (figure 3.4).

[2] *Wadi* and *khor* (Tb) are used interchangeably in Sudan to refer to the dry riverbed of seasonal water courses.

Figure 3.3 **Relief, eastern Sudan**

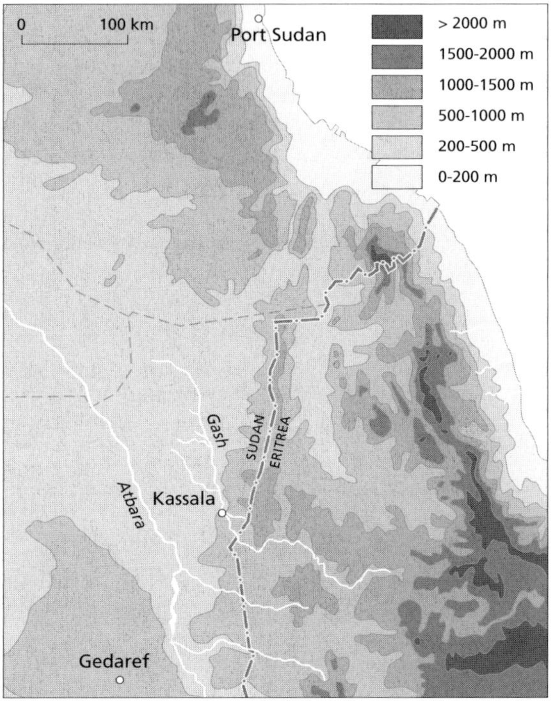

The Border Area plain is cut by some 30 larger khors discharging into the Gash river and its inland delta. These are continuous khors and have almost their entire catchment in the foothills of the Ethiopian Highlands. The smaller and frequently discontinuous ones either spring from the *inselberg* outcrops, or drain the gently sloping plains. In these latter systems, floodwater repeatedly spreads on alluvial flats, and concentrates again on channel slopes. Were a second transect to be made in the direction perpendicular to the slope, the main land units encountered would be (i) interfluve; and (ii) valley-bottom. This sequence repeats again with the interfluve (figure 3.4). These land units and drainage patterns are typical of the Border Area, and of great importance to the SWC techniques locally practised.

Soils and vegetation

The soils of the central Sudan reflect the North-South gradient in climate and associated rainfall and vegetation characteristics. According to the FAO-UNESCO 1:5,000,000 Soil Map of the World (*cf.* Mitchell 1991,13), the main zonal soils are classified as Cambisols, Luvisols, Vertisols, Arenosols, Xerosols, and Yermosols. Local factors also influence soil formation. The azonal types are classified as Lithosols, Fluvisols, and Regosols. Mainly

Xerosols and Regosols can be found in the Border Area. According to the 1988 FAO revised legend, Niemeijer (1993) classified these as Eutric Fluvisols, Eutric Regosols, Vertic Cambisols, and as Vertic, Haplic, and Calcic Luvisols. There were no clear vertisols found, but typical vertic properties and sinkholes have been noticed in the eastern part of the Border Area. These soils were called "proto-vertisols" by Blokhuis (1993). The Border Area soils are generally shallow, suffer from surface-crusting, have a low chemical fertility and organic matter content, and have a restricted depth for plant rooting (Van der Kevie & Buraymah 1976).

Figure 3.4 **Drainage pattern and land units in the Border Area (schematic N.-S. and E.-W. transects)**

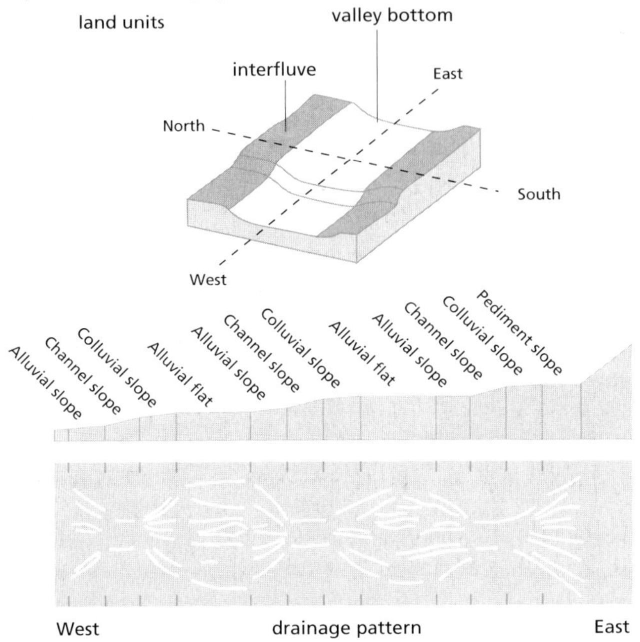

The authorative 1:4,000,000 vegetation map of the Sudan was compiled by Harrison and Jackson in 1958. The vegetation types shown are mainly a function of the distribution of rainfall and soil types. The main divisions relevant to the central Sudan include (i) Semi-desert; and (ii) Woodland Savannah. The relevant sub-divisions for Semi-desert include (i)(a) *Acacia tortilis-Maerua crassifolia* desert scrub; (i)(b) Semi-desert grassland on clay; (i)(c) Semi-desert grassland on sand; (i)(d) *Acacia mellifera-Commiphora* desert scrub. The subdivisions for Woodland Savannah include (ii)(a) *Acacia mellifera* thornland alternating with grass; (ii)(b) *Acacia seyal-Balanites* savannah alternating with grass; (ii)(c) *Anogeissus-Combretum hartmannianum* savannah woodland; and (ii)(d) *Acacia senegal* savannah (Blokhuis 1993,31). Units (i)(a), (ii)(a) and (ii)(b) occur in the Border Area.

Surface and groundwater

The surface waters in central Sudan include, besides a number of larger seasonal wadis and khors, the main Nile, White Nile, Blue Nile, and its tributaries of semi-perennial rivers Dinder, Rahad, and Atbara. The water courses are larger and more densely spaced on the clay plains than elsewhere because less water is lost due to seepage. Several water courses in the dryer central Sudan never reach permanent and perennial rivers. They fade out into inland deltas. The river Gash, for example, springs south in the Ethiopian Highlands of Eritrea. It continues its course in the Sudan for some 100 km, before fanning out in the Gash delta and its spillwater area called Gash Dai. The Gash river was an upper tributary of the main Nile until 2,000 B.C. The river carries an average annual discharge of 950 million m^3 (TNO 1982). This is generated by rainfall in the elevated catchment where the Eritrean Decamere station (2,200 m above msl) receives 635 mm per year, and Adi Ugri (1,920 m above msl) 571 mm per year (Ibrahim 1980). The Gash locally provides favourable irrigation opportunities. These are seized in irrigated horticulture around Kassala town, and in a parastatal irrigation scheme in the Gash delta proper. A similar seasonal river is the Baraka which enters the Sudan to fan out near the Red Sea coast around Tokar. Its level of discharge is not known. The Baraka annually floods an average 32,000 ha and provides irrigation opportunities in the parastatal irrigation scheme Tokar delta (Cole 1989).

Two main groundwater systems can be distinguished for the central Sudan. The first consists of extensive and deep aquifers associated with the formations of Nubian Sandstone and Umm Ruwaba Series. The second consists of more dispersed and shallow aquifers. These are formed in areas where sands overlie impervious clays and rock formations, in deposits of water courses, and in outwash aureoles (Walsh 1991). Storage capacity and water yields are generally less favourable in the dispersed aquifers which can be found in the Border Area (TNO & NCDRWR 1990).

3.2 Socio-economic setting[3]

We refer in this section to the 1981-1991 administrative unit then called "Eastern Region" as a wider socio-economic setting for the Border Area. The government of the Anglo-Egyptian Condominium (1899-1955) is referred to as Anglo-Egyptian Administration. The administration of the Independent Sudan after 1956 as the Government of The Sudan, or shortly GOS. The administrative divisions have always been regional, provincial, by districts, and by councils. The names and geographical extent of these areas regularly

[3] The regional data used in this section were taken from MFEP & UNDP (1990, 1991). These sources also present National Census data of 1955, 1973 and 1983. Other data were taken from DHV & IES (1988), IES (1991) and Cole (1989).

changed[4] (figure 3.5). These areas, moreover, only partially corresponded with the census counting divisions. By consequence, no continuous series of 1955, 1973, and 1983 census data is available at scales lower than the common geographical region. We refer to this common region as eastern Sudan in the next sections.

Figure 3.5 **Administrative divisions, eastern Sudan 1899-1951, 1951-1971, 1971-1981, 1981-1991, 1991-present**

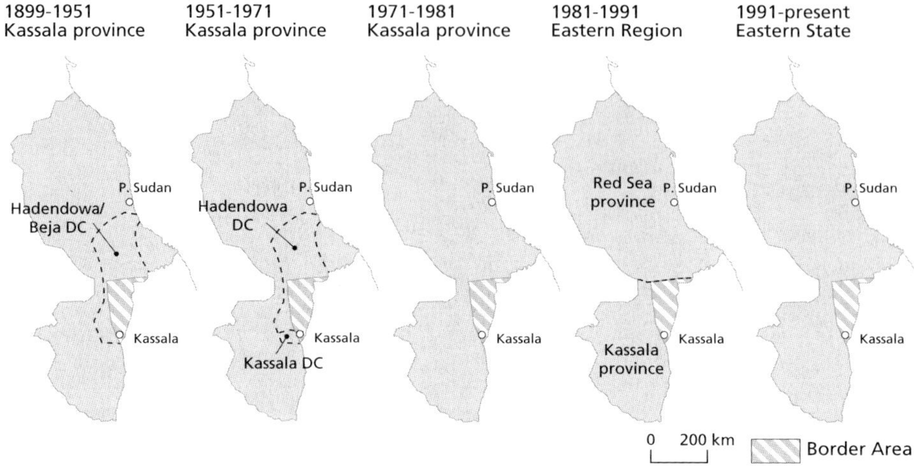

People and settlements

The population in eastern Sudan considered as "sedentary" increased from 149,218 out of 970,449 (15 %) in 1955, to 984,217 out of 2,923,195 (34 %) projected for the situation in 1990. The population considered as "rural settled" increased from 33 % in 1955, to a projected 44 % for 1990. Finally, the population considered as "rural nomadic" declined over this same period from 52 % to 22 % (table 3.1 *cf.* eastern Sudan).[5] The most recent

[4] The Border Area has been part of the following administrative divisions (per period from higher to lower scales). 1899-1951: Beja District (a). After the introduction of District Councils (DC) in 1943, the Border Area is part of Hadendowa DC also called Beja DC (b). 1951-1971: Kassala Province (a), Hadendowa DC, and Kassala Rural DC (b). 1971-1981: Kassala Province (a), Kassala Rural Council (b). 1981-1991: Eastern Region (a), Kassala Province (Area Council) (b), Kassala Rural Council and Border Rural Council (c). 1991-present: Eastern State (a), Kassala Province (b), Kassala Rural Council, Aroma Rural Council, and Border Rural Council (c).

[5] The common reservations with respect to the definition of nomadic people (*cf.* Dietz 1987) also hold here. In the 1973 census, this definition is (quoted by Salih 1982,23): "[..] a group of households following a sheikh without settled agriculture or defined occupation other than cattle, sheep, goat, or camel husbandry [..]". The definition in the 1983 census is "[..] tribes who tend to live in mobile houses (tents, huts). They raise livestock and are continuously moving in search of water and pasture and do not stay in specific areas for long time [..]" (MFNE 1987,12).

1983 population data for the Border Area are given in table 3.2. The population considered as "settled" and "nomadic" in the Kassala Rural Council is 58,625 out of 71,592 (82 %), and 12,967 out of 71,592 (18 %) respectively. The same figures for the Kassala Border Council are respectively 55 % and 45 %. The population of Kassala town is considered as entirely sedentary.

Table 3.1 **Population numbers in eastern Sudan, (selected years)**

	1955/56	1973	1983	1990
KASSALA PROVINCE		1,059,891	1,510,252	1,998,277
– Urban		231,178	394,136	608,228
– Rural Settled		579,695	937,935	1,181,315
– Rural Nomadic		249,018	178,181	208,734
RED SEA PROVINCE		437,490	698,127	914,918
– Urban		157,673	243,644	375,988
– Rural Settled		113,579	74,007	93,211
– Rural Nomadic		166,238	380,475	445,719
EASTERN SUDAN [A]	970,449	1,479,381	2,208,379	2,923,195
– Urban	149,218	388,851	637,780	984,217
– Rural Settled	320,369	693,274	1,011,942	1,274,525
– Rural Nomadic	500,862	415,256	558,656	654,453

Source: National Census 1955/56, 1973, 1983 and projections for 1990 (MFEP & UNDP 1990). Note: [A] The region is named Kassala Province in 1955/56 and 1973, Eastern Region in 1983 and Eastern State 1991.

Table 3.2 **Population numbers in the Border Area, 1983**

	SETTLED		NOMADIC		TOTAL	
	No. of households	Total population	No. of hh	Total pop.	No. of hh	Total pop.
Kassala Town	21,477	142,909	n.a.	n.a.	21,477	142,909
Kassala Rural Council	12,315	58,625	2,924	12,967	15,239	71,592
Kassala Border Council	15,398	113,758	11,551	92,355	26,949	206,113

Source: Third Population Census 1983, Kassala Province by Area Council and People's Councils. Note: Administrative divisions listed are the best-fit equivalents for the Border Area as defined in this study, n.a. means not applicable.

The most important ethnic and tribal groups in eastern Sudan are the Beja and Arabs. The ancestors of the Beja were called "Blemmyes" or "Blemminges". They have occupied this territory for at least the last 5,000 years. The tribes in the Beja group include the Amarar, Besharin, Hadendowa, Kammalab, Morghumab, Sigolab, Melhitkinab, Beni Amer, Arteiga,

Shaiab, Ashraf, Kimmeilab, Hassanab, Halanga, and the Haram (Paul 1971,137-139). The Beja have their own unwritten language called *To Bedawe* (Tb). Several of the sections of Beni Amer also speak Tigray. Recently, other lists of Beja groups based on their political history and linguistic particularities have also been proposed (*cf.* Morton 1989). The Arabs originate from northern and southern Arabia and Yemen, and reached Sudan in the first centuries A.D. The main ethnic groups are arabized Nubians of the Barabra and Ja'ali and the (semi-)nomadic Juhayna (Holt & Daly 1988,3). Other smaller ethnic and tribal groups in eastern Sudan include the Shukriya, Lahawin, Rashyda, Ethiopians, and mainly urban migrant settlers of West-African Haussa, Fula and Bornu and West-Sudanese Nuba. Since the 1960s, the urban settlers also include war and drought refugees from Eritrea and Tigray (Kuhlman 1994).

The National Census of 1983 lists a total of 20 main urban areas in eastern Sudan. Of these, 13 are located in the then Kassala province (Kassala, Aroma, Wagar, New Halfa, El Masna, Kashm El Girba, El Fau, Gedaref, Gala El Nahal, Showak, El Hawata, El Mefaza and Doka). Kassala is the regional administrative capital and most important town. Its population in 1990 was an estimated 217,000. Besides Kassala, there are no main urban areas in the Border Area proper. During fieldwork, we counted a total of some 110 settlements and villages. A small number of these are shown in figure 3.1. The villages are of relatively recent origin. Most were established after 1950 (*cf.* section 4.3.3). There are 7 urban areas listed for the then Red Sea province (Port Sudan, Suakin, Sinkat, Gabit, Tokar, Haya, Gabit El Maadin). The harbour town of Port Sudan is the most important with an estimated population of 308,000 in 1990 (MFEP & UNDP 1990), (figure 3.6).

Organization and gender

The Border Area is the tribal domain of the Hadendowa and Beni Amer. The tribes have been described by Seligman & Seligman (1930), Owen (1937), Nadel (1945), and Paul (1971). The Hadendowa are numerically and politically the most important. The tribe, or *gabela* (Tb), can be subdivided into smaller sections and lineages. At the lowest scale, the minimal lineage called *deewa* (Tb) is more or less comparable with the extended family. Ausenda (1987,66,244) described it as the agnatic residential unit of socio-economic cooperation for herding and farming. The Beni Amer are sub-divided into sections, or *badana*. Traditionally, a differentiation exists between aristocratic and serf sections, although this was formally abolished in 1948. The Beni Amer of the Border Area are of the former serf status. The Beni Amer tribal domain, unlike that of the Hadendowa, extends over Sudan and Eritrea. Approximately one third is located in Sudan in two regions around Tokar and in the Border Area. Beja indigenous chieftainship is vested in the functions of *omda* and *sheikh*. The first oversees tribal sections and the second different lineages contained in these sections. However, the regional variations are great in this respect. The executive power in the village is held by a group of sheikhs and other dignitaries. They have seats in the Village Committee. Leadership is normally inherited, but can also be achieved by election when the natural heir is not acceptable to the community. Birth, particular virtues, mastery of skills, or the occupation of an esteemed profession then favours one's chance of election (Morton & Fré 1986).

Figure 3.6 **Urban areas and infrastructure, eastern Sudan**

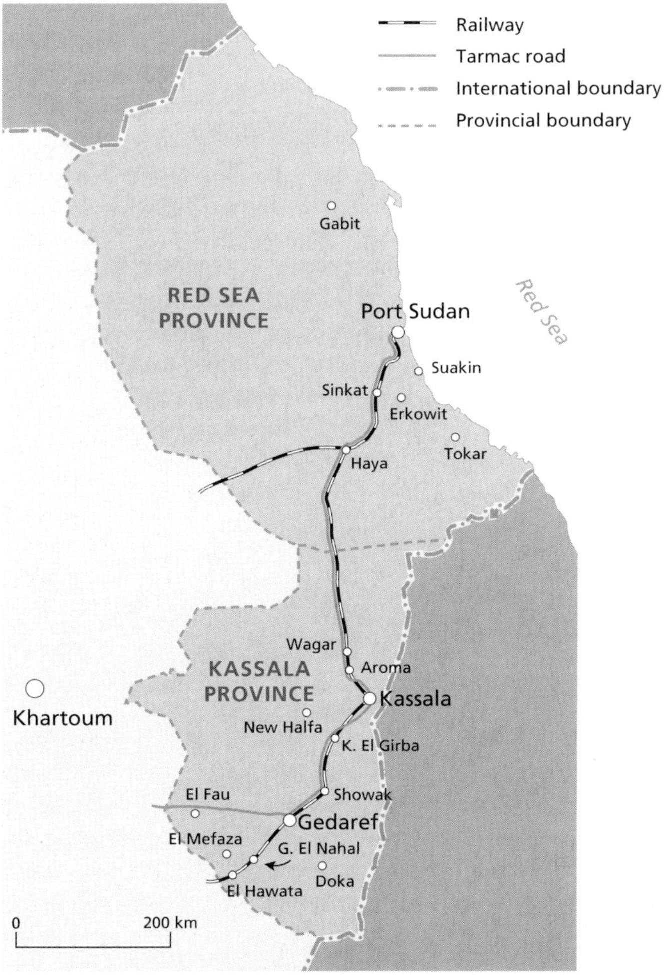

With the exception of mainly Christian southern Sudanese and Eritreans who predominantly live in Kassala town, all ethnic and tribal groups are Muslim. Different Muslim sects are adhered to. Several have established religious centres in the rural areas. The most influential is the *sufi* movement of Ali Betai. It originated in 1951 among the Hadendowa of the present-day village of Hameskoreib. The *sufi* doctrine is highly regional in outlook. It is argued that for this reason it particularly appeals to peripheral groups, like the Border Area Beja. The doctrine emphasises a local ancestor cult, its various religious directives can be afforded even by the poorest, and its style of preaching is simple and comprehensible (Eickelman 1981,225-228). The movements[6] usually have a political connotation and maintain good contacts with religious orders in the Arab Gulf countries. Some sects, including the movement of Ali Betai, receive substantial funding from governments in the Gulf region.

Beja custom is characterized by a particularly strong gender segregation. Women are supposed to confine themselves to the domestic domain. They work in household activities and small-scale indoor handicrafts. Appearance in public is a cultural taboo. Some groups in the Red Sea province however allow their women to provide assistance in weeding, sowing, harvesting, livestock raising, and in the collection of water, fodder, and fuelwood (Cole 1989). Although the orthodox outlook prevails in the Border Area, the rules are less strict among the Beni Amer than among the Hadendowa. Household labour availability in eastern Sudan, by consequence, is generally lower than in other SSA countries where women normally perform the bulk of agricultural and other labour.

Table 3.3 **Gross domestic product of eastern Sudan (in million £s), 1985/1986**

	EASTERN SUDAN	%
PRIMARY SECTOR		
– Agriculture incl. forestry	797.3	30
– Total	797.3	30
SECONDARY SECTOR		
– Mining and quarrying	1.6	0
– Processing, manufacture, handicraft	267.9	10
– Electricity and water	27.9	1
– Building and construction	117.5	5
– Total	414.9	16
TERTIARY SECTOR		
– Commerce and services	598.7	23
– Transport and communication	444.4	17
– Finance and insurance	281.9	11
– Duties	78.3	3
– Total	1403.8	54
GRAND TOTAL	2615.5	100

Source: MFEP & UNDP 1990 (adjusted). Note: In 1985/86, eastern Sudan was administratively called Eastern Region.

Economy

The economy of eastern Sudan is mainly based on tertiary activities, while the primary sector comes next in importance (table 3.3). Agriculture is the main activity in the primary sector and consists of crop production, livestock husbandry, forestry, and some fishing.

[6] Other movements in the Border Area incude *i.a.* the Khatmia, Gadria, Tijania, Ansar Mahdi, Shazhalia, Ansar Sunna (*cf.* Ausenda 1987).

Table 3.4 **Land use in the eastern Sudan by province (in 1,000 fd), 1990**

	KASSALA PROVINCE	%	RED SEA PROVINCE	%
Traditional rainfed cultivation	2,300	7	59	1
Mechanized rainfed cultivation	5,700	19	n.a.	
Irrigated cultivation	1,000	3	20	0
Horticulture	55	1	7	0
Forestry	1,900	6	2,700	5
Range and pasture	19,000	63	24,000	47
Mountain and desert	81	1	24,000	47
Total	30,036	100	50,786	100

Source: MFEP & UNDP (1991,18). Note: In 1990, eastern Sudan was administratively called Eastern Region. Traditional rainfed cultivation includes i.a. indigenous SWC.

Figure 3.7 **Irrigation schemes and main water courses, eastern Sudan**

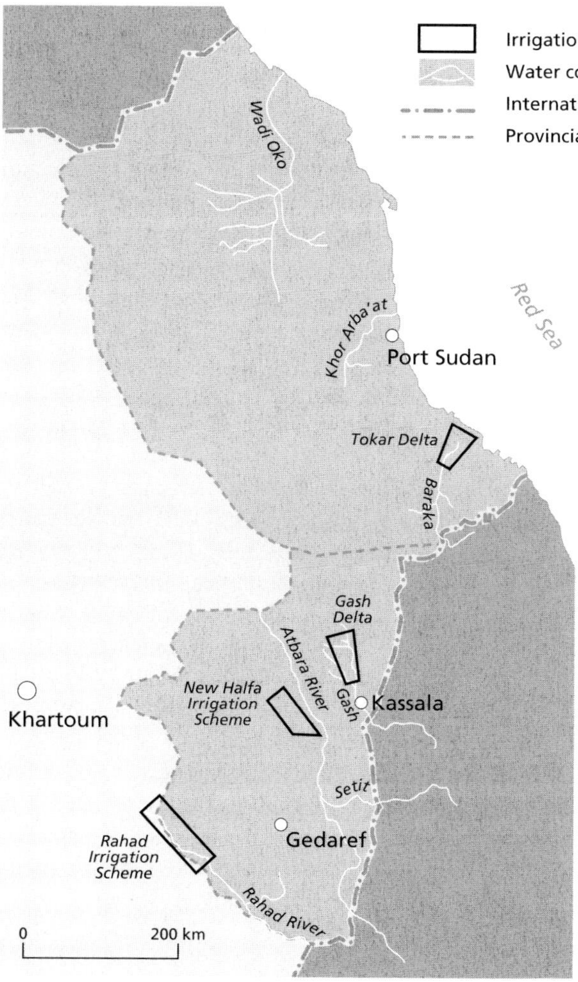

Crop production is based on gravity and pump irrigation. The gravity irrigation schemes in eastern Sudan include (figure 3.7) (i) Rahad (120,000 fd, or 50,200 ha). This scheme mainly produces cash crops of cotton and groundnuts, and some subsistence crops of sorghum, fodder and vegetables; (ii) New Halfa also called Kashm El Girba (330,000 fd, or 138,000 ha). This scheme mainly produces cash crops of cotton, groundnuts and sugarcane, and some subsistence sorghum and wheat; (iii) Gash Delta (300,000 fd, or 126,000 ha although the factually commanded area in the late 1980s reached only 80,000 fd, or 35,000 ha). This scheme mainly produces cash crops of sorghum and watermelon, and some subsistence sorghum; and (iv) Tokar Delta (200,000 fd, or 84,000 ha although the factually flooded area in the late 1980s was only 77,000 fd, or 32,000 ha). This scheme mainly produces cash crops of cotton and sorghum, and subsistence millet. Two more 500,000 fd irrigation schemes are projected by the Government of The Sudan along the Atbara river (Rumela scheme), and Setit river (Burdana scheme), (GOS 1982). These have not been implemented as yet. Pump-irrigation schemes are located on the margins of the rivers Gash, Atbara, Rahad, and along the main seasonal water courses of Wadi Oko and Khor Arba'at in the Red Sea province near Port Sudan. Mainly cash crops of fruits, vegetables, fodder and maize are cultivated here (DHV & IES 1989). Crop production in the region is also based on large-scale mechanized rainfed farming. The main production area is in the 600-1,000 mm rainfall area around Gedaref (*cf.* figure 3.6). The cash crops are predominantly sorghum and sesame grown on individual schemes of 1,000-1,500 fd (420-630 ha) in size. Irrigation and mechanized farming are modern crop production activities. These are predominantly state-controlled and heavily subsidized (Bascom 1990b). Finally, crop production is also based on traditional rainfed subsistence cultivation. This uses systems of perennial cultivation without resting periods, or *bilad*, cultivation in slash-and-burn systems, or *hariq*, flood-recession, or *gerf*, valley-bottom farming, or *seel* (GOS 1986, DHV & IES 1989), and various SWC techniques. The traditional activities receive little or no government support.

Livestock production in eastern Sudan depends on (semi-)nomadic and settled transhumant pastoralism. Livestock includes camels, cattle, sheep and goats which are raised with little or no capital investment. The herd movements are over horizontal distances mainly in north-south directions. These follow the seasonal advance of rainfall. The movements over vertical distances go from the Border Area into the Ethiopian Highlands. For the coastal region, these go into the Red Sea hills (figure 3.8). Herd movements are influenced more in detail by the composition of species, characteristics of location, and seasonal climate (GOS 1986). According to a study of MFEP & UNDP (1991), range production of dry matter varies from some 500 metric tons deficit for the dry season situation in the Kassala province, to a 1,800 metric tons surplus for the dry season situation in the Red Sea province. There is no important livestock migration between the two provinces however (*cf.* figure 3.8).

Forestry in eastern Sudan is important for its production of fuelwood, charcoal, and Arabic gum of the *Acacia senegal* tree. A total of 64 gazetted Forest Reserves have been demarcated with a total area of 209,000 ha in 1990 (Vink 1991). The trend in domestic fuelwood demand will already in the late 1990s result in absolute shortages in the region (DHV & IES 1989).

Figure 3.8 **Livestock movements, eastern Sudan (based on National Livestock Census 1975-1976)**

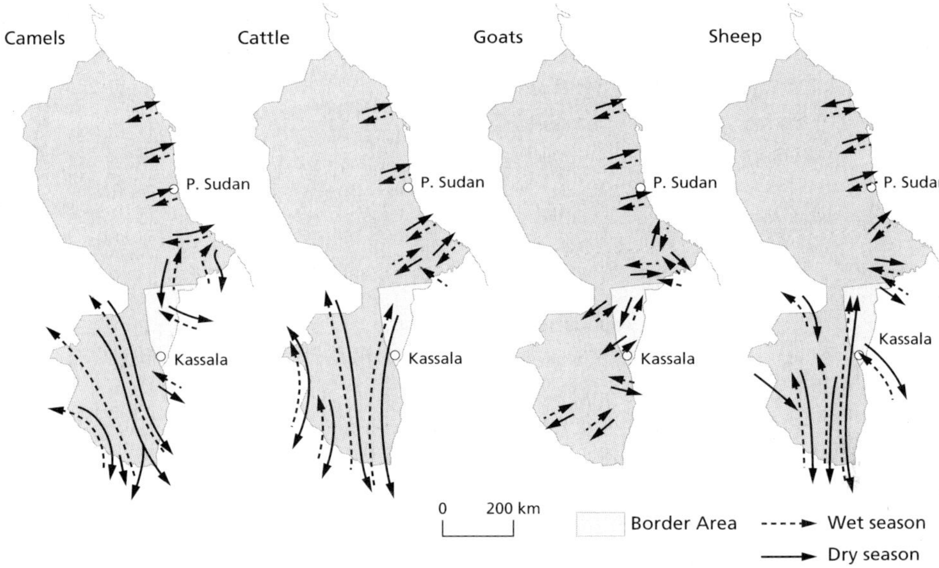

Subsistence fishing in eastern Sudan is done in the Red Sea, in the Atbara and Rahad rivers, and in lake Girba. The trade in fresh and dried fish is declining as a result of overfishing and modern irrigation development.

Eastern Sudan's secondary sector is importantly based on harbour activities and the processing of agricultural produce. The activities are usually located in or near main urban areas. In the area of Port Sudan, chemical and oil-refinery industries can be found. There is also vehicle-assemblying, spinning, softdrink-bottling, and salt production. Industries in Kassala town include fruit-canning and the production of juices and jams. In New Halfa, these include the refining of sugar, the ginning of cotton, and the production of sesame seed-cakes. These activities are associated with the New Halfa irrigation scheme. In El Fau, cotton-ginnery is associated with the Rahad irrigation scheme. In Tokar, there are brick work industries. Rural industries are situated around Kassala and Port Sudan. These are mainly based on granite mining and quarrying of limestone and clay.

The tertiary sector in eastern Sudan mainly comprises commerce, services, transport and communication. Commerce centres in Port Sudan and services in the regional capital Kassala. Sudan Railway and Sudan Airways are in a state of near collapse. The bulk of private and commercial goods is therefore transported by road. The tarmac network consists of one single 1,200 km leg which connects Khartoum with Port Sudan. All main towns in the region, including Kassala, have now been linked to this tarmac road.

The regional financial situation is a major constraint to the economic development of eastern Sudan. There has been a deficit on the balance of revenues and expenditures ever since the first distribution of national budgets was made in 1981 (Farah 1990). The allocation of funds, in addition, is extremely skewed. Some 94 % of the regional budget was allocated to the Kassala province and 6 % to the Red Sea province over 1980-1986 (Meeuws 1990). This inequality is even greater when allocations are considered at district levels. Almost 55 % of this regional budget was allocated during this period to the Kassala District (Kassala Rural Council and Border Rural Council), while this area accounts for only 19 % of the total regional population (*Ibid.* 1990).

3.3 Land and water rights

Access to land

The Land Law of the Sudan codified in 1925 and 1970 is an amalgam of Islamic, customary, statutory, and judge-made laws (El Mahdi 1979). The amount of land officially registered is still a minor proportion. It includes (i) long-term leases to private persons. These mainly occur in private pump schemes, rainfed mechanized farms, and in the form of tenancies in irrigation schemes; (ii) free-holding ownership. This mainly applies to riverain horticultural land. Other officially registered lands in rural eastern Sudan include (iii) Forest Reserves and the Wild Life Reserve of the Dinder National Park. The remainder of the lands is considered unregistered and is government property under the 1970 Unregistered Land Act. This applies to same 90 % of the Kassala province area. The land may be used, however, under (iv) customary rights of usufruct which is based on tribal rules. Because future claims of the government must be respected, the greater part of the land in the traditional sector in eastern Sudan is held under insecure tenure conditions.

The lowest tribal unit of the Beja with access to land is the minimal lineage. The *deewa* was in this context defined by Sörbö (1991b,218) as "[..] territorial units in the sense that each of them has collective rights to land, vegetation and water in a limited length of a wadi bed and the barren upland countryside surrounding it [..]". These tribal rights contain rights of collective ownership of territory and all its natural resources called *asl* (Tb), and rights of usufruct of pasture and arable land including the use of individual trees, wells and campsites called *amara* (Tb). The collective ownership rights are vested in first occupancy gained by military conquest and occupation of virgin land. Beja lineages have access to each others collective territories, but this must be requested and should be materially rewarded with gifts of livestock or grain. The amount of this gift called *gwadab* (Tb) is normally not stipulated. It depends on personal circumstances and may be waived by the recipient altogether. Its importance is mainly symbolic. The usufructary rights control the use of natural resources by the own tribe or others, and have a strong ideological significance. Usufruct is ruled by *silif* (Tb) which was defined by Morton (1989,272) as "[..] the state of indefinite expectation, or reciprocity, where the counter is not stipulated by time, quantity or quality [..]". Access to these natural resources, accordingly, is not governed by hard and fast rules, but mainly by social conduct. The interpretation of this conduct, however, may vary among the various Beja groups (*Ibid.* 1989,102). Tribal land under usufructary use, unlike under collective ownership, need not be geographically contiguous.

Among the Beja, rights to land are exclusively inherited by the male members in the patrilineal line of descent. This largely follows Muslim *sharia'h* law and means that land is divided in unequal shares according to the strength of blood ties of the heirs. Contrary to *sharia'h*, Beja women do not inherit any rights to land (Sörbö 1991,219). Although both collective ownership and usufruct rights cannot be transferred to others, recent selling of cultivated lands has been reported for the Port Sudan area (Morton 1989). It also has been encountered in the rural Border Area (*cf.* section 5.1.2.1).

Access to water

Islamic laws of water use similarly rest on different pillars including the Koran, interpretations of the Koran, and local custom (FAO 1973). The *Sunnite* doctrine applies in the Sudan which mentions (i) Right of thirst. This rule dictates a free access to water for human and animal consumption under all circumstances. (ii) Various irrigation rights. In general, it holds that the number of restrictions upon its use decrease with the increase in the amount of water supply. Lakes, for example, would normally be held in communal use. Water in rivulets would normally be held in private. The use rights in the latter situation accrue from private water development enterprise, like the digging of a well, or the diversion of a stream. When water is scarce, however, this private use also may become subject to rules of good conduct. These refer to equity (no negative effects on downstream users), and urgency of water need (priority to allocations in greatest need of water). The other rules include (iii) Right of wells. This means that a builder of a well becomes the exclusive owner and beneficiary of this water, respecting the Rights of thirst. (iv) Right of springs and rainwater. This means that the title holder of land with a spring and of land which captured rainfall becomes the exclusive owner and beneficiary of this water, respecting rules of good conduct (*Ibid.* 1973,13-24).

Rules in the Border Area

Within this framework of land and water rights, the Beja of the Border Area add several rules of their own. Traditionally, landholdings are demarcated in the direction of the slope only. This is done either on sight by throwing a stick, or by using a special measuring rope.[7] This leaves the area that can be cultivated entirely to the opportunities provided in the terrain. Usually, the watershed crests on both sides of adjacent interfluves factually demarcate the land in this second direction perpendicular to the slope. This practice reflects the early Beja preoccupation with the cultivation of mainly valley-bottom lands (*cf.* section 4.1.1). It is also a fair practice in the dynamic environment of the Border Area, where khors regularly change their course. It maximizes the chance that all land users receive some floodwater, and minimizes the chance that some get all. Cultivated lands on the interfluves and at locations away from seasonal water courses are usually demarcated in two directions. The allocation of cultivated lands is usually to the minimal lineage. Further sub-divisions over individual members in this group are left to their responsibility. Local

---

7 The indigenous measures used and their metric equivalents are given on page vi.

village leaders have the powers to temporarily allocate cultivated lands to individuals who hold no landtitles. This is called *akul gom* and literally means "eat and go". It is a practice used to support destitutes, for example in drought years. Land users in the Border Area eventually may have access to different farming zones, and within a given farming zone, to different landholdings.[8] This is a purposeful action to reduce the risk of crop production under variable water supply conditions. Similar rules for land and water use have been described for the northern Beja of the Red Sea province by Morton (1989,87-88).

3.4 SWC techniques

The different types of indigenous and project-introduced SWC techniques applied in the Border Area are listed in table 3.5 by type of water source, implementation context and organizational level of use.

Table 3.5 **Indigenous and project SWC techniques in the Border Area, eastern Sudan**

	PRINCIPAL RUN-OFF SOURCE			
	Floodwater		Rainwater	
TYPE OF TECHNIQUE				
Indigenous SWC	earth bunds	P	earth bunds	P
	brushwood panels	P, (C)	brushwood panels	P
	(inundation) channels	(P), (C)	rills	P
Project SWC	earth dams	C	earth bunds	P
	earth embankments	C		
	stone lines	C		

Note: P=private use, C=communal use; brackets indicate that use is not common.

The indigenous SWC techniques include:

(i) Earth bunds called *teras* (Ar) (plural *terus*) and *odruk* (Tb) (plural *odirib*, *durkab*). The system consists of low earth walls generally up to 0.5 m high which collect run-off from adjacent ground-level catchments (rainwater harvesting), or from spate flows in the seasonal water courses (floodwater harvesting). *Teras* is usually the name given to the entire landholding. It includes the cultivated area which is bunded on three sides and the catchment area which is located on the upslope unbunded fourth side. There are many variations in the form and function of the basic system. The *teras* is the most elaborate indigenous SWC technique used in the area and the most important in terms of acreage cultivated in the Border Area. The technique is discussed at length in the next section 3.5.

[8] Landholdings and farming zones were defined in section 2.1.1.

(ii) Brushwood panels called *libish* or *shok* (Ar). Panels of wattled branches of local tree and scrub species are used to collect run-off in much the same manner as *teras* bunds do. The panels are seasonally erected and generally up to 0.5 m high. The technique is sometimes also called stick bunding (Millington 1987a,b).

(iii) Rills called *boka*, *hashu*, *shaweer*, *serap* (Ar), and *teboka* (Tb). Low earth ridges 0.2-0.5 m in height are built on the cultivated lands to manipulate small flows of the run-off collected. The technique is also used in combination with SWC techniques such as earth bunds and brushwood panels, but also in combination with non-SWC techniques, such as rainfed and valley-bottom cultivation. The elevated rills and depressions in between in fact make a micro-scale water harvesting system in itself. The ridges are normally a side effect of weeding, but their use to manipulate run-off is a purposeful activity (*cf.* section 5.1.2.1).

(iv) Channels called *shayote*, *sagia*, *tura* (Ar), and *fedit* (Tb). Channels of varying widths, but generally not deeper than 1 m, are cut to divert floodwater away from the main water course. Two techniques are used. The first is based on inundation. The channel has its bed just above the level of normal flow and taps only floodwater from seasonal high spates. The second is based on common water diversion. The collected floodwater may be used at the level of individual landholdings in combination with earth bunds, brushwood panels, and rills.

New SWC techniques have been introduced by the Department of Soil Conservation in the Border Area. These are to some extent based on indigenous principles. The introduction of earth embankments and stone lines was supported by the Sudanese-Netherlands MFEP/KADA programme. The main project-introduced techniques are discussed in more detail in sections 4.2.2. and 4.3.

(i) Earth bunds. The indigenous *teras* is the starting point for earth bunds introduced by projects. The mode of construction of these bunds is mechanized, instead of manual in their indigenous application. This usually results in larger overall dimensions of the structures.

(ii) Earth dams called *torabi* (Ar) and *tareet* (Tb). High earth walls generally up to 2-3 m are built by using machinery across the stream-bed of the water courses upstream of the cultivated lands. Three dam variants are used. The first regulates floodwater by blocking smaller khor branches in order to route the water into a favoured main branch. The second directly diverts floodwater from the main branch onto the cultivated lands. The third regulates the distribution of captured floodwater on the cultivated lands.

(iii) Earth embankments. A series of broad-based earth embankments with a design height of 0.35 m and channels to the same depth are built by means of machines on the near-contour. These embankments divert floodwater away from the water course. Various designs have been tested in three pilot schemes in the Border Area. However, unlike the project-introduced techniques of bunds and dams, this technique remains experimental, and has not been widely applied.

(iv) Stone lines. The technique resembles the earth embankment system, but stones packed to some 0.2 m high are used instead. Another difference is that no floodwater is diverted, but only local sheetflow run-off is harvested. The construction is mechanical with respect to the transport of stones, but their placement is manual. One pilot scheme was implemented, and the technique is still being tested.

Other techniques used for crop production in the Border Area are of a non-SWC type. These include rainfed and valley-bottom cultivation. The first is uncommon in the area. The second is not, and is of great importance to the household economies. The technique is called *seed* (Ar), or *ogwad* (Tb). It is also locally referred to as *moya tatla baraj* literally meaning "free flowing water". We refer in this study to this technique as wildflooding.

3.5 The *teras* system

The *teras* is the most elaborate indigenous SWC technique applied by households in the Border Area. The *teras* is a construction which is built once. It usually needs maintenance and repair in subsequent seasons of use.

The system

The main *teras* characteristic is the base bund which is built approximately on the terrain contour (figure 3.9). It impounds the captured run-off, and allows it to infiltrate into the soil. The outer collection arms partly have the same function, but also act as a conveyance structure. The arms are placed at a right angle to the base, and direct the run-off to the cultivated area. The shorter inner arms divide the land into smaller basins to effectively impound smaller supplies of run-off. In certain areas, the internal circulation of run-off is manipulated by changing the length of some of these inner arms. The individual basins can thus be given a thorough wetting before the spill is routed to neighbouring compartments. The same principle is behind the use of the smaller "child" *teras*. This is sometimes built in the catchment of the main structure, which is called "mother". Its dual objective is to benefit from small run-off volumes generated by light early rains. It also reduces the velocities of flow to less erosive magnitudes when the rains swell later in the season. The construction of earth bunds leaves shallow channels in the field. The excavation material is usually taken from the inside face of the bunds. The resulting ditch is used to support the conveyance and circulation of run-off. Excess water is normally drained along the tips of the outer arms. These are reinforced for this purpose with different materials including

Figure 3.9 **Typical elements of the teras**

1. Base contour bund
2. Outer collection arm
3. Inner collection arm
4. Shallow channel
5. Basin
6. Cultivated area
7. External catchment
8. Internal catchment
9. 'Mother'
10. 'Child'

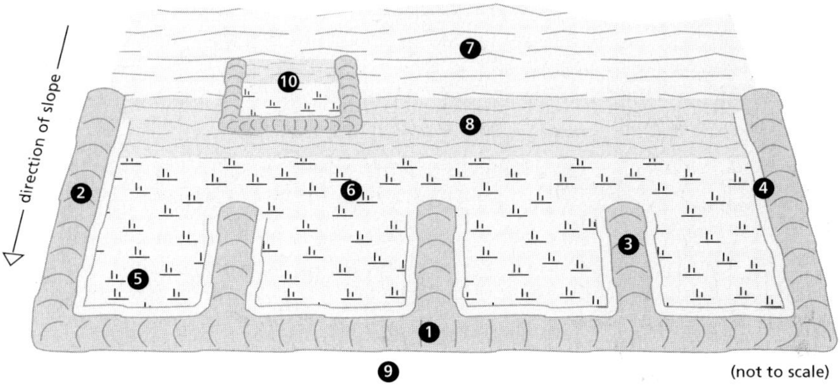

brushwood, small stones, and worn-out tires. The contour bund is deliberately breached in case of severe flooding to avert the more devastating effect of an uncontrolled burst. Common bund heights and base widths used in the Border Area are 0.5 m and 2 m respectively. However, these dimensions vary greatly with the slope and amount of run-off expected in a given area. Maximum recorded bund heights in the Border Area reach over 2.5 m, with corresponding bases in the same area of some 25 m. The lengths of base bunds are between 50-300 m, with recorded maximum lengths up to 700 m, and arms usually 20-100 m (up to 200 m).

The ratios of the base-to-arm dimensions are not fixed. Base bunds with lengths under 100 m usually have more constant ratios 1: 0.6 or 1: 0.7. This makes the *teras* relatively deep. Base bunds over 100 m length generally have ratios around 1: 0.3. This makes the *teras* relatively shallow. The maximum size of the area that can be cultivated normally ranges between 0.2 - 3 ha. The catchment is external to this cultivated area. Any lands not under crop, however, add an additional internal catchment area. The total catchment, furthermore, is to a large degree defined by the configuration of different *terus* in a given farming zone. A regular spacing is common in the Border Area. The base bunds in this situation are aligned with the contour, and are slightly staggered in the direction of the slope. When there are no upstream land users, run-off can even be collected from an unbounded area. Inside a given configuration of *terus*, the bounded external collection area usually has a *c: ca* ratio (*cf.* section 1.4.2) of 3: 1 to 2: 1, and up to ratios of 4: 1. The internal catchments vary in size according to the area cultivated. Frequently, these cover not more than 50 % of the total bunded area.

Variations and functions
As a result of the complex Border Area drainage pattern, there is a wide range of *teras* designs. These can best be described in terms of shape (rectangular/irregular), dimension (shallow/deep), and basin characteristics (single/multiple basin). The available 1986 aerial photographs of the Border Area show that the base and outer collection arms usually make a rectangular bunded area. However, also *terus* in the form of irregular semi-circular hoops are used. These differences seem mainly related to the micro-topography, characteristics of water supply, and local environmental dynamics. In relatively stable environments with an evenly distributed supply of run-off over level land, rectangular *terus* prevail. These are shallow (base bunds longer than collection arms), and of the multiple-basin type, but single basins also occur. Where drainage lines, run-off volumes, and erosion and sedimentation processes are more subject to change, irregular hoops prevail. The structures are usually deep (collection arms longer than base bunds), and more often of the single-basin type. Another type of *teras* in the Border Area is a rectangular basin with one concave outer collection arm. Different explanations for this design have been given locally. Some land users model the arm in this shape to catch the maximum amount of run-off with a minimum of breach hazard. Others refer to it as spillway to drain excess water.

The *terus* of the Border Area are either built on the alluvial flats of the valley-bottom, or on the interfluves (figure 3.10). In the latter situation, the technique is entirely rainwater dependent. In the first, also additional floodwater is collected. The catchment of a floodwater-using *teras* is usually smaller because the rainfall multiplier-effect is of lesser importance. The configuration of different *terus* is in this case adjusted to the drainage lines. Their positions are not strictly aligned and staggered. The slopes on which the bunds are built are generally under 1 % in the Border Area. The average *c: ca* ratio calculated for landholdings sampled in this study is 2.3: 1 for largely rainwater-dependent structures. This average is 2.1: 1 for largely floodwater-dependent structures.[9]

Table 3.6 **Average labour demands for manual building, repair and catchment cleaning of a one hectare teras (in man-days), Border Area, eastern Sudan**

	MAN-DAYS
CONSTRUCTION	6-16
SEASONAL MAINTENANCE	
– Repair	8-12
– Catchment cleaning	3-6

Source: L&E Survey. Note: the original data are in man-hours per fd. Recalculations are based on 7-hour working days.

[9] Um Safaree and Ilat Ayot research villages, total catchments including all internal and external collection areas, averages over 1983 and 1988.

Figure 3.10 **The teras in the landscape of the Border Area**

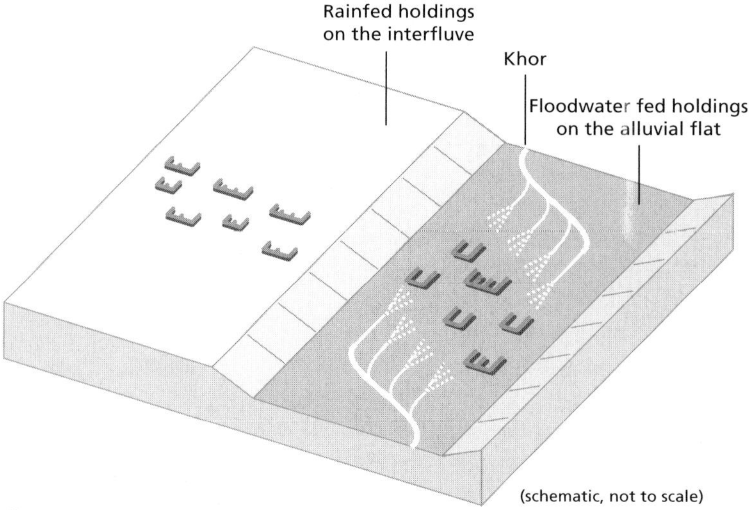

Construction, maintenance and repair

Teras bunds are built of the local alluvial and colluvial material. They are invariably positioned in the field on sight. A detailed knowledge of the terrain allows *teras* users to orientate the structures in such a way that commonly the most profitable use can be made of run-off. Building sites on soils with vertic properties called *badobe*, and on soils with gravels or hard pans, are usually avoided. The construction is manual. Only simple tools are used, like a hoe, spade, axe, and basket. At several locations, land users have their bunds raised naturally by wind action. They only build an initial 20-30 cm high ridge and place brushwood on the windward side to capture the dust and sand. The labour requirements for *teras* building mainly vary with the soil characteristics and the period of construction. The highest demands on labour are made in the dry season when the soils are hard, and on heavier clay soils. The labour demands for maintenance depend on the type of bund damage, erosion occurrence, and the degree of noxious plant growth in the catchment area. *Teras* building is usually carried out in different phases. This allows the necessary adjustments in its design by trial and error experience. An average *teras* would need a total of 6-16 man-days per ha for construction. Annually, between a minimum of 3 man-days per ha (cleaning of the catchment only) and a maximum of 18 man-days per ha (average repair and cleaning of the catchment combined) are required for maintenance (7-hour working day, table 3.6). The labour demand for general cultivation is not included in this. This figure is approximately equal under SWC and non-SWC systems in the Border Area (*cf.* section 5.1.2).

From the 1960s onwards, tractors are increasingly being hired by land users in the Border Area for construction, repair, and maintenance of *teras* bunds. This takes about 2.5 hours per ha and includes ploughing. These services are offered by urban contractors. Regular seasonal maintenance is largely manual. It usually starts in May or June shortly before the onset of the rainy season. Maintenance activities include the application of protection means at the tips and weak parts of the bunds, like former breaches, low sections, and gullies. The materials used include brushwood, stalks, crop residues, sand bags, and small stones. A serious menace to the system of *teras* SWC is deposition by wind and water. Windblown sand can still easily be cleared from the land. Sedimentation by collected run-off inside the bunded area is more difficult to deal with. Firstly, the bunds need to be regularly heightened to keep pace with the rising level of the cultivated lands within. Secondly, the efficiencies in water distribution over the cultivated lands decline. These problems are structurally solved only by rebuilding parts of the *teras*, or the entire structure, on upstream land.

4

History of crop production, SWC interventions and livelihoods

The following sections discuss the historical developments in crop production under indigenous SWC and government SWC interventions in the Border Area. They also address the changes in local Beja livelihoods from more to less nomadic types. We do not pretend to be exhaustive. We merely aim to provide a broad background against which the local situation of the 1980s and 1990s is to be appreciated. Crop production in the Gash delta irrigation scheme, and the SWC activities of the MFEP/KADA programme are dealt with separately. The Border Area is our main area of reference. However, the Beja tribal lands in the Red Sea province and Eritrea are sometimes also referred to. This wider area is then called the "Border Area region".

4.1 Crop production and indigenous SWC

The historical data used in this chapter have been collected at the National Records Office (NRO) in Khartoum, and the Sudan Archive of the Durham University (DSA) in England. The archival references to the cultivation techniques of the Border Area region are not consistent. For example, "raingrown cultivation" is known to have included in several cases early techniques of rainwater harvesting such as the *teras*. Equally, "khor cultivation" may have included techniques of floodwater harvesting. Accounts of first use of techniques can only be dates of first observation by government officials who knew what they were reporting about, and knew what name to give it. The early descriptions of *teras* use in central Sudan provide a good illustration. According to Hill "[..] simple bunds to trap rainfall and to increase the amount of water infiltration into the soil [..]" were already used around the 1820s (*cit.* in Trilsbach 1991,181-182). The technique was described by Wingate in 1913 as: "[..] land is cultivated when heavy rains fall on the higher land, and run off and collect at the lower end of the *hadab* (rain bund) [..]" (DSA Wingate 112/3). The technique of "[..] light earthen training banks [..] in which rainfed (*sic*) sorghum is grown [..]" was called *teras* by Davie in 1924. Some degree of standardization in reporting can be assumed to have begun after the publication of the Sub-*Māmūr*'s handbook (Sudan Government 1926). This book discusses indigenous SWC and describes the *teras* technique in detail. Its contents, being the handbook of civil servants, must have been widely known. It is for this reason that consistent field-reporting on *teras* use in the Sudan can be expected to have only started from 1926 onwards.

4.1.1 Earliest cultivation

The first evidence of land use in the Border Area is based on archaeological finds in a Neolithic settlement located due east of the present-day town of Kassala (Fattovich 1990). Pastoral people, possibly cultivating cereals, have lived in the Gash delta region since 3,000 B.C. There is proof that they were growing sorghum for the period starting at about 1,500 B.C. (*Ibid.* 1990,13-14). A first eye-witness account of land use is given by Ibn Hawqal [988](1975)[1] who travelled in the area in 955 A.D. Sorghum and millet was grown by "[..] the Nuba and the sedentary Beja [..] in the mountainous territory extending from the sea (Red Sea) to Dukn (present-day Gash delta) [..] the river Dujn (probably the Gash river) waters the district and its cultivation [..] the people of Bazin (present-day western Eritrean lowlands and Border Area) dwell in straw huts [..] they breed cattle and small animals and cultivate the land [..]" (*Ibid.* [988] 1975,162-166; Director of Department of Archaeology, pers. comm. Khartoum, Feb. 1992). The next evidence of local land use is dated almost one millennium later. This comes from early 19th century explorers. Burckhardt visited the region in 1814. He describes the Beja as a predominantly nomadic group. They are involved in cultivation in seasonal water courses and "[..] after the harvest is gathered the peasants return to their pastoral occupations [..]". Crop production on a notably larger scale was practised in the Gash delta. This area was already in Burckhardt's time known as the granary of eastern Sudan. It exported a part of its production of sorghum to Jeddah (*Ibid.* [1819](1987,378-416). James [1883](1969) travelled in the region in 1881-1882. He describes the sorghum fields at the northern outskirts of the town of Kassala which had been established in 1840.

The public servants of the Anglo-Egyptian Administration were the first to make more systematic expeditions into the Border Area. These started from the late 1890s onwards. Occasional "khor cultivation" is reported in the area of khor Togan (N.E. of the present-day Telkook). Apart from the headquarters of the Hadendowa, and one market town in the Gash delta, no other settlements and cultivated areas are mentioned for the northern part of the Border Area. In its southern part, however, villages of straw huts and

---

[1] For reasons of historical clarity, in this chapter square brackets indicate the year of first publication of the source.

cultivation are already common (Gleichen 1898,271-288). "Raincrop cultivation" is reported for the area between Kassala and Sabderat in 1902 (DSA Savile 427/5). This area coincides with the present-day Um Safaree and Hafarat village domains. Other crop production was reported in 1904 around the well centres in the Gash delta, the Hadendowa headquarters, and in the khors of Eilabillei and Fagada (present-day Gash Dai area) (DSA Savile 427/6). Numerous khors are cultivated between Kassala and the Gash Dai area in 1916. Contiguous areas of cultivated lands amounted up to 172 fd (73 ha) each (DSA Thomson 402/6). A Beja land use sketch-map in Thomson's diary suggests that in these years there was a close association of seasonal camps, wells and khor-based cultivation (*Ibid.* 402/6/43). In 1929, the Gash riverbed was cultivated by means of a technique nowadays called flood-recession (*cit.* in: NRO Allen 1929).

The first production figures for the Border Area have been found in the 1931 Hadendowa District Annual Report (DSA Barter 448/1). Some 25,500 fd was planted in this year. This produced in the northern part of the Border Area a total of 108,500 *ardeb* (217,000 sacks). This is an average production of about 1,865 kg/ha. In the southern part including the Kassala area, Maman, Gash and Um Adam (present-day Gash Dai), the production was 14,500 ardeb (29,000 sacks) from 5,000 fd. This is an average 1,270 kg/ha. The latter figure also includes an unknown area in the Gash delta irrigation scheme, however. "Rain cultivation" was reported to expand around Kassala between 1930-1937 (NRO Blackley KD/20.H.10, KD/8.B.2). However, the population and the area between Girgir and Togan in the present-day area of Ilat Ayot and Telkook are still described in 1949 as "[..] not well provided with grain land, having only a few wadis and otherwise depending on scanty rain cultivation [..] they are mainly animal owners [..]" (NRO Carlisle BD/19.B.4). In the handing-over-notes of 1951, usually the most detailed archival document on the regional state of affairs, Carlisle mentions that in this area only "[..] narrow wadis in which cultivation is practised in good years [..]" can be found (DSA Carlisle 725/3). The first explicit mention of *teras* application in the Border Area is by Richards (1951). The first evidence on this use comes from vertical aerial photographs dated 1953 (Khartoum Survey Department, 1:15,000, series 138).

The absence of early references in the archival sources, obviously, can not imply that certain techniques were not being used. Still, we consider it unlikely that the adoption of the *teras* in the Border Area preceded the year 1950. Firstly, such early reports do mention *teras* applications in other areas of central Sudan (*cf.* NRO Lea GRDC/22.C.1 on *teras* use in the Gedaref area). This indicates that the technique was known and was also being reported on. Secondly, such early reports were also made for the Border Area. These however only mention other techniques, some of which are less conspicuous than earth bunds. This indicates that extensive field visits were being made which could have revealed early *teras* application if it had existed.

4.1.2 Indigenous techniques

Non-SWC techniques

A *maya* is a natural depression that can be cultivated (*cf.* section 1.4.3). Such depression is indicated west of the Gash delta on the first topographical map of the Border Area region (1:250,000, sheet 56-A Kassala surveyed in 1900-1902). Another reference comes from the Gash Dai area, where "[..] pond type of irrigation occurs naturally [..]" (Bell 1956,37). Expectedly, Gleichen had also observed a *maya* when he described in 1897 the Hadendowa use of "[..] desert-grown grain [..] cultivated in [..] inland desert basins flooded by rains [..]" (*Ibid.* 1898,331). We do not consider the use of *maya* as indigenous SWC in this study (*cf.* section 1.4.2). Grass-firing cultivation, or *hariq*, is an indigenous technique in central Sudan. It was introduced in the Gedaref District by the Anglo-Egyptian Administration. It was later autonomously adopted and tried for three seasons by the Beja on the Clay Plains west of Kassala town. A total of 120 fd was burned and planted by them in the last of these three years, 1947. The yield forecast of sorghum was 180 *ardeb*. This would equal a production of 650 kg/ha, which was half the Gedaref level forecasted in the same year (NRO Dingwall BD/2.F, BD/2.F.1, BD/2.F.6). The technique was after three seasons rejected by the Beja for reasons not documented. Its modest returns are likely to have played a role. "Khor cultivation" in the Border Area yielded in the 1930s with less effort already double or triple this yield forecast (*cf.* section 4.1.1). The Beja had all tillage work under *hariq* carried out for them by hired West-African labourers (NRO Dingwall BD/2.F.1). *Hariq* too is not an indigenous SWC technique.

Muslim West-Africans have been travelling through the Sudan on their pilgrimage to Mekka since the 14th century (Al-Naqar 1972). In 1900, a new route gained in importance which ran through Kassala town. Many settled permanently in the area, especially after the establishment of the Gash delta irrigation scheme in 1924. They where allocated land in the scheme because they "[..] are the best labourers to be obtained [..]" (NRO Cumming 57.A.9). The main groups of West-Africans living in Kassala in the 1930s included the Fulani, Hausa and Barno from northern Nigeria, and the Bergu from Chad (NRO *Ibid.*). The West-Africans proved highly successful in crop production. Besides their involvement in occasional *hariq* experiments, the Beja also employed them on other lands. They were subletted Beja tenancy titles in the irrigation scheme (Salih 1980), and were given access to Beja tribal lands in the Gash Dai (DSA Owen 414/5/15-20). West-Africans have also been important SWC users in the Kassala area until the present day.

Origins of the teras

Two possible source areas of the *teras* technique as used in the Border Area region can be considered. The first is West and Central Africa, the second the Nile region.

There are two arguments in favour of West and Central Africa. The first is that the Africans interviewed in different parts of central Sudan themselves claim *teras* use since what they called the "Turkish time" (1820-1880). This date precedes our period of documented *teras* use in the Border Area by the Beja. The second argument is that a highly similar technique is also used in the Ouaddai region of Chad (Sommerhalter 1987).

The resemblance is striking and pertains to elaborate aspects of design. These include the shape of the structure (open and closed basins), engineering directives (base bunds on the contour, staggered configurations, sub-divisions into smaller internal basins), materials used (earth bunds and brushwood cores), and finally its applications in the field (combined rain and floodwater use, deliberate breaching to avoid a burst under conditions of excessive run-off supply) (*cf.* section 3.5). There are also two arguments in favour of the Nile region. The first is based on archaeological finds in the "Project for Systematic Archeological Survey of the Mahas Region" of the University of Khartoum. Stone relicts of an estimated age of 4,500 B.C.-1,500 B.C. have recently been discovered in the Nubian stretch of the Nile valley. These are likely to have served for the purpose of run-off diversion (Osman, pers. comm. Khartoum, Feb. 1992). Several of these structures can still be identified on aerial photographs. The second argument is linguistic in character. Certain components of the *teras* are being referred to by different ethnic groups, including West-Africans, with words of a Nubian origin. The word *teras* itself is not derived from the English "terrace" as Cole (1989) suggests, but is Arabic. It is also used in the Beja language of To Bedawee, but has a clear Nubian origin. In the Dongolese Nubian lexicon (Armbruster 1965,197) *tērar* means "sowing", "tillage cultivation" and "to grow a crop". In the To Bedawee lexicon of Almkvist (1884,65), *téra* means "half" and *teráb* or *térib* means "to divide". The more specific Beja meaning given to a general Nubian concept is striking (Nubian Nile-dwellers do not have to divide their cultivated lands to reserve part as a catchment as the Beja are compelled to do in their dryland habitat). There are also references to the symbols of the later Nile-based Funj Kingdom[2], where *teras* was still in use in the first two decades of the 19th century (Trilsbach 1991,181-182).[3] The Beja have always maintained good contacts with the Nubian and Funj cultures (Ibn Hawqal [988] 1975, Crawford [1951] 1978, Mokhtar 1990). The Beja, finally, are also known to have adopted their technique of basin irrigation from the Nile region (*cf.* section 4.1.3).

We consider the evidence in favour of the Nile region more convincing. In addition, claimed early (*teras*) land use, as the Africans do, is over much of the continent a recognized strategy to safeguard or improve entitlement to land. These statements can therefore be deceptive. Finally, the technique now in use in Chad may well have originated in Sudan in the first place, and have been introduced locally by pilgrims on their return journey.

Other indigenous SWC

The historical use of brushwood in SWC techniques is undocumented for the Border Area. Brushwood panels are difficult to identify on early aerial photographs. They are seasonally

[2] The *teras* base bund in the area of New Halfa area, for example, is called *kakar*. This is the name of the six-legged coronation stool of the Funj kings (Osman, pers. comm. Khartoum, Feb. 1992). When this stool is depicted upside-down in two dimensions only, exactly the groundplan of a *teras* arises. The *teras* outer and inner arms are then formed by the legs of this chair.

[3] The Funj Kingdom (1504-1820) itself was greatly influenced by the Nubian culture (Mokhtar 1990).

erected just before the start of the growing season, and are small in size. However, brushwood was used to protect the river banks of the Gash before stone-pitching was introduced. It remained in use for this purpose until at least 1975 (El Berier 1982). Brushwood wattling was applied on earth dams in the Red Sea province already in the 1930s (DSA Barter 448/1). It was also used here in the form of stake and brushwood dams in the late 1940s (NRO Dingwall BD.19.B.2). These applications have all been introduced by the Anglo-Egyptian Administration. Brushwood was autonomously adopted by the Beja around Khor Arab in the Red Sea province for SWC purposes in the mid 1950s.

4.1.3 Irrigation in the Gash delta

Basin, controlled spate and groundwater-based irrigation have been in use in the Gash delta area from the mid-19th century onwards. The first two traditionally make important contributions to the Beja household economies in the Border Area. Groundwater-based irrigation is encroaching on Beja tribal lands in the Border Area since the mid 1980s.

Basin irrigation
Fleming (1922), Stone (1955) and Ausenda (1987) give details of the oldest types of irrigation in the Gash delta area. This is called *shayote*. The system is known to have been in use by the Hadendowa in two main variants before 1840. In the first type, water was diverted by a dam from the Gash river and its branches. Irrigation water was then led by gravity through a canal to the cultivated lands. According to Ausenda, the diversion dam widths were 5-20 m, the canals were of 250-800 m length. The areas under irrigation command measured 50-100 fd (21-42 ha) each. These areas were divided into smaller 8-12 fd basins, which were irrigated in turn. Most *shayotes* were located on the Gash east margin in the Border Area (*Ibid.* 1987). In the second type, no such dam was built. Irrigation water was tapped by a canal from areas naturally flooded. Preferably, these canals were dug to convey water into the "rain land region". Bush-clearing was less demanding here than in the thickets of the delta proper (Fleming 1922,73). The canal and commanded lands were administered by a *sheikh*. He seasonally allocated the irrigated landholdings over the members in his group. This commonly resulted in 2 fd-landholdings per member (expectedly representing one household, but this is not detailed). The main crop grown was sorghum, but also vegetables like *bamia* and *lubia* were cultivated. Cotton was introduced by the Turco-Egyptian Administration as a cash-crop in 1860-1880. The documented average sorghum production was 8-12 sacks per fd (1,754-2,630 kg/ha). The total area cultivated under *shayote* at any one time was estimated by Ausenda to have ranged between 4,500-6,000 fd. The system disappeared with the establishment of the Gash Delta irrigation scheme on these lands.[4] Indigenous basin irrigation in the Gash delta originated in the Nile region (Stone 1955).

---

[4] Certain present-day practices in the Gash Dai area, as well as irrigation at locations south of the town of Kassala, can still be called *shayote* however (Ausenda 1987).

Controlled spate irrigation
The first pilot farms in the Gash delta, in preparation for the construction of the irrigation scheme, date from 1905. Stone weir diversions were built by the Anglo-Egyptian Administration in the Gash river. New designs of basins and new crops like castor, sesame, wheat and barley were tested (DSA Tottenham 185/1). An area of 1,920 fd was thus irrigated in 1910 (El Karrar 1976). Scheme construction proper started in the 1920s. The first canals became operational in 1924. The delta is annually flooded by the Gash between July and September. The total domain under judicial supervision of the management board is 700,000 fd (295,000 ha). Only half this area has been brought under irrigation command in 6 blocks by an equal number of main canals. Irrigation water is led into the system on the west bank of the Gash. The water is distributed by gravity through unlined main and secondary canals, and finally led into basins of varying size. Normally, two irrigation cycles can be run per season. Parastatal organizations named Kassala Cotton Company (1924-1928), Gash Board (1928-1967) and Gash Delta Agricultural Corporation GDAC (1967-1994)[5] have always supervised and controlled the scheme. The GDAC ma-nagement resides under the Ministry of Agriculture and Natural Resources in Khartoum.

Figure 4.1 **Number of tenants in the GDAC irrigation scheme, 1926-1991 (approximate). Source: various, see text**

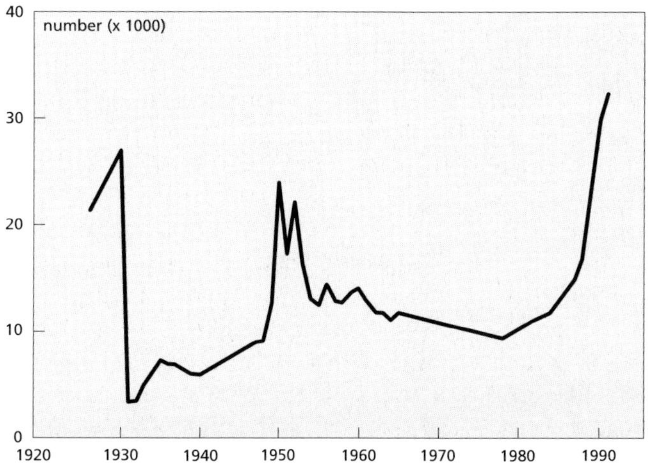

The management board allocates land to registered tenants. Initially, the Beja tribe of Hadendowa held 65 % of all land titles, the remaining was held by West-Africans and Northern Sudanese (Euroconsult & Newtech 1987). The current division over groups is not

[5] At present, the organization is called Gash Delta Scheme. We refer in this study only to the period under GDAC management, and earlier, however.

known. The tenancies are inherited and held for life-time. However, they change hands in the form of gifts. New tenancies can only be allocated with the consent of the management board and national Ministry. The seasonal allocation of land over a group of tenants takes place by means of a lottery system. The division over individual tenants is made by informal leaders, as under *shayote*. The share of the entitlement which factually can be cultivated is annually prorated per tenant to the total irrigated area in the scheme. Normally, this is about half the size of the formal title. This tradition of "collective allocation" has resulted in a situation where the precise total number and size of tenancies are unknown. This has been the situation since at least the late 1950s (Ahmed 1966).

Figure 4.2 **Total cultivated area in the GDAC irrigation scheme (in fd), 1944-1991 (approximate). Source: various, see text**

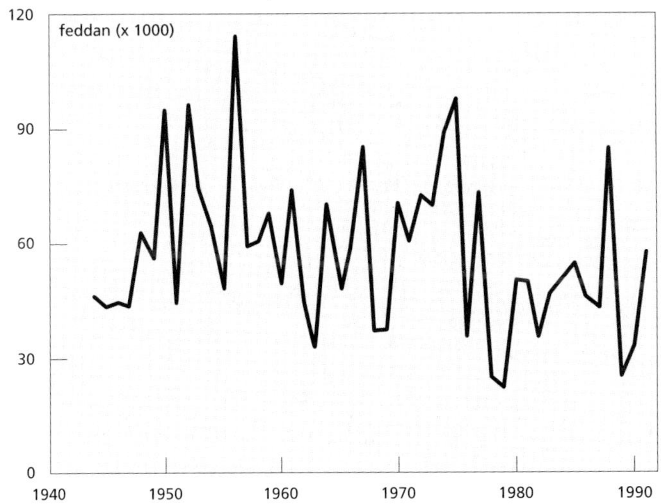

Historical land use in the scheme is referred to by Richards (1951), Ahmed (1966), El Karrar (1976) and Ausenda (1987). The main cash crop until 1970 was cotton. After this year, it was castor bean and from the mid 1980s onwards sorghum. The tenants have always been permitted to grow subsistence sorghum in marginally watered areas and at the end of irrigation canals using spillwater. These sometimes covered 20 % of the scheme domain. For the Hadendowa, this meant a doubling of the areas under sorghum when compared to *shayote* (Ausenda 1987). Cotton growing was initially welcomed by them because the returns were favourable. This enthusiasm diminished in the 1930s when the world market prices for cotton fell. Many Hadendowa deserted the scheme to take up their traditional livelihoods in livestock husbandry and indigenous cultivation. Their place was largely taken over by the West-Africans (DSA Owen 414/5/15-20). When cotton production was stopped in 1970 because of pests and unfavourable prices, castor bean became the main cash crop for the next 14 years. In the first year of 1984-1985 drought, the national ministry ordered the GDAC to grow staple sorghum on all its irrigated lands.

From that year onwards, the choice of crop has been left free to the tenants. The Hadendowa still in the scheme spontaneously introduced new crops like watermelon, pulses, cowpeas and various beans from the mid 1980s onwards. Others took great efforts to acquire tenancy titles (see below).

Figure 4.3 **Annual discharge of the river Gash (in million m3), 1907-1984. Source: various, see text**

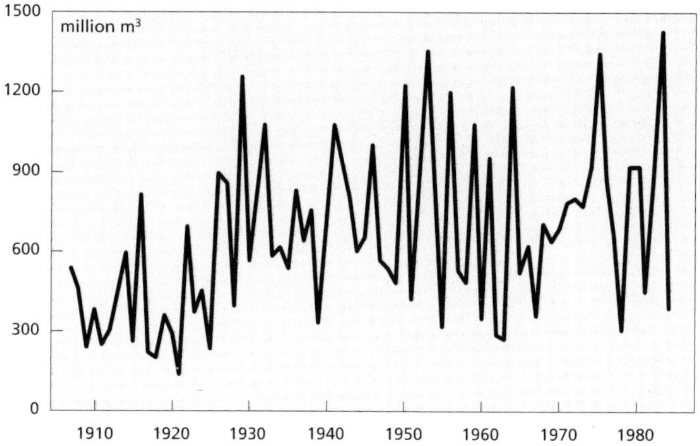

Until 1980, the different management boards have directly taxed their tenants over the value of the cash crop. The rate was 20 % under the Anglo-Egyptian Administration. This was raised to 50 % under the Kassala Cotton Company and Gash Board, and again lowered to 45 % under the GDAC. Direct taxation was replaced by a system of water fees in the early 1980s. Ahmed (1966) calculated the returns to the tenants for the late 1950s and early 1960s. These figures indicate that the cash crops only started to generate a positive return when 30 % of the tenant labour would not be priced. Castor and sorghum still produced a seasonal cash loss in the early 1980s. Tenants financed this by the sale of livestock and by engaging in non-farm work (Euroconsult & Newtech 1987). Recent crop budgets under the system of water fees are more favourable, however, largely because of subsidization.[6] The 1989/1990 sorghum returns per feddan in the irrigation scheme are the second highest on a list of regional crop production systems (table 4.1).

---

[6] GDAC water-fees are underrated and do not reflect true irrigation costs in the scheme (NEI 1991).

Table 4.1 **Sorghum crop budgets in 1989/1990 (net returns in £s per fd, 1989 prices) for different production systems in the eastern Sudan**

	Net return in £s per fd
Mechanized rainfed farming, Gedaref area	66
Traditional rainfed farming, eastern Sudan	180
Permanent gravity irrigation, Rahad scheme	260
Permanent gravity irrigation, New Halfa scheme	867
Controlled spate irrigation, Gash delta scheme	1,096
Controlled spate irrigation, Tokar delta scheme	1,451

Source: MFEP & UNDP (1991,104-130).

The number of tenants greatly increased over the last decade. At the same time, the annual Gash discharge remained variable, and the cultivated area declined (figures 4.1-4.3).[7] Two notable outliers in tenant numbers occur in the late 1920s and early 1950s. These are not explained in any of the sources. Probably, these relate to new registration procedures (as are known to have been introduced in the 1920s) and incorrect data (this is likely to be the case for the outliers in the mid 1950s). The growth in tenant numbers since the mid 1980s is almost entirely on account of new allocations to the Hadendowa of the Border Area. These have been succesfully negotiated by the leader of the movement of Ali Betai (*cf.* section 4.3.3). An estimated 20,000 households and 36,000 fd of land titles have thus been made available to them between 1984 and 1990 (Deputy Director GDAC, interview Aroma, Jan. 1992). The greater part went to the village of Hameskoreib (just outside the Border Area study region, *cf.* figure 3.1) which is the religious centre of the movement. New allocations were made in 1990, this time also including households in the research villages Ilat Ayot and Telkook.

The main cause of the declining cultivated area in the scheme is the deterioration of the irrigation infrastructure. A total of 50,000 fd is annually cultivated at present. This was only more in the wet year 1988. The GDAC scheme was nominated for a large-scale upgrading in the mid 1980s. This programme was to be financed by international donors. After initial phases of management support, only a modest rehabilitation programme was finally approved in 1991. This was called the Gash Delta Rehabilitation Plan and was reformulated as Gash Delta Project in 1992. The project objectives are in the field of improved irrigation management, food self-sufficiency, forestry and women in development (HVA 1990, 1992, 1993, GOS & GON 1992).

[7] The data were taken from various sources, including: DSA Gash Board 499/1, Richards (1951), Salih (1971), Paul (1971), Euroconsult & Newtech (1987), MFEP & UNDP (1990) and GDAC Annual Reports 1980-1990. The data on the number of tenants and cultivated areas must be considered as indicative. Frequently, contradicting data were presented in different reports. Gash discharge gauging was terminated in 1984.

Groundwater-based irrigation
Groundwater is largely confined in the Border Area region to a narrow strip of alluvial deposits in the Gash basin (Saeed 1969). The greatest extent of horticultural land irrigated by groundwater is located south of Kassala town. Horticulture development is now expanding in northern direction. Groundwater was first raised from shallow wells by means of ox-drawn water wheels. The wells were equipped with engine-driven pumps from the mid 1940s onwards. The first deep wells were introduced in the 1950s. Lands for horticultural development were rented and from 1947 onwards also purchased from the government. Mainly urban entrepreneurs of northern Sudanese origin engaged in these activities (NRO Cummings 57.A.9). Initially mainly sorghum was grown, but more profitable horticulture and fruit trees were introduced in the later decades. At present, some 1,200 farmers cultivate about 17,000 fd. An estimated 4,000 fd was developed without government consent. Illegal horticulture presently takes place within the GDAC irrigation scheme and on the eastern Gash margin in the Border Area. Several Beja groups reportedly sell traditional land rights to urban investors for this purpose. This practice started at the time of the 1984-1985 drought (Director Kassala Horticulture Department, interview Kassala, Dec. 1991).

4.2 Government SWC interventions

The Anglo-Egyptian Administration and the Government of the Sudan have been engaged in various activities in the field of water resources development in the Border Area. We only discuss in this study the SWC interventions. This will exclude the activities in domestic water supply, like the construction of *hafirs*, shallow wells and deep tube-wells. It also excludes other crop production interventions, like groundwater development. The main institutions involved are the regional Ministry of Agriculture (MANR) and its Departments of Soil Conservation (SCLUWP) and Range and Pasture Management (RPM). In the period 1966-1979, this also involved the Rural Water and Development Corporation.[8]

---

[8] The names of the ministry and its departments have regularly changed. MANR was earlier named Ministry of Agriculture; Ministry of Agriculture, Food, Natural Resources and Irrigation. SCLUWP was earlier named Land Use Department; Rural Development Department; Soil Conservation, Land Use and Rural Water Programming Department. The present-day NCDRWR (National Corporation for the Development of Rural Water Resources) was earlier named Rural Water and Development Corporation; Rural Water Corporation. It was part of different bodies, including the Ministry of Agriculture, Ministry of Animal Resources, Ministry of Cooperation and Rural Development, Ministry of Agriculture, Food, Natural Resources and Irrigation, Ministry of Energy and Mineral Resources, and the Ministry of Housing, Water and Public Utilities.

4.2.1 National policy and earliest interventions

National developments

The Anglo-Egyptian Administration had shown until about 1940 little developmental interest for areas located away from the Nile and its large-scale Gezira irrigation scheme (Sheperd *et al.* 1987). The only exception to this formed the Gash and Tokar deltas in eastern Sudan. These had already been visited by officials of the Egyptian Irrigation Service in 1892 and 1904 to assess their potential for irrigation development (DSA Tottenham 185/1, Trilsbach 1991). From the 1930s onwards, a series of SWC experiments in floodwater diversion was started in the Border Area region following earlier promising experiments in the Red Sea province (see below). The activities outside the Nile region and inland deltas of eastern Sudan remained incidental for the period of another decade, however. In the meantime, the great anti-erosion campaigns were gaining ground in the USA. These gave rise to the general idea of soil conservation (Jefferson 1952). The Colonial Development and Welfare Acts passed in England between 1940 and 1945. This facilitated the intensification of government interventions in the British territories of Africa, including in this new field of soil conservation (Adams 1992). A Soil Conservation Committee was set up in the Sudan in 1942. Its report published two years later (Sudan Government 1944) is a milestone. For the first time, the development of low-potential regions in the Sudan is being addressed. The Committee's Five-Year Plan contained 651 different projects in a variety of fields. These included the development of floodwater harvesting and rainland cultivation: the *hariq* experiments mentioned in section 4.1.2. However, the Border Area is not on this list (*Ibid.* 1944,10-12). It illustrates that the region still holds a relatively peripheral position in national planning of the 1940s.

After the publication of the Soil Conservation Committee report, two active periods of government intervention in low potential drylands can be discerned. These are 1947-1954 and 1960-1971 (Sheperd *et al.* 1987). The year 1954 demarcates a clear policy change. The government decides to confine all agricultural and pastoral development to those regions receiving "[..] over 400-500 mm annual rainfall in order to [..] fetch the fastest possible returns to the country [..]" (*Ibid.* 1987,71-72). Dryland interventions gained new momentum in the independent Sudan in 1960. The first cadre of Sudanese civil engineers had now been trained. Ambitious national programmes had been launched called "Anti-Thirst" and "Anti-Hunger Campaigns" (*Ibid.* 1987,51-55). However, continuous institutional reforms (*cf.* footnote 8) in what had become known as the "Water Organization" announced the end of also this second active period. The negative effects which resulted from the separation of the planning and executive bodies in the Water Organization are mentioned to have had a major disruptive effect (*Ibid.* 1987). Great international dryland development programmes were initiated in the Sudan in the 1970s. One of the first was the "Desert Encroachment Control and Rehabilitation Programme" (DECARP) supported by UNEP, UNDP and FAO. Its working area was in northwestern Sudan. Gradually, the regional scope of international programmes was expanded to cover other parts of the country, including eastern Sudan. The long neglected Border Area started to attract national and international development attention from the mid 1980s onwards. The devastating effects of the 1984-1985 drought have clearly served as a catalyst in this respect.

SWC interventions in the Border Area

The first government SWC interventions in the Border Area date from 1930. These projects were called "water schemes", "diversion banks" and "cultivation banks". The District Commissioner (DC) Barter on a trip from Kassala to Mamman reported on "[..] excellent sites for diversion banks [..]" in the Fagoda and Togan area (DSA Barter 646/2). The assistent DC of Sinkat, touring in the northern Border Area between Mamman and Odi in 1933, however reported that he is "[..] constantly on the look-out for cultivation-bank sites throughout the area, but the country does not lend itself to such operations [..]" (NRO Owen DCH/66.K1). The first experiments in the Red Sea province area date from 1929, where "[..] rough banks are made to get water out of the khors where these are incised and do not form a cultivatable level [..]" (DSA Barter 646/2). Dams were built of earth, sometimes reinforced with sand bags and brushwood wattle. One report mentions a dam of some 70 m long, 2 m high, with a 3 m base and a channel cut of 3,500 m^3. Another dam reportedly irrigated an area of 1,500 fd which produced 4,500 ardeb of sorghum (1,315 kg/ha). Several of these early dams were co-financed by the government and the Beja. One project in 1930 mentioned that the first paid 65 % of the construction costs, while the Hadendowa contributed the remaining 35 %. The total costs of the dam amounted to £ 61. It was laid down as a policy that banks should be tribal and not individual, because "[..] small banks are more of a nuisance than a profit [..]" (DSA Barter 448/1). Another policy in certain areas was that land users who refused to contribute to repair and maintenance were denied access to the irrigated lands. In other areas, however, the Hadendowa repaired damaged banks themselves, and built new dams without outside support (*Ibid.* 448/1). The average production in the 1930s under such techniques was about 6 sacks of sorghum per fd (1,315 kg/ha). It has not been mentioned whether or not the early interventions were forcefully sanctioned by the Anglo-Egyptian Administration.[9]

The first dams in the Border Area were built one decade later. In the area of Um Safaree and Hafarat, a cultivation bank was constructed in khor Aderjewab in 1941. Another one was built in Khor Obilit near the present-day village of Awad (NRO Sandison KD/2.N.1). The Land Use and Rural Water Development Department formulated a special programme for the northern Border Area in 1962. This was called "Togan-Mamman Waterspreading". However, no funds were available for intervention activities at that time. Another project was proposed in 1964 in khor Mindoweet near Telkook. Also this work was never executed because the required heavy-duty equipment to be transported from western Sudan by rail arrived out of season (Baasher, interview Kassala, Nov. 1991). The Rural Water and Development Corporation executed two projects for floodwater diversion in the areas of Hameskoreib and Tahadai in the mid 1970s. The first supports the centre of the Ali Betai movement. The project includes a 270 m stone-masonry dam of 3.2 m height and a 29 million m^3 water reservoir covering an area of 40 km^2. Two years after its construction, however, the feeder khor changed its course and put the dam out of service

[9] This was a regular practice in many SSA countries at that time (*cf.* IFAD 1986). No reference of this was found in the archival sources and field interviews for the situation in eastern Sudan however.

(Baasher, interview Kassala, Nov. 1991, files of the NCDRWR office in Kassala). In 1988, an area of 60 km² was again flooded by this dam for the first time in seven years (OXFAM 1988). The second project in Tahadai includes a series of 3 earth dams with dry stone-pitching protection and spillways. These were built by the Rural Water and Development Corporation in 1977. The dams are positioned on the approximate terrain contour and measure 1,550 m each. They were still in place and in use in 1990, despite some damage incurred as a result of breached sections (files NCDRWR Kassala). Until the initiation of the Border Area Earth Dam programme by the Kassala Soil Conservation Department in 1983, a total of four SWC projects had officially been registered in the then Kassala province and a total of eight in the then Red Sea province (Salih & Khadam 1982).

4.2.2 The Border Area Earth Dam programme

Regionalization in 1981 (*cf.* section 3.2) provided entirely new opportunities for SWC interventions in the Border Area. Firstly, the regional Ministry of Agriculture could from now on prioritize its own budget allocations. Secondly, the change of power had placed Hadendowa politicians at high ranks in public administration. The outcome was that the Soil Conservation Department in Kassala was given the task to formulate a specific programme for the Border Area. This is based on the earth dam technique earlier used. The implementation in the field, however, does not include stone-pitching protection and spillways. The objectives of the Ministry of Agriculture for this period have been stated as follows (GOS/GON 1988,22):

> *(i) to increase food production; and (ii) to combat desertification.* At the end of 1988 a few more specific objectives were formulated in the Four-Year Plan of the ministry which include: *(i) "to stimulate agricultural production for increased food security; (ii) to stimulate exports of agricultural products; (iii) the rehabilitation of drought-affected areas; (iv) to stimulate agricultural settlement of nomads and refugees; (v) to encourage agro-industries; (vi) the promotion of sound animal husbandry techniques and veterinary services; and (vii) to increase the productivity of the range lands by improved management techniques"* (GOS/GON 1990,25).

The objectives of the Border Area Earth Dam programme of the Soil Conservation Department have been formulated in rather general statements such as (Ahmed 1987,2):

> *(i) "to increase the area of irrigated land; (ii) to increase the area of cultivated land; (iii) to increase crop yields; (iv) to increase farmer incomes; (v) to contribute to the growth of the national economy; (vi) to improve natural range land; and (vii) to reduce land degradation".*

The Soil Conservation Department with its headquarters in Kassala has five more offices in eastern Sudan. These are located in New Halfa, Gedaref, Sinkat, Port Sudan and Tokar. The total executive staff is 33, of whom 11 were employed in Kassala in 1992. In the next discussion, we refer only to the Kassala office. The available equipment for projects includes three tractors and one loader.[10] The project budget is discussed in section 7.1.1 Efficiency. The office includes the sections of Soils and Air Photographs, Land Use and Water Programming, Socio-economic Surveys, Projects and Planning and Follow-up. This organization reflects the department's planning status in the former Water Organization (*cf.* section 4.2.1 *National developments*). This background expectedly is also the reason why the department continued with the simple earth dam concept without more elaborate stone-pitching and spillways. These had always been implemented in the field by the executive body in the Water Organization: the Rural Water and Development Corporation. Figures 4.4-4.6 show the working areas and performances in the Border Area over the 1982-1990 period.

Earth dams[11]

The working procedures of the Soil Conservation Department in the Border Area Earth Dam programme include the following phases. (i) Potential project locations are identified from a list of valid requests. A valid request is made by a group of land users with access to cultivated lands suitable for floodwater harvesting. In practice, this means that the area is flooded by a khor with an adequate seasonal discharge. (ii) SCLUWP staff visit the proposed location. They assess whether the following two conditions apply. Firstly, there must be reason to believe that these lands are already being cultivated by the group of applicants. Secondly, there must be reason to believe that the area is free of tribal conflict. However, there are no hard and fast rules to form a judgement. Practically, the procedure followed is one of consulting as many different people and groups as possible locally. (iii) When these conditions are met, the location is approved as a future project and included in the following year's work plan for field survey. (iv) The first activity at the approved project location is to execute a land use and socio-economic survey. The precise contents may differ per location. However, the survey always contains details on population, livelihood activities, cultivated crops and local rules for floodwater use. The claimed entitlements to land and alleged absence of conflict are also field-checked in this phase. (v) After the completion of this survey and writing of a report, a grid survey is made of the area. The output is a 1: 2,500 map indicating terrain contours at 0.5 m intervals. (vi) According to information from the survey and map, the standard design for earth dams is adjusted where required. After this, the location is included in the following year's work plan for construction. Per year, a cluster of such locations is identified by the technical

---

[10] The department purchased a loader in 1985. A first tractor was locally bought in 1987. Two new tractors financed by MFEP-KADA became operational in 1988.

[11] The following account is based on interviews (Director and staff of the Department of Soil Conservation, Kassala, Dec. 1991-Jan. 1992) and the department's Annual Reports 1982-1993.

Figure 4.4 **SWC projects of the Department of Soil Conservation in the Border Area**

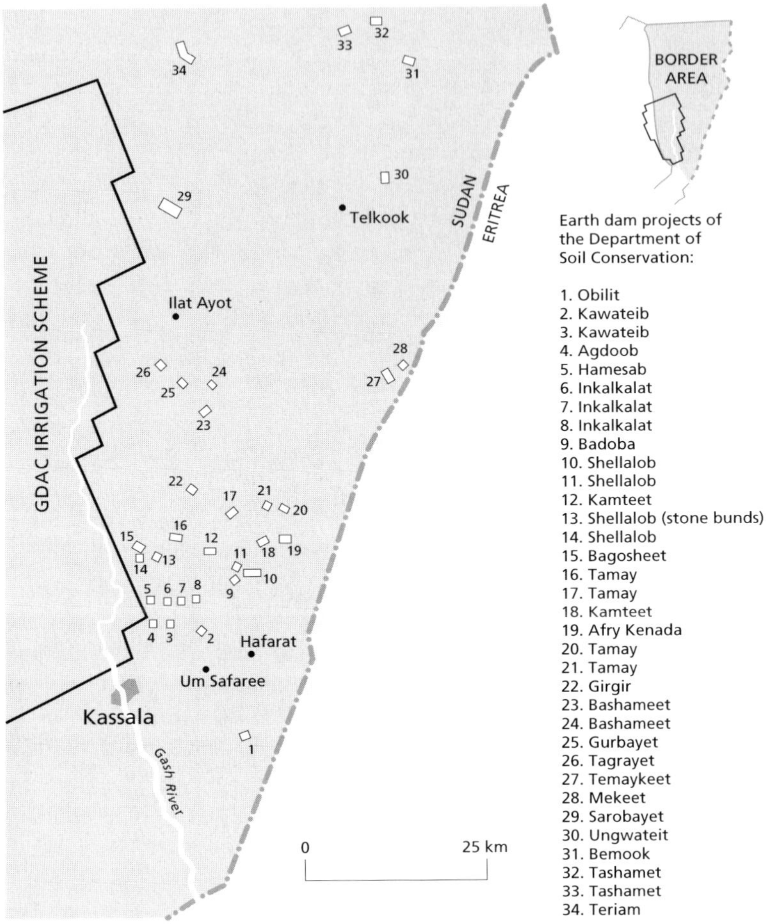

field staff for development. (vii) The construction activities start with the pegging of contour lines. The dam itself is raised mechanically by tractor or loader. There are no costs involved for the group of project applicants. (viii) The execution of protection works with brushwood and sand bags is considered to be the entire responsibility of the applicants. However, the Ministry of Agriculture did proclaim a new policy in this respect in the drought year 1984. The policy is still effective, and directs the Department of Soil Conservation to provide also all required materials for protection, repair and maintenance free of charge. The department, in addition, is to pay daily wages for casual labour [..] in order to generate incomes for the local people [..]" (Ahmed 1987,1). (ix) Since 1985, grain seeds and tree-seedlings are being distributed free of charge as part of the programme. The seeds are mainly local and hybrid short-maturing and drought-resistant sorghums. Also seed of millet, okra, watermelon and rosella have been distributed. Tree-seedlings distributed include: *Acacia mellifera*, *A. nilotica*, *A. seyal*, *Prosopis juliflora*, *Parkinsonia acuelata* and *Hyphaene thebaica*. Since 1987, mechanical preparation of the cultivated lands by tractor is an optional activity in the programme. Land users must pay for this service, but the rates are subsidized. The envisaged future activities of the department in the Border Area include the distribution of manure use to improve the soil structure and fertility. Tentative formulations of future policy refer to making manure use conditional to the continued receipt of free SWC intervention support.

Figure 4.5 **Construction and maintenance of earth dams in the Border Area (in m^3), 1982-1990. Source: Annual Reports SCLUWP 1982-1990**

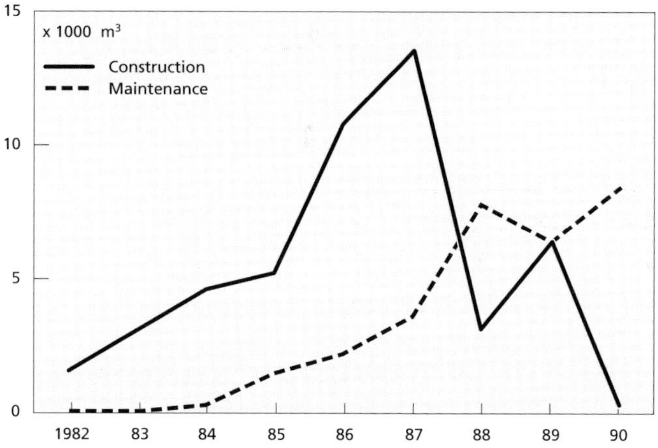

Figure 4.6 **Number of earth dam projects in the Border Area, 1982-1990. Source: Annual Reports SCLUWP 1982-1990**

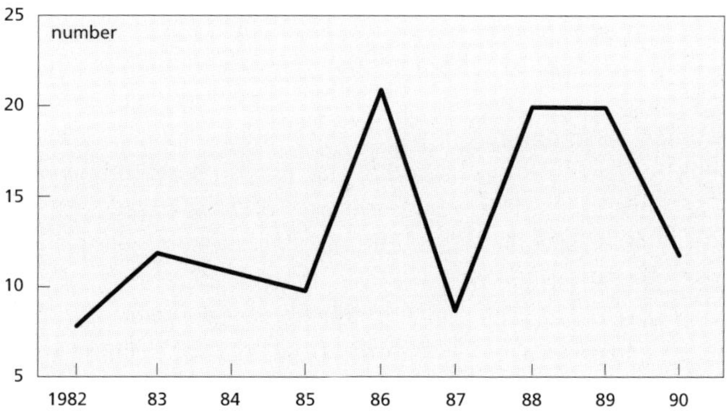

The technical details of the Border Area earth dams are given in Ahmed (1987). Three basic types are in use in accordance with the main dam functions. These are either the blocking of smaller khor branches, floodwater diversion, or regulation of captured water on the cultivated lands (*cf.* section 3.4). In the first two situations, the dam is raised in the water course. These dams have a standard design of 3 m base width, 1 m top width and a height of 2 m with 1: 2 slopes. 1 metre dam length, accordingly, equals approximately 2 m^3 dam volume. These structures are straight and usually built diagonally across the khor bed. The length of the dams in the Border Area varies between 50-200 m. This is mainly determined by the khor width. The third type of dam on the cultivated lands is largely similar, except for its shape. These spreader dams are built in series. They are frequently winged at 45°-60°-angles, on one or both ends, pointing either upslope or downslope. The shape is mainly determined by the terrain characteristics. The position in the direction of the slope is exactly one dam after the other, or staggered. The intervals between subsequent dams in series is variable and depends on local topography. Most project locations developed by the department in the Border Area are almost flat. The gradients vary between 1:300 to 1:400 (about 0.3 % slopes). The length of spreader dams varies between 100-300 m. By rule of thumb, a package of one diversion and one spreader dam of 500 m length each can be built by the department in one month. This includes supervision by SCLUWP staff and the application of protection materials by local daily paid labourers.

Observations in 1990 by the MFEP/KADA backstopping mission (Mulder 1990) indicated that the above formulated design criteria are usually not adhered to. The mission found both oversized and undersized dams. It also commented that monitoring by the Department of Soil Conservation was generally poor. The only records kept concern total lengths or volumes of dams constructed and maintained. These are assessed once after completion of the activity. There is no follow-up to monitor the intervention performances. Data collected on the total areas irrigated and cultivated are obtained by driving around the circumference by car. This is a crude measure and is also not performed regularly. Finally, the input of diesel, oil, lubricants and spare parts of the fleet of cars, tractors and the loader is recorded. However, the data cannot be linked to the physical output of single projects. The project costs therefore remain unknown. Several observers had already reported earlier on the generally poor performances in the Border Area of these types of dams (ILO 1986, Musa *et al.* 1989). This mainly concerned technical shortcomings and these were also outlined by the mission.[12] The problems are acknowledged by the Department of Soil Conservation. The department, in addition, also raises the logistical problem. The poor performances partially derive from the limited capacities of the department to carry out adequate maintenance. This point is illustrated by figure 4.5. This figure also suggests that no more new dams can be built in the Border Area, because most existing ones need maintenance.

---

[12] (i) The earth dams are too small and too weak. (ii) The crests are not at a constant height and therefore liable to overtopping by floodwater. (iii) Erosion protection measures are inadequate especially at the dam ends. (iv) Loose earth is an inappropriate building material. (v) No proper site preparation is performed to prepare the foundation and anchoring of the structures (Mulder 1990,18).

Teras rehabilitation

The Department of Soil Conservation in Kassala had already given occasional support to *teras* users in the Border Area since 1988. This new field of activities received the formal approval of the Ministry of Agriculture in September 1991 as part of a new programme called "Traditional Cultivation".[13]

Table 4.2 **Teras construction and maintenance by the Department of Soil Conservation in the Border Area, 1988-1990**

YEAR	NO. OF SITES	NO. OF TERAS	BUND LENGTH MAINTAINED (m)	TERAS AREA MAINTAINED (fd)	TERAS AREA PLOUGHED (fd)
1988	7	94		270	
1989	1	33	13,200		
1990	4	28	93,904		53

Source: SCLUWP Annual Reports 1988-1990.

The main activity is the rehabilitation of existing *terus* by means of tractor and loader. Land users are charged for the work. The demand for this service is high. Accordingly, a maximum was set to the total length of bunds treated per applicant. The prices were not fixed in the first year 1988. In 1989, maintenance of bund lengths up to 300 m costed £s 100 when carried out by tractor. This was £s 200 when carried out by loader. In 1990, these prices were respectively £s 0.5 and £s 1 per m of bund length maintained. Land preparation by tractor was performed in this same year at £s 75 per fd. New *terus* have been raised by the department outside the Border Area on the Clay Plains west of Kassala town from 1988 onwards. Table 4.2 shows the available data on *teras* rehabilitation work by the department in the Border Area only, however.

4.2.3 The MFEP/KADA pilot scheme

The Department of Soil Conservation received support from MFEP/KADA in the period 1987-1991 to introduce new SWC techniques in the Border Area. The new techniques were to solve the main difficulty with earth dams, namely their vulnerability to breaching. The MFEP/KADA objectives with respect to the specific task of support to SWC interventions were formulated as (Kraayenhagen 1987,19 and largely based on Green Mission 1985):

> *(i) "To restore the ecological balance [..] arresting the deterioration by erosion and desertification of the Border Area [..] caused by mismanagement of natural resources; (ii) to upgrade the quality of life of the community, including in the fields of nutrition, water, health and education; (iii) to increase agricultural production, particularly in rain-dependent agriculture, forestry and livestock husbandry, to overcome the malnourished state of much of the population and*

[13] Which, in turn, is part of the 1990-1993 National Economic Salvation Programme.

> *livestock, and for the country to become self-sufficient in food supplies; (iv) to reach an integrated rural development programme based on sound agricultural development through the regional administration, considering the socio-economical conditions and technical viabilities and its replicability for alternative areas elsewhere in the Eastern Region, with emphasis on people's participation and motivation".*

The short-term goals of MFEP/KADA support were formulated as (Kraayenhagen 1987,8):

> *(i) "To develop waterspreading systems that can utilize additional run-off [..] to supplement rainfall thereby increasing agricultural production; (ii) to extend the present agricultural areas by providing economically viable waterspreading systems [..] to demonstrate the optimum land use possibilities to include [..] fodder trees and grasses, and [..] village woodlots; (iii) to develop a [..] system which permits surface and drainage run-off water to penetrate the soils between the khors and crest boundaries located at the higher lying ridges of the surrounding sub-watersheds, and which can be used elsewhere considering its replicability; (iv) to provide technically trained staff for the planning and surveying of waterspreading systems; (v) to stimulate self-reliance through active people's participation within the target population, to promote self-help schemes; (vi) to monitor and evaluate the proposed waterspreading systems for its impact; and (vii) to carry out applied field research in order to improve and modify initially introduced waterspreading systems".*

These MFEP/KADA activities were separately managed in the Border Area Pilot Project (BAPP). BAPP received support for research and monitoring from the Sudanese National Council for Research (NCR) in the WARK project in the period 1989-1991. These activities were part of the NCR programme "Field Studies in Sudan's Rainlands". Not all activities of the Department of Soil Conservation were supported in BAPP. The department at the same time continued the implementation of its own Earth Dam programme. Several other SWC projects have been proposed for the Border Area, but did not come into effect. These include a labour-intensive public works programme to be implemented at five locations under ILO and UNDP auspices. The proposed intervention was based on improved earth dams similar to the type built by the Rural Water and Development Corporation in the 1970s (*cf.* section 4.2.1 *Border Area SWC interventions*). The total costs to develop 28,000 fd (12,000 ha) in the Border Area would amount to $ 1.2 million, which is $ 100 per ha (ILO 1986). The second proposal concerned a spate irrigation scheme to be developed under auspices of a Chinese *ad hoc* group of experts. Only one of the five locations proposed is in the Border Area. The total costs to develop 10,000 fd (4,200 ha) here would amount to $ 1.6 million, which is $ 380 per ha (Ad Hoc Chinese Experts Group 1989).

The Department of Soil Conservation, together with MFEP/KADA and BAPP, introduced two new SWC techniques in the Border Area. These are floodwater harvesting by using a system of (i) broad-based earth embankments; and (ii) stone bunds. The first was tested in a pilot scheme built in 1987 in the Hafarat farming zone of Hedadeib. This intervention will be referred to in the following as the "Hedadeib pilot scheme". The second was tested in a pilot scheme built in 1990 in Shellalob some 20 km north of Kassala town. In addition, the activities in Hedadeib had been preceded by trials at the Mokram experimental farm of the Ministry of Agriculture on the northern outskirts of Kassala town. These Hedadeib activities were also followed by trials in a second pilot scheme located downstream of the first in an area called Kawateib (*cf.* figure 3.1). In this study, we only refer to the government interventions made in the Hedadeib pilot scheme, for these are the best documented of all (Hashim 1987, El Mosbah 1987, Kraayenhagen 1987, KADA 1987a, 1987b, Cosijn & Van Dijk 1989, Musa *et al.* 1989, WARK 1989, Van Dijk 1990, 1991, Van Dam & Houtkamp 1992).[14] The following sections mainly draw from these reports.

Initial design

The basic water harvesting elements (*cf.* section 1.4.2) in Hedadeib are a rock outcrop catchment, and a system of earth embankments and shallow channels on the cultivated area. The impervious rock outcrop intercepts rainwater. This reaches the pilot scheme as floodwater run-off through a shallow natural khor. The embankments and channels subsequently distribute this water over the cultivated lands. Here, the water infiltrates into the soils and contributes to crop production. Most run-off discharged in the khor is intended to flow over the embankments. Only a small amount is intended to be intercepted in the system of channels for diversion away from this khor. The channels are built for this purpose along the embankments with a gradient of a 1: 1,000 fall towards the watershed crest. These "near-contour" embankments are built in series on a sloping plane. This causes the captured water not only to spread laterally via the channels, but also to move upslope until it meets the crest and the next upslope embankment (figures 4.7-4.9). The advantages when compared with earth dams would result from the use of small run-off amounts. This was expected to reduce the breach hazard of the structures. The advantages when compared with the existing indigenous SWC and non-SWC techniques would result from the increase in the area that could be cultivated (Kraayenhagen 1987).

The basic dimensions of the structures used are given in table 4.3. The optimal spacing of embankments (horizontal interval *hi*) is determined from the terrain slope and the design height of the embankments (vertical interval *vi*) according to equation 1 (figures 4.8 and 4.9):

---

[14] The experiments at Mokram, Shellalob and Kawateib are documented in the work reports of the Kassala Department of Soil Conservation, the progress reports of MFEP/KADA and El Faki (1991).

$$hi = vi \times 100 / s \quad (1)$$

where:

- hi = horizontal interval, or distance measured from the embankment crest to the foot of the next upslope embankment
- vi = vertical interval, or embankment height
- s = percentage slope

Figure 4.7 **Schematic pattern of embankments and channels**

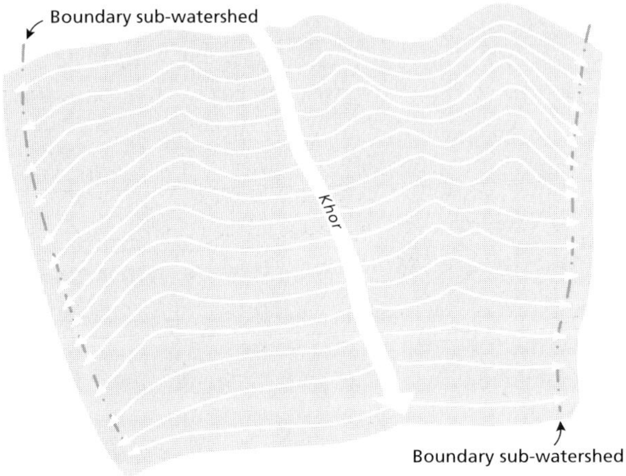

Figure 4.8 **Measurements of embankments and channels**

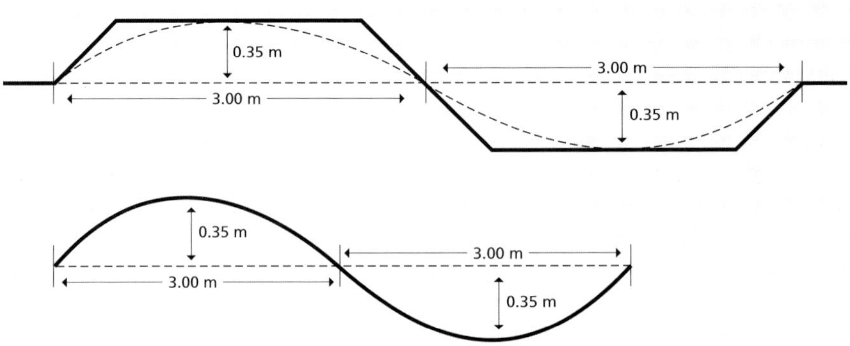

Figure 4.9 **Throw back of run-off between two embankments**

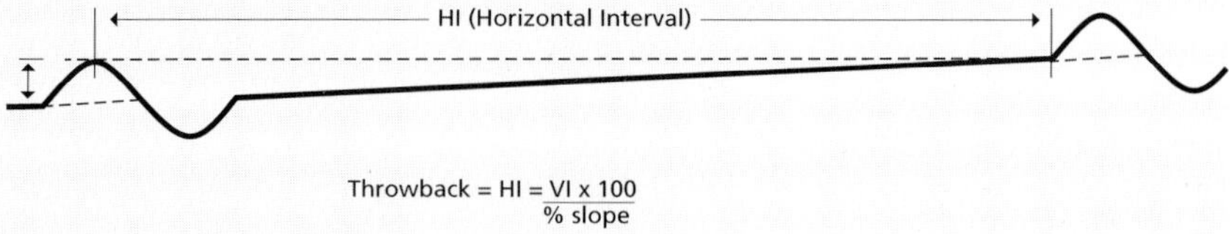

The optimal conditions for floodwater harvesting were never realized in the Hedadeib pilot scheme for three main reasons: (i) several components were purposefully not built according to the masterplan design (*cf.* Kraayenhagen 1987); (ii) several other components built according to design were of poor standard; (iii) necessary measures to protect earth embankments against floodwater did not materialize. The arguments for abandoning the design have never been properly documented by the Soil Conservation Department or MFEP/KADA (Musa *et al.* 1989). With respect to the latter two reasons, it was later acknowledged that the technique is too sophisticated for implementation by the Department of Soil Conservation (Mulder 1990) and that the harshness of the Border Area environment had been substantially underestimated (Van Dijk 1991).

Table 4.3 **Basic dimensions of broad-based embankments and prevailing terrain characteristics in the Hedadeib pilot scheme**

Slope of terrain	0.5 to 0.8 %
Vertical interval (embankment height)	0.35 m
Horizontal interval (embankment spacing)	40 to 70 m
Base width of embankment	3 m
Channel depth	0.35 m
Channel width	3 m
Average embankment length (1990)	625 m
Channel gradient	0.1 %

Source: Van Dijk (1991).

Implementation in Hedadeib

A first proposal to implement the pilot technique concerned the Idi Kedib area in the village of Um Safaree. The activities started early 1987. However, the population refused to co-operate further after the completion of the first four embankments (Musa *et al.* 1989). The official argument was that the project would negatively affect the water supply to the village *hafir*. Unofficially, informants expressed the fear that the project would eventually lead to land alienation by the government. The BAPP activities also comprised a forestry component and according to 1932 Central and Provincial Forestry Acts, all afforested land is government property. A new location was selected in the Hedadeib farming zone of Hafarat. According to Kraayenhagen (1987,32) reporting on the situation in Hafarat in March 1987 [..] it became clear that the proposed approach would be acceptable to most people [..]. We can only speculate why the same fears of land alienation have not been expressed here. A first reason could be that Hedadeib is of smaller overall importance to the village economy than the proposed area in Um Safaree. This is shown in the outcomes of the L&E Survey (*cf.* sections 6.4, 6.5). A second reason could be that the expected gains of intervention, which we argue in section 7.1.3 involve considerable fringe benefits, were perceived in Hafarat to outweigh any chance of loosing (relatively unimportant) land. The local BAPP forestry activities at any rate did not result in land alienation in Hafarat.

Figure 4.10 **Initial design of the pilot scheme in the area of khor Hedadeib. The area of actual implementation is shown in the frame**

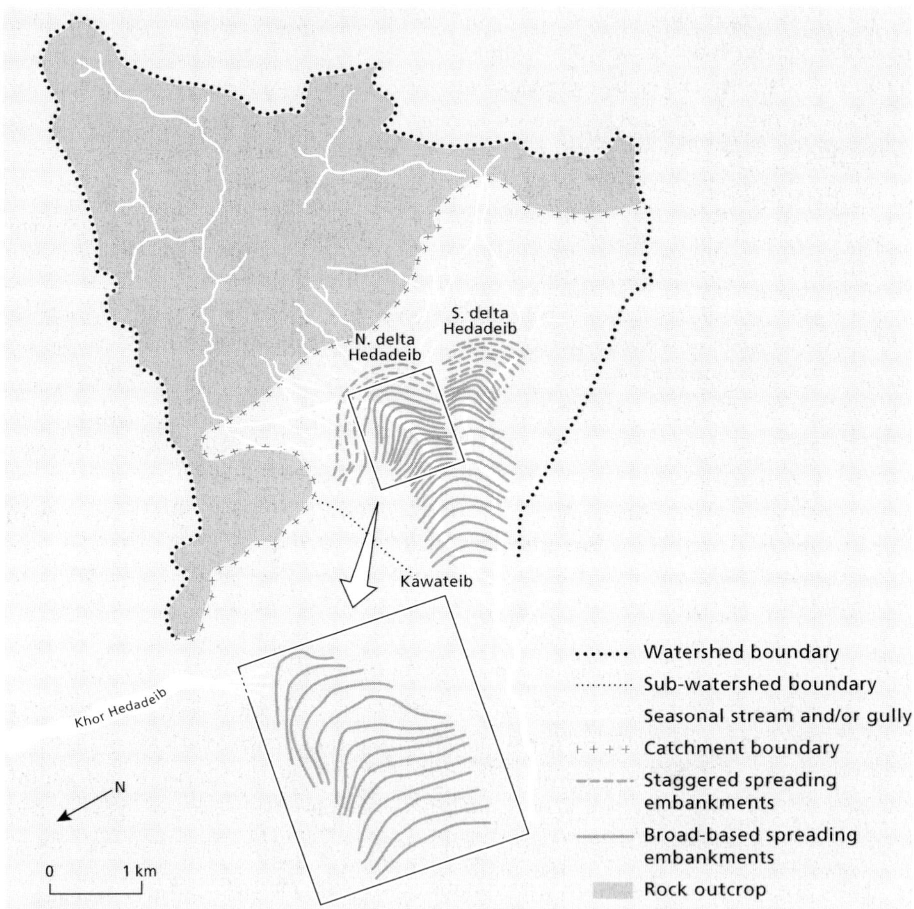

The area demarcated for development is located in the upper tributary of khor Kawateib. This section is called khor Hedadeib (figure 4.10). Physiographically, the pilot scheme covers the upper alluvial fan and part of the alluvial slopes of khor Hedadeib. The soil textures are highly variable loamy to clayey. Vast areas are covered by a sandy top layer. Socio-economically, the area is one out of a total of eleven farming zones cultivated by the population of Hafarat. Mainly wildflooding techniques and some *terus* were used here before the establishment of the scheme. The Department of Soil Conservation initiated the work according to its normal schedule. However, in the case of Hedadeib, the socio-economic and land use survey was made after construction work had already begun. Time pressure due to delays in Um Safaree is the most likely explanation for this. In September 1987, a field report of Department of Soil Conservation mentions that in the downstream area of Hedadeib "[..] conflict arose between farmers and the technician in charge of land preparation [..]" (Hashim 1987,1,9). Only at this phase of development, the area proved to be cultivated by a relatively heterogeneous group of land users of rivalling tribal sections. This caused regularly returning disputes over land (*cf.* sections 6.5, 6.5.3).

A total of 93 ha including the disputed downstream zone had been developed and put under earth embankments by the end of 1987. However, these downstream areas were excluded from further activities in 1988. In this year, also, the embankment lay-out in the upstream section was adjusted. By the end of 1988, 75 ha of cultivated land was under command of 14 earth embankments. The total length of the embankments reached just over 10 km. These 75 ha receive floodwater from a total catchment of 875 ha. An area of 710 ha (81 %) of this is located in the Jebel Haura mountain range and 165 ha (19 %) is on almost bare pediment slopes.[15] The *c: ca* ratio (*cf.* section 1.4.2) is about 1: 12. When this is compared to indigenous SWC in the Border Area (*cf.* section 3.5), the catchment is 4-6 times larger and mainly includes rock outcrop. Substantially greater amounts of floodwater are therefore used. This was considered beneficial for the expansion of the total cultivated area.

Land users in the Hedadeib pilot scheme cultivate their "own land". No policies of re-allocation of land titles, or procedures for admission have been applied. The total number of households in Hafarat entitled to cultivated land in the Hedadeib area is 59. A total of 30 hold titles in the upstream lands developed from 1988 onwards.

Deviations from initial design

A number of deviations from design can either directly or indirectly be attributed to the land dispute. Most importantly, this dispute resulted in a partial relocation of the pilot scheme. The embankments were built further upstream along the khor than proposed in the plan. Probably, this was to compensate for the downstream "loss of land". The consequences were twofold. Firstly, the conditions of run-off supply changed. Secondly,

[15] The total is 520 ha when only the sub-catchment of khor Hedadeib is considered. Of this, 484 ha (93 %) is rock outcrop and 36 ha (7 %) is on bare pediment slopes.

the qualities of local soils for the building of embankments changed.[16] The type of embankment proposed, however, was not adjusted to any of these new conditions. In addition, also a range of other complicating factors were faced at this new location. Floodwater simultaneously arrives in this area from different directions. However, the gradient of embankments and channels was designed to distribute water arriving from only one source: khor Hedadeib. As a consequence, all run-off supplied from the eastern pediment slopes is collected in the first upstream channel. After this, it is directly routed to the watershed crest without making any contribution to irrigation in the pilot scheme (figure 4.10).[17]

Imperfect implementation

Besides deviations from design, some work executed in the field was also of a low standard. To some extent, this is related to the characteristics of the less favourable upstream location. Embankments cannot properly be built in more sandy areas. Small and localized deviations in their dimensions turned out to have great negative effects on the functioning of the entire system. It was estimated that some 15-20 % of the cultivated lands developed was not under irrigation command of earth embankments as a result of imperfect implementation.[18]

---

[16] In the initial design, irrigation water is to be supplied by overland flow which is evenly discharged over the alluvial flats by a naturally spreading khor. At the new upstream location, contrary to design, the scheme was built on the alluvial slopes. Locally, more than one water course occurs, these water courses are more pronounced, and the bulk of run-off arrives in a concentrated flow. In the initial design, the building material for embankments is taken from the alluvial flats. These consist of loamy and clayey soils. After relocation, contrary to design, embankments were built with more sandy material from the alluvial slopes and locally more common khor beds.

[17] More detailed design deviations include the following. (i) A proposed system of staggered embankments to be located upstream of the pilot scheme was not built. Expectedly, no land was left for this after the expansion of the scheme area into this direction. The staggered embankments were meant to maximize run-off spreading, to reduce the run-off velocities, and to induce siltation before run-off would arrive at the first embankments in the scheme proper. (ii) 5 out of 14 embankments were built on coarse-textured alluvial slopes instead of on heavier soils of the alluvial flats. (iii) Sections of earth embankments which cross the smaller drainage lines were constructed without the application of the additional protection measures proposed. (iv) The total area developed in Hedadeib in 1988 (75 ha) is smaller than the area proposed. The southern delta of Hedadeib and the area of Kawateib were never included in the project (figure 4.10).

[18] More in detail, imperfect implementation includes the following. (i) Embankment and channel dimensions deviated from the design. (ii) The construction difficulties had built up a time pressure. Construction work in Hedadeib was therefore partly executed by road grader instead of tractor. This lowered the credibility of techniques which, after completion, were to be maintained locally with simple means.

Table 4.4 **Materials tested in the Hedadeib pilot scheme for the protection of earth embankments, 1988-1991**

BIOLOGICAL MATERIALS	TYPE	PROPAGATION
Cenchrus ciliaris	grass	seeds
Cenchrus setigerus	grass	seeds
Paspalidium desertorum	grass	seeds
Panicum tergidum	grass	seedlings
Panicum coloratum	grass	seedlings
Cynodon dactylon	grass	seedlings
Vetiver zizanioides	grass	seedlings [A]
MECHANICAL MATERIALS	TYPE	APPLICATION
ENKAMAT 7010	fabric	covers
ENKAMAT 7010 with		
– C. ciliaris and C. setigerus	fabric and grass	covers and seeds
GEOTEXTILE HF1200	fabric	covers
Sand bags	bags	piles secured with brushwood
Acacia tortillis, Prosopis juliflora	brushwood	wattled panels

Source: Cosijn & Van Dijk (1989), Van Dijk (1991). Note: [A] not surviving in the Kassala FNC nursery.

These new upstream conditions made the development of suitable techniques for the protection of earth embankments urgent. This became one of the main research efforts of the project in the subsequent years. Table 4.4 lists the different materials tested for this purpose. The low-cost option of biological protection measures was part of the initial design of the pilot scheme. However, these measures never reached the operational stage. Trial plots in Hedadeib were either washed away after flooding, or withered in the dry years.[19] The best alternative in terms of performances, costs and social acceptability proved to be the use of brushwood wattled panels and sand bags. These are indigenous techniques locally applied for the purpose of protection of earth SWC structures (Van Dijk 1991). The imported industrial fabrics listed in table 4.4 were only used as an emergency solution for the sake of research and monitoring. Their performance was mediocre and their costs are beyond any viable application in the Border Area (Cosijn & Van Dijk 1989). The problem of embankment protection has never been solved in Hedadeib. The average damage caused by breaches remained a high 20 % over the 1987-1989 period (*cf.* section 6.5.3). From 1989 onwards, the embankments were no longer repaired after breakage, and more attention was given to indigenous techniques as an alternative for the purpose of repair. An evaluation of the original pilot scheme design is therefore only

[19] In addition, these trials indicated that biological protection is not feasible locally. Cheap grass seeds are easily disseminated by hand and remain dormant in dry years. However, they easily disperse over the cultivated lands. Land users dislike the technique because it raises the weeding-demands. Grass seedlings are less likely to infest the cultivated lands. However, they incur higher costs for nursery treatment, planting, and initial irrigation when transplanted on the embankments.

possible for the years up to 1989. These performances are discussed in section 6.5.3. The overall effect of these government interventions on the household economies in Hafarat is evaluated in chapter 7.

4.3 Changing livelihoods of settling nomads

Regional studies of eastern Sudan and the Border Area (*cf.* Tothill [1948] 1951, Barbour 1961, Lebon 1965, IES 1991) acknowledge that Beja livelihoods have entered a process of transformation from mainly nomadic types, to more semi-nomadic and agro-pastoral livelihoods. This development gained momentum in the 1950s. It started somewhat earlier for selected groups, such as the Beni Amer. They took up more sendentary lifestyles from the 1920s onwards (*cf.* section 4.3.3). Such newly emerging livelihood profiles are usually defined according to the criterium of either household mobility, or household involvement in non-livestock husbandry activities (Toulmin 1983, *passim*). The first provides no meaningful distinction for the case of the eastern Sudan. The nomadic groups do not move very far and sedentary crop producers also depend to a great extent on mobile livestock (Sörbö 1991). Agro-pastoralism is, for this region, better defined according to the second criterium. Strictly speaking, most households in the Border Area would (temporarily) not classify as agro-pastoral when the 1980s-1990s period is considered. They would be called mixed-economy households according to excepted definitions (*cf.* section 6.6).[20]

We refer to them in this study as "multi-resources economy" households. This underlines the variety of livelihood activities the households of the Border Area were found to engage in during this particular period. In the next sections, we discuss for the Border Area the historical developments in livestock and non-farm activities (section 4.3.1, 4.3.2), the factors commonly mentioned to have influenced sedentarization in this area (section 4.3.3), and factors commonly mentioned to play a role in the adoption of SWC techniques in the drylands in general (section 4.3.4).

4.3.1 Livestock developments

No archival records have been consulted for the purpose of documenting historical developments in the livestock sector of the Border Area region. The official data available are at the provincial level and cover the period 1976-1990. These expectedly are of low

---

[20] Households with gross income contributions from livestock and related activities of 50 % or more are called "pastoral". Households with 10-50 % of livestock and related contributions, and 50 % or more crop production contributions are called "agro-pastoral". Households with livestock (related) contributions of less than 10 %, but more than 50 % contributions from crop production are called "agricultural". Finally, households are called "mixed economies" in all other combinations where non-agricultural incomes contribute more than 50 % (Wilson 1986,14-15; Dietz 1987,15; Rutten 1992,13-14).

reliability.[21] Table 4.5 therefore only shows selected years based on this source. Other data can be found in Harrison (1955) and in the 1975 National Livestock Survey. The most recent figures for the study region were collected in 1986 as part of the BAPP activities (MANR 1988).

Table 4.5 **Livestock numbers in the Kassala province, (selected years)**

	No. of camels	No. of cattle	No. of sheep	No. of goats
1976/1977	618,971	726,657	1,558,980	1,003,571
1983/1984	672,875	881,620	1,956,783	1,152,788
1984/1985	605,587	528,972	1,565,426	1,037,509
1985/1986	612,854	543,783	1,617,085	1,058,259
1986/1987	620,209	559,009	1,670,449	1,079,424

Source: Animal Resources Administration Kassala. Note: years are fiscal years. For example, 1976/1977 covers the period from July 1976-June 1977.

Figure 4.11 **Livestock numbers in the study region. Source: Harrison (1955), MAFNR (1976), MANR (1988). Note: the area "total Beja" is about 20 times larger than the area "total Border Area"**

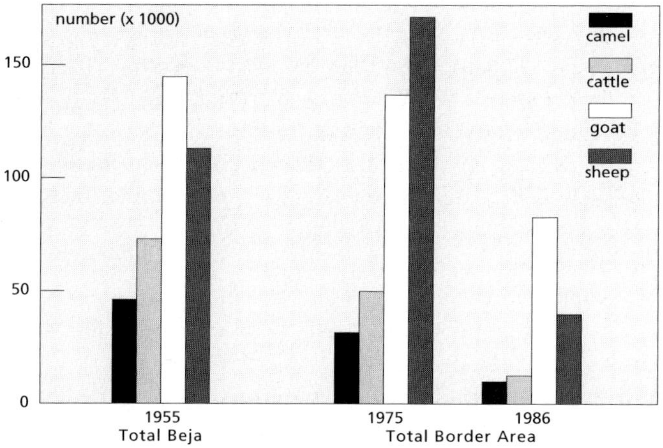

The data given by Harrison are for 1955. They refer to the total livestock of the Beja tribe. The area of reference is the 248,000 km² zone of the Hadendowa (Beja) District and Kassala Rural District combined. The 1975 and 1986 data are selected for the 12,000 km²

[21] Official livestock numbers regularly increase over the years until the drought years 1984 and 1985 when they drop with exactly 10 % for camels and goats, 20 % for sheep and 40 % for cattle.

area of "strata 41, 42, 45, 46, 56" in MAFNR (1976b), and "strata II and III" in MANR (1988,6). This area approximates the Border Area study region. Two conclusions can be drawn from them (figure 4.11). Firstly, livestock numbers in the Border Area are likely to have substantially increased between 1955-1975. The two levels of livestock totals are comparable, but the Border Area is much smaller than total Beja. Secondly, significant losses of livestock were suffered between 1975-1986. These are known to have been caused by the 1984-1985 drought (see below).

4.3.2 Non-farm livelihood categories

Non-farm refers in this section to all livelihood activities other than crop production and livestock husbandry. The earliest references to non-farm activities of the rural Beja are given in the accounts of European explorers. Burckhard mentions their trade of handicraft which included: "[..] a variety of mats and baskets [..] pots for cooking and ablution [..] camel saddles, ropes of reeds, hides, water skins, a few fowls, dried camel flesh, bark of the *tama* tree, gum arabic, *Acacia* pulses for tanning leather, salt, and black ostrich feathers [..]". Artisans were working as "[..] blacksmiths [..] mending knives [..] lance heads and camel gear [..]". Trade connections of the Kassala region, then named *Taka*, extended to the Red Sea coast and the Nile valley (Burckhard [1819](1987,398-400). James travelled in the Kassala area in 1881-1882. He mentions trade in wild animals. Lions, leopards, wild cats, antilopes, giraffes, ostriches and ant-eaters were locally being hunted for European zoological societies (*Ibid.* [1883] 1969). Although the Beja are not mentioned by name, they are likely to have assisted on these expeditions made on their tribal land.

The Beja have engaged in commercial camel porterage since at least the early decades of the 20th century. Morton & Fré (1986) mention that 15,000 camel loads were carried between Kassala and the Red Sea coast every year. The commodities included *senna mecca*, which are pods of *Cassia* sp. used as a herbal medicine, colocynth fruits *Citrullus colocynthus*, milk, hides, charcoal, clarified butter and palm fibre (*Ibid.* 1986,23). Porterage started to decline after completion of the Kassala-Port Sudan railway in 1924. At about this same time, however, another trade in dom-nuts of the branch palm *Hyphaene thebaica* gained importance. The kernel of these nuts is referred to as vegetable ivory. It was used in the European industries for the production of buttons. Female trees yield an average 74 kg of nuts. The average annual production in the Beja district amounted to 2,205 tons (1 Eng. ton equals 1,016 kg) over the 1925-1931 period (DSA Barter 448/1). The average annual production in the entire Kassala province was 6,036 tons over the period 1927-1940 (NRO Sudan Government K.D./57.A.1 and various Annual Reports). Around 1930, the Hadendowa "[..] appear to have had a healthy and diverse economy which was based on subsistence agriculture, cotton in the Gash delta and Tokar delta irrigation schemes and dom-nuts, alongside pastoralism [..]" (Morton & Fré 1986,41). Barter mentions that *Cassia* pods were an important cash crop in the 1930s (DSA Barter 448/1). Early types of networking incomes among the Beja in the mid 1930s are referred to by Owen as "[..] the pocket of the *nazir* (tribal leader) is always open [..] no solicitation went unsatisfied [..] (the nazir is) [..] the distribution centre of the wealth of the tribe [..]"

(DSA Owen 414/5/21).

Smuggling had become an important source of income of Beja living in the Border Area in the late 1940s. According to the 1948 Annual Report of the Kassala province, the commodities of grain, sesame, oil, rice and hides were crossing the borders into Eritrea and Arabia "[..] on a fairly large scale [..] the trade is so lucrative that confiscation of goods is no deterrent [..]" (NRO Sudan Government K.D./57.A.1). Cotton smuggling was one of the chief reasons for the management board to decide on a reduction of its acreage in the Gash delta irrigation scheme in the late 1950s (Ahmed 1966). The dom-nut trade which had started to develop in the 1920s declined in the late 1960s as a result of the introduction of plastics in the European industries. Only 367 tons of vegetable ivory were collected in the Kassala province in 1968/69 (Morton & Fré 1986). The National Livestock Census of 1975 also includes data on rural livelihoods by tribe. The Beja of the Border Area were at this time engaged in the collection of fuelwood, wild fruits, pods and honey, the production of charcoal, and in hunting and fishing (MAFNR 1976, table 28.10b).

In the meantime, Kassala town was rapidly growing especially since the 1970s.[22] This generated employment opportunities, including a wide range of informal jobs (Post 1987). The rural Beja directly benefitted from this through daily and temporal labour migration to Kassala. The jobs typically include employment as trader, middleman, street vendor, porter, messenger boy, coffee and tea seller, and night watchman. The growth of Kassala expectedly also indirectly generated work in the rural Border Area. Wood cutting, charcoal production, and the collection of stones, gravel and sand is performed here as off-farm employment. These products are delivered to the urban industries. A number of small-scale industries emerged in the Border Area from the 1970s onwards. A complex of 16 lime kilns is located near Shellalob. Granite is quarried at Jebel Mokram, northeast of Kassala town. The Gash clays provide the raw material for brick-making industries located at the river margin. Most of these rural industries are managed by urban entrepreneurs. However, part of the labour employed is recruited in the surrounding Beja villages (Kuhlman 1990). Several reports on the Border Area mention that the Beja have become increasingly dependent on non-farm incomes to sustain their livelihood after the 1984-1985 drought (Hubach & Marouf 1985, Westhoff 1985, De Leeuw 1987, Kuhlman *et al.* 1987, Hashim 1987).

4.3.3 Sedentarization in rural areas

The first sedentary communities in the Border Area were encountered in the first decades of the 19th century by Burckhard [1819](1987). These were located along the caravan route from Kassala to the Red Sea coast. In 1882, the Beni Amer reportedly still lived in tents (James [1883](1969,64)). The first topographical maps of the Border Area (1:250,000, sheet 56-A Kassala) surveyed in 1900-1902 display for this period only four settlements, including our research village Hafarat. The collection of tribute by the Anglo-Egyptian

[22] The annual urban population growth rate was 4.5 % over 1955-1964; 4.7 % over 1964-1973; 7.2 % over 1973-1979 and 9 % over 1979-1986 (Kuhlman 1990). The 1990 population of Kassala town was estimated at 217,000 (MFEP & UNDP 1990).

Administration was in these years carried out at the well centres. These were among the few places where the still largely mobile Beja and their herds could be found (DSA Savile 427/3). The Beni Amer are usually mentioned for being the first Beja group to have adopted more sedentary lifestyles from the 1920s onwards. Certain sections of Eritrean descent, firstly, have always been more inclined to crop production than livestock husbandry. Their involvement in subsistence cultivation and cotton growing in the Tokar delta, secondly, is usually given as another explanation for their early sedentary ambitions (Cramer Roberts 1923, Nadel 1945, Paul 1950). Sedentary communities gathering around a Koran school, or khalwa[23], become more common in the Border Area from the 1930s onwards (DSA Barter 448/1).

Several scholars discussed the factors which they consider to have been important in the process of nomad sedentarization in the Border Area region (table 4.6). The sequence of our presentation in this table is chronological with respect to the date of initiation of the development. This does not necessarily coincide with the period of greatest impact in the area. However, it is shown that usually different factors play a role over different periods of time. All are external to the Beja livelihood, with one notable exception of the religious movement of Ali Betai.

Table 4.6 **Main factors related to the sedentarization of nomadic groups in the Border Area region mentioned in the literature**

(i) Infrastructural development and market integration – roads, railway, urban development and expansion	Lebon (1965)
(ii) Irrigation development Gash and Tokar deltas – economic opportunities	Newbold 1935, Spencer [1948](1984)
(iii) Pacification Beja tribal domain – Colonial Administration, Indirect Rule	Paul (1950), [1954](1971)
(iv) Ali Betai religious movement – explicit call to sedentarize and study Koran	Fadl (1988), El Hassan (1991)
(v) Government sedentarization policies – national and international agricultural projects	Yasein (1967), Ahmed (1976)
(vi) Restriction on livestock movement – irrigation schemes, mechanized farming, civil war in Eritrea	Bascom (1990a), Morton (1993)
(vii) Adverse natural conditions and drought – late 1940s and mid 1980s livestock losses, forced selling and slaughter, unfavourable terms of trade	Khogali (1980), Dahl (1991)

[23] Khalwa has different meanings of state, place of solitude for worship, place of religious and social functions, "men's house", and place where Koran is taught and which is used in combination with any of the other meanings (El Hassan 1991,1). The latter broad meaning of *khalwa* holds for the Border Area.

(i) Infrastructural development and associated processes of market integration are mentioned to have either induced, or forced, the adoption of new lifestyles. This developed through the increased exchange of goods and information. Kassala was initially only connected with other regions through a network of tracks. These follow the old caravan routes. One of the most important led from Khartoum, via Goz Regeb and Kassala, to Suakin and Massawa in present-day Eritrea situated on the Red Sea coast. Telegraph services were introduced in 1871, and its wireless variant in 1914. The railway from Khartoum to Port Sudan reached Kassala town in 1924. A bridge connecting Kassala town on both sides of the Gash margin was built in 1949. This is a significant event because the riverbed cannot be crossed after flooding. Without a bridge, the entire Border Area used to be seasonally cut off for several weeks from its surrounding regions, except in northern and eastern directions. Kassala was linked to the tarmac road network in 1978. In the late 1970s, an airport was constructed with a 2,500 m runway and facilities which allow international air traffic. However, the bulk of transport in the region is still by lorry truck. In the Border Area itself, motorized traffic depends on dirt tracks. These become impassable for several days after heavy rains and khor spates. Another bridge built in 1990 at a strategic point just outside Kassala town has recently improved the town's all-season access from the direction of the Border Area. The growth of Kassala town is to some extent part of this same process of infrastructural development and market integration. Ausenda (1987) estimated that at the end of the 19th century, Hadendowa households in the Border Area region were for some 10 % dependent on the market. Kuhlman (1990) found for the late 1980s that the rural-urban wage differentials in the same area had virtually disappeared. This can be considered as indicative of a greatly increased orientation on commercial markets of the households in the Border Area since.

(ii) Irrigation development in the Gash and Tokar deltas generated the same infrastructural change and forces of market integration as discussed under (i). Irrigation schemes, however, also came to occupy Beja tribal lands used for traditional cultivation and grazing. They came to block the seasonal trekking routes of the Beja and of other nomads in the region (*cf.* vi). Sedentarization of the Hadendowa, in addition, was an explicit objective of the management board of the Gash delta scheme in the mid 1930s (*cf.* v). Although the developments in the Gash delta are of greatest direct importance to the Border Area households, similar developments in the Tokar delta have excerted indirect effects. This region is, after the Border Area, the second Beni Amer enclave in the Sudan. These enclaves have acted, so-to-speak, as "communicating vessels" because the remainder of their land in Eritrea had been unsafe during the previous three decades as a result of civil war and drought (*cf.* vi and vii). The Beni Amer expectedly also first adopted subsistence cultivation in the Sudan in this area of Tokar. This was induced by the early introduction between 1850-1860 of irrigated cotton by the Turco-Egyptian Administration (Cramer Roberts 1923).

(iii) Pacification policies of the Anglo-Egyptian Administration in the first decades of this century reduced the general socio-economic insecurity and facilitated settlement. The Beja were difficult to control because of frequent skirmishes between different tribal sections, and the harshness and remoteness of their habitat (Cumming 1937, 1940). The pacification objective was therefore turned to in a dual approach. This included the introduction of administrative divisions and the delegation of executive powers over these areas to Beja leaders (Indirect Rule). Khalaf (1965) distinguishes the epochs of British Administration (1899-1920), Change to Indirect Rule (1921-1925), Indirect Rule (1926-1936) and Change to Local Government (1937-1942). Tribal rivalry remained vivid, however, as the records of armed confrontations in the Border Area over 1943-1946 witness (Paul [1954] 1971).

(iv) The movement of the late Sheikh Ali Betai summoned the Hadendowa of the Border Area to sedentarize and study Koran. In 1951, Ali Betai had a revelation. He was ordered by the prophet Mohammed to guide his tribesmen away from what a Koran teacher in Telkook would later describe as "[..] the undisciplined life of wandering people [..] who were pagan and aggressive [..] and were constantly on the look out for opportunities to assault people and raid their livestock [..]" (Ahmed, interview Telkook, Nov. 1991). Ali Betai established a first centre for education and study in Hameskoreib. Domestic water and health care was provided for the community of his followers. Crop production was promoted to sustain the livelihoods locally. This group of followers steadily increased over the years. The movement received the support of various national administrations who regarded it as a valuable vehicle to check the unstable border region. Since the death of Ali Betai in 1978, the movement has been led by his son Suliman. The NGO Hameskoreib Area Development Agency (HADA) was established and substantial financial support is now received from patrons in Saudi Arabia, Quatar, Oman, Kuwait and Yemen. The annual donation in the late 1980s from the first country alone was in the order of 1 million Saudi riyals, or US $ 280,000. Hameskoreib is now a place of worship attracting students from Sudan, Chad, Eritrea and Somalia. The current activities of the movement in the Border Area consist of the establishment of schools of mixed Koranic and secular teaching. It succesfully lobbies with the GDAC management to increase the number of tenancies for its followers in the Border Area (*cf.* section 4.1.3 *Controlled spate irrigation*). The movement especially appeals to the Hadendowa. However, similar forms of organization also can be found among the Beni Amer with connections to religious sects in Saudi Arabia (Liverenz 1989).

(v) An early attempt of the Anglo-Egyptian Administration to sedentarize the Hadendowa and to make use of their labour was made in 1936. A new key for land allocation in the Gash delta irrigation scheme was proclaimed and a batch of new titles was allocated to them for this purpose (Mohamed 1983). The attempt was unsuccessful partly as a result of their declining interest caused by the low cotton prices. A clear policy framework for nomad sedentarization was first formulated in the Independent Sudan of the 1960s and 1970s. This was linked up with the national Anti-Thirst and Anti-Hunger campaigns (*cf.* section 4.2.1). One such proposal of 1964 for the Border Area region mentions the construction of "[..] 37 barrier dams, 1 underground dam, 13 *hafirs*, 100 deep wells and 70 shallow wells in the Beja country to run a 130,000 fd irrigated area so as to settle 8,000 families [..]" (Khogali 1980,242). This project was never implemented, however. At this time, also pilot settlement schemes co-financed by international organizations were established in different parts of the country (Khogali 1981). A similar project was proposed in the Border Area by the Ministry of Finance and Economic Planning in 1982. This "Mixed-farming Project Salem Aleikum" was approved in the same year. It was funded as part of the MFEP/KADA programme. The short-term project goal was the settlement of 210 families of semi-nomads on 2,100 fd of irrigated land (MFEP 1982). The project was terminated six years later as a result of technical failures in the development of groundwater for irrigation. However, its general objectives were still deemed viable by all parties involved (GOS/GON 1988,23). The 1988 Four-Year Plan of the Ministry of Agriculture in Kassala includes the objective "[..] to stimulate agricultural settlement of nomads and refugees [..] (*cf.* section 4.2.2.).

(vi) In the course of the 20th century, agricultural schemes were developed in eastern Sudan. These occupy the Beja grazing grounds and block the routes for seasonal migration of livestock (Morton 1993). If one were to take a position in the centre of the Border Area, these obstacles are situated in the north in the form of the 84,000 ha Tokar delta irrigation scheme. In the west, this is in the form of the 126,000 ha Gash delta irrigation scheme. In the south, large-scale mechanized farming estates have been established since the mid 1940s. These cover some 2.5 million ha in the Gedaref area alone. The estates continue to expand and the corridors through which livestock can move between dry and wet season pastures become narrower, and in some places become obstructed altogether (Bascom 1990a). Finally in the east, the havoc of civil war in Eritrea and Tigray has endangered the passage into this area for three decades (Cliffe 1989) until the fall of the Mengistu regime in 1991.

(vii) Household losses of livestock as a result of adverse natural conditions and drought include extinction, and off take for forced selling and slaughter. Drought is the shortage of water or moisture for a specific purpose. Rasmusson (1987) defined it as implying an extended and significant negative departure in rainfall relative to the regime around which a society has stabilized itself. This negative departure can be assessed by relatively arbitrary criteria applied to a series of data, for example by considering the standard deviation below the long-term mean rainfall. This could be called a meteorological definition of drought. It can also be assessed by considering the impact of the event on society, for example in terms of deviations from normal crop production and forage availability. This could be called an agro-economical definition of drought. Figure 4.12 illustrates both such meteorological and agro-economical drought events for the Border Area region since 1881. The data for the meteorological drought years were taken from Cole (1989). These were updated with recent information from the Meteorological Department in Kassala. A drought year is defined as a situation where rainfall is at least 1 standard deviation below the mean rainfall. A severe drought is defined as the consecutive occurrence of two such drought years (*Ibid.* 1989,83). The data to determine agro-economical drought events were taken from regional references of food stress and livestock loss in Khogali (1980,237). These data are based on official reports of pasture failures in the Kassala province and his own fieldwork among the southern Beja. Other references used include those in Dahl (1991,189). These data are based on archival records and her own fieldwork among the northern Beja of the Red Sea province. Furthermore, also reports in Salih (1971), Morton & Fré (1986), De Waal (1989) and in the monthly bulletins of the

Sudan Early Warning System Unit of the National Relief and Rehabilitation Commission were consulted for this purpose. Individual drought years are indicated by a single symbol. Figure 4.13 shows the same data, but with more detail for the 1980-1990 main research period.

Figure 4.12 **Meteorological and agro-economical drought years, Border Area region, 1881-1991. Source and definitions are given in the text**

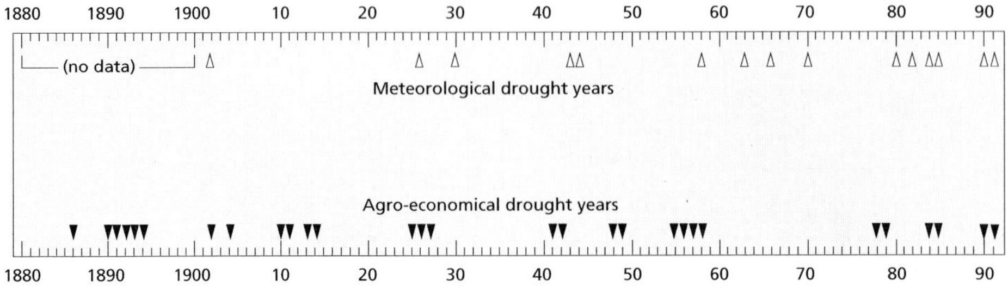

Figure 4.13 **Same as figure 4.12, Border Area region, 1980-1991**

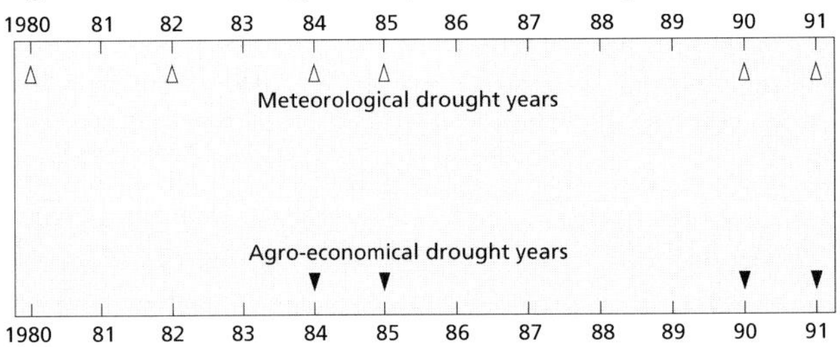

The meteorological and agro-economical drought years in figure 4.12 do not perfectly match. This is especially not so around 1910, the late 1940s and 1960-1970. Two reasons may explain this. Firstly, the rainfall gauging station in Kassala is a point location, while the drought impact covers a much wider area. Secondly, the indicators of drought impact are possibly also influenced by other phenomena. These include political instability, unfavourable economic terms of trade, high taxation, epidemic diseases and pests affecting livestock and crops (*cit.* in: Cole 1989). The 1984-1985 drought is reported to have caused levels of livestock loss of about 90 % in the northern Beja lands (Fré 1991). In the Border Area, this was between 80-85 % (FAO 1986c). This figure was further specified by the Ministry of Agriculture as loss percentages of 82 % for cattle, 92 % for sheep, 82 % for goats and 86 % for camels (MANR 1988).[24] Losses of livestock for the very dry years 1990 and 1991 (*cf.* section 5.1) were expected to reach the same 1984-1985 levels according to surveys based on remotely-sensed rangeland resources in eastern Sudan (Pflaumbaum 1991).

---

[24] *Cf.* section 4.3.1 and footnote 21 for estimated levels of provincial livestock losses.

4.3.4 Adoption of indigenous SWC

The developments which contributed to Beja sedentarization in the Border Area still do not explain why they decided to adopt SWC for crop production. Other authors suggested in this context that for central and eastern Sudan, nomad sedentarization is a necessary condition for SWC adoption, but not a sufficient one (Abu Sin 1970, Born *et al.* 1991). In regions such as the Border Area, where crop production can be pursued either with or without SWC, usually a better relationship between labour costs and production benefits exists for techniques of the non-SWC type. The argument is that other inputs being largely equal, SWC will always demand additional attention for construction and maintenance of its structures. It is also for this reason that usually more compelling factors are suggested to explain the adoption and use of SWC. A central role is played by a constellation of factors referred to as "pressure of population on scarce resources" (IFAD 1992). The adoption process itself is frequently interpreted in terms of the theory of Boserup (1965, 1981). This theory suggests, for a given area, a strong positive relationship between critical population densities and the process of agricultural intensification. Although critics pointed out that usually also causes other than population pressure contribute to intensification, and that usually also consequences other than intensification result from rising population densities, the core argument of the theory still stands. The processes in question were discussed by Dietz (1987,142-146) for dryland West Pokot in Kenya. He also refers to the dynamics of the household in other livelihood activities. In an initially pastoral society, increasing population densities may result in land use intensification in either livestock production, or in combined livestock and crop production activities. When after adjustments new critical densities are reached in later phases, land use intensification may occur in one sector only, for example in crop production. The adoption of SWC techniques would fit in such a sequence of events. A more economical dimension to population density has lately been stressed by Tiffin *et al.* (1994). They illustrate for Machakos in Kenya that higher population densities also tend to induce more traffic, information exchange and, in the long run, more infrastructural development. These increase the exposure of communities to market forces. This can be beneficial for the adoption and improvement of SWC as was the case in Machakos. The techniques were locally fine-tuned to newly emerging economic conditions. However, also a long list of case studies exists where the same market forces were found to have largely undermined the application of indigenous SWC and the SWC techniques introduced by projects (Reij 1992).

The first references to population densities in the Border Area were found in the archival sources. Owen reports in 1933 for the Maman-Odi area in the northern part that "[..] remoteness still exists [..] you may ride for hours and then come on one house hidden in a narrow khor [..] ride for hours more and come on two [..] I never saw more then three houses together [..]" (NRO Owen DCH/66.K1). In 1935, other tribal groups were granted permission to cultivate on Hadendowa land. This indicates that locally still little pressure existed on the cultivated lands (NRO Owen BD/66.K).

Table 4.7 **Population census data and average population densities for selected areas of the Border Area region**

YEAR	ADMINISTRATIVE AREA	POPULATION NUMBER	AREA km²	AV. POPULATION DENSITY (no./km²)
1955	Kassala Rural	73,152	1,792	40.8
	Hadendowa	248,573	245,875	1.0
1973	Kassala DC	69,951	12,097	5.8
	Aroma DC	272,721	60,390	4.5
1975	Strata: 41, 42, 45, 46 and 56	92,514	11,594	8.0 [A]
1983	Kassala Rural	71,600	5,248	13.6
	Border Area	206,100	9,524	21.6

Source: Ministry for Social Affairs (1958), MFPNE (1976), MAFNR (1976), MFNE (1987). Note: [A] the range over the five individual strata is 2.7-26.8 persons per km². The calculation of superficies and population densities is based on the area on official maps of the census districts (cf. section 3.2). The areas are best-fit sub-sets of the Border Area defined in this study (8,600 km²).

The first listing of population density figures is in the 1944 report of the Soil Conservation Committee. A map indicates densities of 5-50 persons per square mile (2-20 per km²) for the rural areas immediately surrounding the town of Kassala, and densities of under 5 persons per square mile (less than 2 per km²) for the remainder of the Border Area (Sudan Government 1944,145). The data listed in Tothill [1948](1952,699) indicate the equivalent of some 32 persons per km² for the Kassala District in 1940, but this includes Kassala town. The same source mentions a density of 0.6 persons per km² for the Beja District, which equally includes smaller towns. Average population densities calculated from figures of four National Census campaigns are presented in table 4.7. The variation in the areas where population was counted makes a direct comparison over years impossible. In general, it holds that the population densities decrease in northern direction, except in the 1983 data. None of the counting areas, in addition, exactly match the Border Area as defined in this study. The best coverage is provided by the 1975 Livestock Survey. This source also allows the most elaborate breakdown into smaller areas. In 1975, the average population density in the Border Area was some 8 persons per km². This ranged from some 27 persons per km² around Kassala town, to some 3 persons per km² in its northern parts (MANFR 1976, table 14.11). The population increase was, until 1967, mainly based on natural growth. Early immigrants, like the northern Sudanese and West-Africans, predominantly settled in Kassala town, the Gash delta and on the western Gash margin. After 1967, population growth became locally dominated by Eritrean refugee immigration. Kuhlman *et al.* (1987) and Kuhlman (1990, 1994) report that the first refugees were mainly lowlanders of different Beni Amer sections. They settled on the Sudanese side of the border among Beni Amer kin. The refugees arriving in 1978-1979 and 1982-1983 consisted of more mixed groups. Smaller numbers of people found their way to the settlements in the Border Area. Another heterogeneous group of drought victims fleeing from Eritrea and Tigray entered the Sudan near Kassala in the period 1983-1985. Most were intercepted at the border and directed to large refugee camps run by the United Nations and international aid organizations. The latest influx, this time of Ethiopians, took place after the fall of the Mengistu regime in 1991. The total number of refugees who have sought shelter in the Kassala area is enormous. This amounted to 153,000 in 1986, representing 20 % of the total population in the research area of Kuhlman *et al.* (1987). Almost all except the lowland Beni Amer settled outside the rural Border Area, either by choice or force. In the latter group of rural settlers, in turn, almost all settled in the southern Border Area around Kassala town.[25] However, no significant refugee populations can be found in the research villages Um Safaree and Hafarat which are located in this same southern part. The main reason for this is the poor domestic water supply condition in these villages (*cf.* sections 6.4, 6.5).

4.3.5 The role of government SWC interventions

Sedentarization of nomadic peoples is a policy objective of the regional Ministry of Agriculture (*cf.* section 4.2.2). It also figured as subsidiary objective in international programmes because these were based on the official Sudanese policy directives.[26] Still, nomad settlement never seems to have been a chief argument for implementing SWC interventions in the Border Area. More likely, this chief argument is the region's enormous hydraulic potential. One look at the map shows that seasonal water courses, springing from the foothills of the Ethiopian Highlands, are abundant. This is in stark contrast to much of the remainder of central and eastern Sudan. After early, albeit modest, government SWC interventions and the rise of indigenous SWC in the Border Area since the first half of this century, these interventions mainly developed in the early 1980s when the political tide was right. The 1984-1985 drought did much to attract international attention and rural development projects to the Border Area region. This further propelled the implementation of introduced SWC techniques in the framework of projects on a larger scale.

---

[25] Kuhlman (1990,11) gives a 40 %-refugee population in the research area he calls "small-scale rainfed agriculture". This largely coincides with the Beni Amer tribal domain in the southern part of the Border Area. He found no refugee population in the research area he calls "agro-pastoral", which largely coincides with the Hadendowa tribal domain in the northern part of the Border Area (*cf.* figure 3.1).

[26] See for example the problem-tree and logical framework of the MFEP/KADA project BAPP in KADA (1987b,21-23).

5

Natural resources and economic sectors

The current status of land and water resources in the Border Area is discussed in more detail in the next sections. The general characteristics of household and village economies and government interventions are presented, and the format of selected variables used in the research is discussed.

5.1 Land and water resources

The physical setting of the Border Area was discussed in section 3.1. More detailed attention is given here to village areas and research years. The climatological references to the long-term mean are based on Kassala data of 1961-1990, unless otherwise indicated. The four agro-climatologically defined growing season months are July to and including October. This follows definitions in Van der Kevie & Buraymah (1976).[1]

Rainfall and evaporation

The 1901-1992 rainfall data of the Kassala station indicate a mean of 299 mm. However, this is only 250 mm when the 1961-1990 mean is considered. As in central Sudan (*cf.* figure 3.2), these last three decades were relatively dry in Kassala (figure 5.1). Insights into the variability of rainfall over time and place in the Border Area can be gained from

[1] The growing season is defined as months with rainfall and soil moisture depth exceeding 50 % of monthly Et_p. In the Kassala area, this situation arises in the period of 4 months July-October. In chapter 6, also economically-defined growing season months will be used. This period is defined from the labour-demand perspective and covers a period of five months, including June as an additional month for land preparation.

Figure 5.1 **Annual rainfall (in mm), Kassala station, 1901-1992**

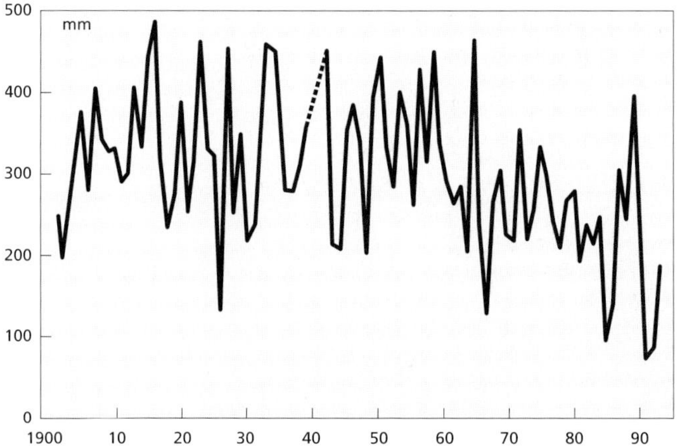

1988-1991 data collected at the project locations of the Department of Soil Conservation. Telkook is the most northern location and Kassala the most southern in this 60 km by 70 km area (table 5.1). The rank numbers for annual rainfall vary over the years. They also indicate considerable spatial variability in this relatively small area. Finally, a clear north-south gradient is shown when the mean ranks of the 5 stations are compared. This gradient gave us reason to consider this matter further. Rainfall distributions had initially been assumed to be uniform at the scale of the Border Area. This is for example shown by Whiteman (1971), which figures are based on 1931-1960 data. However, this proved not to be the case when more recent data were consulted. The, in this respect, best data were found in Baasher (1992). He presents rainfall isohyets for the Border Area region based on 1964-1972 data. The northern research villages Telkook and Ilat Ayot are located in the approximate 250-300 mm rainfall zone. The southern villages Hafarat and Um Safaree in the approximate 300-350 mm rainfall zone (*Ibid.* 1992,10). The research implications of this finding are discussed in section 6.6.

Figure 5.2 shows the monthly distribution of rainfall over the main research years 1983 and 1988, as also the 1961-1990 mean. Figure 5.3 shows data for other selected years referred to in this study: 1989-1991. Of greater importance to crop production, however, is the distribution over the year of raindays and dry periods. A rainfall event of 10 mm is normally considered as the threshold at which soil moisture storage and run-off generation may become effective (Tauer & Humborg 1992). A rainday, accordingly, can be defined as a day with a total fall of 10 mm or more. The time span between two consecutive raindays can be called a dry period. The latter is important when it occurs after sowing. Such "mid-season breaks" are a notorious cause for harvest failure. Raindays and dry periods for the growing season months in 1983 and 1988-1991 are presented in table 5.2. The 1987 data used in the GIS application (*cf.* section 2.2.2) have been added.

Figure 5.2 and 5.3 **Monthly rainfall (in mm), Kassala station, selected years**

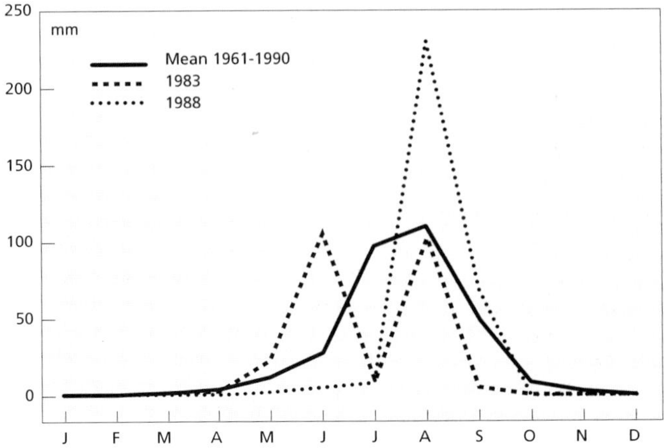

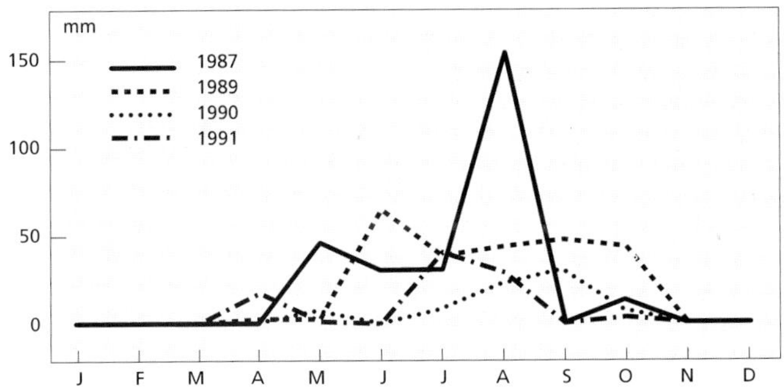

Table 5.1 **Annual rainfall (in mm) and rank number R for selected stations in the Border Area, (1988-1991)**

	1988	R	1989	R	1990	R	1991	R	mean rank
Telkook	231	5	177	4	4	5	na		4.7
Aroma	353	2	166	5	32	4	na		3.7
Dablaweit	332	4	194	2	58	3	23	3	3.0
Hedadeib	335	3	188	3	97	1	33	2	2.3
Kassala	396	1	218	1	76	2	91	1	1.3

Source: SCLUWP Kassala, Meteorological Department Kassala.

The Kassala total 1983 rainfall was 249 mm (table 5.2). This is comparable to the 1961-1990 mean. The monthly distribution was less favourable, however, with two separate peaks in June and August (figure 5.2). The average and maximum dry period length in 1983 was 3 days only because of a relatively late first rainday on 7 August

(31.1 mm) followed by a second on 11 August (50.6 mm). No other raindays occurred in this year. Sowing in the Border Area was mainly done in June, but no exact dates were recalled by the land users interviewed.

The total rainfall in 1987 was 271 mm, which is just above the Kassala mean. The first rainday in a series of five occurred on 5 July (14.6 mm), the last on 24 August (90 mm). The maximum dry period length was 21 days, and the average length 11.5 days. We have no data on the date of sowing in the Border Area for this year.

Table 5.2 **Selected rainfall characteristics of the four growing season months July-October, Kassala station, 1983 and 1987-1991**

	A	B	C	D	E	F
1983	249	2	7 August - 11 August	5	3	3.0
1987	271	5	5 July - 24 August	50	21	11.5
1988	396	11	3 July - 12 September	71	16	6.2
1989	218	8	7 July - 19 October	104	28	13.4
1990	76	2	8 August - 6 September	29	27	27.0
1991	91	2	22 July - 22 August	31	29	29.0

Source: Meteorological Department Khartoum (adapted). Note: A = total rainfall Kassala (mm), B = number of raindays July-October (days), C = date of first and last rainday July-October (days), D = number of days between first and last rainday July-October (days), E = maximum length of dry period July-October (days), F = average length of dry period July-October (days). Raindays and dry periods are defined in the text.

Total annual rainfall in 1988 of 396 mm was some 60 % above the long-term mean. The monthly distribution approaches the normal. The first rainday was early, on 3 July (10.3 mm) with another ten more following distributed at regular intervals over the season. The maximum dry period length was 16 days between 11 July and 27 July. The average dry period length was only 6.2 days. Sowing in the southern and northern parts of the Border Area commenced mainly from 15 July.

Total rainfall in 1989 was under the long-term mean. The rains came early in this year and June was the wettest month. The first rainday was on 7 July (15.7 mm), but the other seven followed at relatively long intervals in between. The maximum dry period length was 28 days. The average length was 13.4 days. Sowing in the entire Border Area commenced mainly from 10 July.

The year 1990 is the dryest on this list of six, with a total rainfall of only 76 mm. There were 2 raindays. The first occurred relatively late on 8 August (15.8 mm), the second 27 days later on 6 September (28.7 mm). Many land users did not sow and those who did mainly started on 10 August. New attempts were made from 9 September onward.

Finally, 1991 received a total of 91 mm rainfall. The first rainday was on 22 July (18.4 mm). The second and last in this year on 22 August (11.6 mm). The average and maximum dry period length was 29 days. Sowing was only done at a few places in the Border Area, mainly in the last week of July.

These figures show that although total rainfall in 1983 was close to the Kassala mean of 250 mm, the distribution was highly unfavourable. Only two raindays occurred in the growing season months, one shortly after the other. 1983 is therefore referred to in this study as a "normal-to-dry year". The 1987 rainfall was above the Kassala mean. It had an average number of raindays and dry period days when compared to other years on the list. This year will be referred to as a "normal year". Rainfall was relatively abundant in 1988. The number of raindays was relatively high and the number of dry periods relatively low. This year will be referred to as a "wet year". The remaining years 1989-1991 all show a decreasing number of raindays, and an increasing number of dry period days. The number of 8 raindays in 1989 is still at about the Kassala long-term mean of 7.6. This year will therefore be referred to as a "normal-to-dry year". The years 1990 and 1991 will be called "very dry years" in this study.

Rainfall intensities were recorded in the Hedadeib pilot scheme in the period 1988-1990. According to data of three years and 28 recorded events, the highest frequency of occurrence is in class 5-10 mm per hour (8 out of 28 recorded events, or 29 %). Events with a 30-60 minutes duration prevail (11 out of 28, or 39 %). The peak intensity recorded was 120 mm per hour measured over a discrete 10 minute-storm segment on 2 August 1988. The largest amount of rain over three years fell in class 10-20 mm (180 out of 555 mm, or 33 %). The largest number of individual events was in class 0-5 mm (40 out of 68, or 59 %). The number of raindays in the Hedadeib pilot scheme over 1988-1990 was 17 (25 % of all recorded rainfall events). This equals an average of 5.7 per year (all data taken from Van Dijk 1991).[2]

Figure 5.4 **Monthly potential evaporation and rainfall (in mm), Kassala station, 1941-1970**

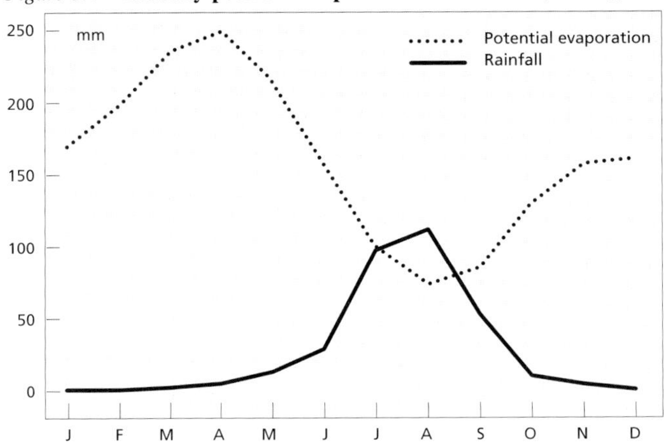

[2] Rainfall intensities were registered by an automatic rainfall recorder. Calculations were made according to the intensity-segment analysis (FAO 1981).

Data on the long-term mean evaporation for the Kassala station are expressed in different measures. The mean is 1,935 mm after Piche evaporimeter measurements (multiplied with a correction factor 0.5 to obtain potential evaporation). This is 2,900 mm after Class-A pan evaporation measurements. Finally, the potential evapotranspiration after Penman for the Kassala station is 2,970 mm (TNO 1982, Shanin 1985, Walsh 1991). A simple water balance based on monthly means of potential evaporation (Piche x 0.5) and rainfall is given in figure 5.4. Evaporation exceeds rainfall in all months, except August. Measurements of Class-A pan evaporation in the Hedadeib pilot scheme in 1990 indicate daily averages for the growing season months June to October of E_0=7.4 mm, 8.6 mm, 6.9 mm, and 12.5 mm respectively. The total pan evaporation in Hedadeib in this year was 2,629 mm.

Physiography, soils and run-off

The Border Area was described in section 3.1 as a pediplain where braided seasonal rivers drain the mountain ranges of the Ethiopian Highlands. The catchments of main khors which supply floodwater to the farming zones vary between some 4-205 km². Their share under rock outcrop is between 0-100 % and shows a great variation for individual drainage areas (*cf.* chapter 6). An indication of the value of run-off coefficients for such type of rock outcrop catchment is provided by three years of measurements in the Hedadeib pilot scheme. For a small 0.28 km² catchment located in the Jebel Haura mountain range north of Hafarat, the run-off coefficient *rc*= 0.6. For a larger 5.2 km² catchment, *rc*= 0.2. Measurements of discharge from this catchment in khor Hedadeib (1988-1990) indicate total volumes between 36 m³-11,040 m³ per year (Van Dijk 1991). The drainage system of the Border Area, particulary its feature of regularly changing water courses, is the main driving force behind a process of continuous redistribution of material. This results in areas alternating in character between predominantly erosional and depositional. On a smaller scale, also sheetflow run-off and sand-laden dust storms redistribute the clays and silts. Soil textures, by consequence, cover the entire range from sand to clay. They are highly variable in distribution, both in horizontal and vertical directions. Soil formation is in weathered Basement Complex, or in this transported material of alluvial, colluvial and eolian origin (Saeed 1972).

The soils in the Border Area were classified by Niemeijer (1993) (*cf.* section 3.1 *Soils and vegetation*). Typical soil-physical properties are micro-lamination, which is the development of thin layers of sediment of different texture, the sealing by micaceous minerals, and the crusting-capacity of their surface. These phenomena are partly interrelated and negatively influence the soils' hydraulic conductivity (*Ibid.* 1993,89-94). The combined effect at locations where soils are undisturbed is a high generation of surface run-off and a low rate of infiltration. In the catchment area of landholdings, this is a favourable property for SWC. For the cultivated lands, this is highly unfavourable and places additional demands on land preparation.

The soils in the Border Area are generally rich in mineral-derived nutrients (*Ibid.* 1993). The content of total phosphorus and minor nutrients is high. The soils can be expected to be rich in potassium. The organic matter content is low, as are total nitrogen and carbonate contents. The surface soils are non-saline, but these become slightly to moderately saline at greater depths. Soil reaction is relatively high with a pH of the top-soil of 8.2-8.8. The detailed physical and chemical properties of the soils studied in Ilat Ayot and Hafarat are presented in Niemeijer (1993, appendix c1, c2).

Run-off is mainly generated in the Border Area in the case of rainfall intensities which exceed rates of infiltration, not by effects of shallow water tables.[3] Infiltration measurements at Hedadeib and selected other locations provide the following outcomes. The mean basic infiltration rate measured with a double-ring infiltrometer in undisturbed soils is 1 cm/hr for light clays, 7 cm/hr for loams, and 13 cm/hr for loamy sands. Under ploughed soils of light clay and loamy textures, this is in both cases 10 cm/hr (Cosijn & Van Dijk 1989). Niemeijer (1993,95-98), using a rainfall simulator, measured high basic infiltration rates of 12-24 cm/hr on undisturbed silt loam to sandy soils. This was mainly induced by horizontal instead of vertical infiltration as a result of the typical soil-physical properties. Also a malfunctioning of the rainfall simulator contributed to this finding (*Ibid.* 1993).

Teras SWC characteristics

A classification of land use in the Border Area was made in a GIS (*cf.* section 2.2.2). The studied regions cover 1,824 km² in the northern part, and 1,460 km² in the southern part. These exclude areas west of the Gash river, the GDAC irrigation scheme and lands in Eritrea. The GIS maps of unsupervised classification, selected land units and best-fit *teras* areas are shown in figures 5.5 to 5.10. The superficies under best-fit *teras*, and the level of biomass production in the normal year 1987 and wet year 1988 in the areas under best-fit *teras* are listed in table 5.3.

The following limitations must be acknowledged. The best-fit *teras* areas are approximate. The GIS procedures applied include both supervised and unsupervised procedures for identification. The spectral reflectance of *teras* areas used for this purpose is based on a combination of characteristics. These include (i) vegetation (mainly crops, but supposedly also grasses and scrubs); (ii) soil moisture (mainly from run-off captured by bunds, but supposedly also from other ponding water); and (iii) SWC structures (mainly earth bunds, but supposedly also other ridges, and soils disturbed by land preparation). Several characteristics, such as vegetation occurrence, in addition, are not unique to land use under *teras*. As a result of approximate identification, both underestimations and overestimations in superficies of best-fit *teras* are made. Underestimations also result from improper spectral resolutions (the smallest resolution is 20 m by 20 m in SPOT). Overestimations also result from the incorrect classification of land known not to be under *teras* from field knowledge. This was the case for a horticultural zone south of the town of Kassala. These lands resemble the conditions under *teras* in all characteristics (cultivation, earth bund parcel boundaries, irrigated land). This total horticultural area is 8 ha, which is 0.5 % of the total best-fit *teras* area of the southern Border Area map. Due entirely to limitations of graphic presentation, the best-fit *teras* area in the northern Border Area (total

---

[3] Tauer & Humborg (1992,44-45) refer to these two processes as the Horton and Dunne models of run-off generation respectively. In the Border Area, groundwater can be found locally only at depths greater than 10 m (TNO & NCDRWR 1990). Therefore, generally the Horton model applies here.

74.35 ha, *cf.* table 5.3) cannot be identified visually in figure 5.10. Figure 5.11 shows a blow-up of the Aderjewab area near Hafarat to illustrate this underlying resolution problem.

Table 5.3 **The distribution of best-fit teras areas (in ha) and biomass occurrence in best-fit teras areas (based on dimensionless scores of NDVI), Border Area, 1987 and 1988**

	NORTHERN Border Area	%	%	SOUTHERN Border Area	%	%
Total area of imagery	182,389.78	100		145,995.61	100	
Best-fit teras area	74.35	0.04	100	1,527.07	1.05	100
1987 biomass occurrence						
– high	10.01		13	547.43		36
– low	64.34		87	979.64		64
1988 biomass occurrence						
– high	51.90		69	762.71		50
– low	22.45		31	764.36		50
Multi-temporal 1987-1988 biomass occurrence						
– high	2.23		3	221.77		14
– medium	48.33		65	868.24		57
– low	23.79		32	437.06		29

Source: Landsat TM imagery 171-049 of 12 October 1987 and SPOT XS imageries 130-318, 130-319 of 24 August 1988.
Note: Normalized Difference Vegetation Index NDVI is a dimensionless indicator of biomass. The multi-temporal 1987-1988 score is high when biomass occurrence is high in two years. The score is medium when this occurrence is one year high and one year low. The score is low when this occurrence is low in two years. Best-fit teras area and NDVI are defined in section 2.2.2.

Taking these important limitations into account, it can still generally be concluded that the *teras* distribution is more important in the southern part of the Border Area than in the northern part. The distribution of *teras* is mainly associated with the land unit "High soil moisture content area". This coincides with the physiographic unit defined by Van der Kevie & Buraymah (1976) as "Khor alluvium". Medium scores of 1987/1988 biomass occurrence prevail in both the northern (65 %) and southern (57 %) part of the Border Area covered (table 5.3). This represents a situation of a one year high and one year low biomass occurrence per best-fit *teras* location (*cf.* section 2.2.2). This is another indication of the high degree of variability of processes governing land use in the Border Area.

5.2 Economic sectors

The villages in the Border Area are described in KADA (1986, 1987) as remote and inaccessible during the rainy season. They are characterized by significant annual fluctuations in the number of inhabitants. The rural population has a very low standard of living. There is a high prevalence of illnesses, as well as a high level of malnutrition. A Sudanese Red Crescent field visit report of August 1991 indicates the most common diseases. These include tuberculosis, diarrhoea, trachoma, night-blindness, and malaria. The level of medical facilities and primary health care is equally low. The livelihoods in the Border Area were classified as agro-pastoral in DHV & IES (1989). However, at the household level a range of different livelihood sectors could usually be discerned in the 1980s and 1990s. These include crop production, livestock husbandry and processing of its produce, labour migration, local off-farm activities and networking. The village economies do not differ greatly in this respect from those found in the adjacent Gash delta area (HVA 1993).

5.2.1 Crop production systems[4]

Domains and techniques

A distinction is made in this study between cultivated lands located inside and outside the village domain. These are referred to as local and non-local lands respectively (table 5.4). The use of indigenous and project SWC is confined to local lands. Also several other techniques are used for the production of subsistence crops here. The non-local lands of the villages include lands in the GDAC irrigation scheme and on the eastern Gash margin. Gravity irrigation is applied in the scheme, while techniques of flood-recesssion and wildflooding are common on the Gash margin. Only a few households in the Border Area have access to non-local horticultural lands along the Gash. These are irrigated by pumped groundwater. Typically, households have access to different regions of local and non-local farming zones, where usually also different cultivation techniques are being applied.

The growing seasons in the Border Area and Gash delta region do not overlap. Cultivation in the first area is dependent on local rainfall. This mainly occurs between June and October (*cf.* figure 5.2). The season starts with land preparation after the first rains have softened the soil. Predominantly short-maturing sorghum varieties are planted. This results in harvest times for the Border Area between August and October. At this time, the season in the Gash delta starts. The supply of irrigation water depends on rainfall in Eritrea. The Gash usually flows in intermittent spates in the same period between June-October. However, land-allocation procedures of the GDAC normally do not allow access to the irrigated lands before October. Long-maturing Aklamoy sorghum is the prevailing crop. The harvest time for the Gash delta is between December and February. Sowing on the Gash margins, equally, can only proceed late in the season when no new spates are being expected. Approximately the same calender for cultivation as in the Gash delta proper applies to these lands.

---

[4] The following account is largely based on reports produced in the framework of MFEP/KADA and WARK (Green Mission 1985, Bokkers & Dabloub 1986, Cosijn & Van Dijk 1989, Van Dijk 1991, Van Dijk & Ahmed 1993), and fieldwork carried out as part of the L&E research programme.

Table 5.4 **Irrigation and SWC techniques in the Border Area and Gash delta area by regions of local and non-local lands**

	LOCAL LANDS	NON-LOCAL LANDS	
		Gash margin	GDAC irrigation scheme
INDIGENOUS SWC	teras brushwood panels	na	na
PROJECT SWC	contour embankments earth dams	na	na
IRRIGATION	wildflooding	flood-recession wildflooding pumped groundwater	gravity irrigation
OTHER	rainfed	na	na

Note: na means not applicable.

Land titles
Individual entitlements to the local cultivated lands are held under customary rights of possession and utilization (*cf.* section 3.3). These rights can normally not be transferred, but some have been in the area of the eastern Gash margin. An estimate of thousands of feddans were sold to urban investors for horticultural development in this area after 1984 (TNO & NCDRWR 1990). However, no such transactions have been reported in any of our research villages. Several households hold tenancy titles in the GDAC irrigation scheme. Cultivation in this scheme is also regularly practised under sharecropping arrangements. Agreements of equal sharing between tenant and land user prevail, which is called *sahib el nus*.[5] Sharecropping is uncommon on local lands. The households in Ilat Ayot and Telkook received through mediation of Suliman Ali Betai (*cf.* section 4.1.3) 900 fd and 2,000 fd of titles in the GDAC scheme respectively. These were distributed among individual households by local leaders. The maximum title allocated per household was 5 fd. More commonly, each received only 2 fd. These titles are still subject to the GDAC lottery procedures, which generally means that the acreage that can be cultivated is about half this size. The first season of use of these lands in Ilat Ayot and Telkook was 1991. The new allocations, for this reason, do not figure in the L&E Survey.

Cropping patterns
In the following sections, only crop production activities on local lands, including lands under indigenous SWC, will be discussed. The dominant land use in the Border Area is

[5] The *sahib el nus* sharecropper provides his labour and shares equally in all production costs of the tenant. For this, he receives 50 % of the production in return. Under another agreement called *sahib el tilit*, the tenant also supplies certain services or goods, such as seeds. The sharecropper in this case receives one third of the production in return.

single-cropping. Occasionally, also some intercropping is practised. Sorghum, *Sorghum vulgare* and *S. bicolor* (coll. *dura* or *esh*), is the main crop of which 11 varieties are locally known. The most common types are 70-days Feterita, 85-days Wad Feraj, 110-days Hagartai and 150-days Aklamoy. Less common varieties include: Mugud, Saffra, Wad Akar and the hybrids Um Bilbul, Korakolo, Agab Seedo and Raas el Girit. The latter three are strains of Feterita. Millet, *Pennisetum typhoides* (coll. *dukhn* or *biltoop*), is another main crop in the area. It is normally only grown on sandy soils. There are two varieties used distinguished by colour. These are named red and white millet. Other crops cultivated include okra *Hibiscus esculentus* (coll. *bamia*), rosella *Hibiscus sabdariffa* (coll. *karkadeh*) and watermelon *Citrullus vulgaris* (coll. *betig*). Less common are fodder bean *Dolichos lablab* (coll. *lubia*), rocket cress *Eruca sativa* (coll. *girgir*) and sesame *Sesamum indicum* (coll. *sim sim*). Where groundwater is available near well-centres, also other other vegetables, fruits and herbs are grown in small gardens. The cropping patterns are presented in the next chapter by technique. The share of the individual landholding cultivated annually, or cropping intensity, is usually between 50-75 %. True crop rotations for the purpose of soil fertility maintenance are exceptional. However, heavy weed-infestation may be a reason to change the location of sowing between years. The irregular run-off supply is another reason why factually different sites are being sown over the years. This is sometimes called "psuedo-rotation".

Cultivation

Camel and oxen-drawn ploughs were in use locally in the Border Area until a decade ago. Hired tractors are in use since the 1960s. The main reason to consider mechanized land preparation, if cash is available, lies in farm economics.[6] However, land preparation in the greater part of the Border Area is still carried out by hand. This starts in June with uprooting and cleaning of cultivated land. Sowing is after the first favourable rains or spates which normally arrive from mid July onwards. Cropping densities are variable with normal row-spacings of 0.6-1 m, and usually more fixed plant-spacings of a "step-length" 0.8 m. The plant populations are high with some 5-10 plants per hole. Thinning is not a regular practice and is normally only carried out in dryer years. Gap-filling (re-seeding) is more common, especially under wildflooding techniques because spates usually remove parts of the seedbed. Various weeds grow. Nut grass *Cyperus rotuntus* (coll. *se'id*) and Bermuda grass *Cynodon dactylon* (coll. *nagil*) are common. Weeding by hand is normally carried out in two or three rounds with simple tools. The first round is usually skipped when land has been ploughed by tractor. Plant residues are left on the field. Common pests and diseases include the prowling by porcupines, rats, mice and birds, plagues of insects, and the occurrence of smut and root parasites particularly witch-weed *Striga hermonthica*.

---

[6] Tractor ploughing at the commercial rate costed £s 450 per feddan in 1991. This usually allows land users to leave out one round of weeding by hand. This benefit is worth £s 200 of labour costs. Besides labour economies in time and money, benefits also result from timeliness (land preparation is on command) and improved infiltration (soils crusts are broken). These factors both contribute to higher overall levels of crop production (WARK field surveys).

Besides the favourable effects on pest control of crop selection and psuedo-rotation, also bird-scaring by young boys placed around the fields is practised, and occasional use is made of seed-dressings against smut. Mineral fertilizers are not used in the Border Area. Organic fertilizers are, albeit not consciously, applied. Crop mulch, weeding residues and animal droppings all add nutrients to the soils.[7]

Cultivated lands are usually collectively guarded against animal trespass. Harvest time is usually between August and October. This depends on the season and crop varieties grown. The entire sorghum and millet plant is cut with a knife, hoe, or axe. It is left in the field to dry, after which the heads are cut. Threshing and winnowing are carried out near the cultivated lands. Substantial losses of grain of 12-20 % may be suffered in these latter stages of production (Van Dam & Houtkamp 1992, Niemeijer 1993). After sacking and bundling of grain, stalks and stover, the produce is transported to the village by donkey, camel, or hired truck. The grain, stalks and stover are stored either inside the residence on a scaffold, or in the village warehouse where room must be rented. In areas of suitable clay soils, the grain is also stored in underground pits called *matamir*. The average production of sorghum grain is between 250-650 kg/ha (*cf.* chapter 6). This is well below the Border Area level of the 1930s of 1,300-1,900 kg/ha (*cf.* section 4.1.1) and the SSA average of 1,000 kg/ha (IBRD 1992).

Labour demands

The specific labour demands in *teras* use were discussed in section 3.5. We discuss in this section the demands placed on labour for general cultivation, irrespective of the technique applied. Crop production is predominantly carried out by the male members of the household. They usually receive support from other men of the lineage, neighbours and friends. Young unmarried women of the Beni Amer also participate in sowing and harvesting, unlike Hadendowa women. Frequently, additional labour is hired. This is allocated to all activities, but usually to the less appealing or physically most demanding activities such as bird-scaring, weeding and harvesting. Payment of hired labour is in cash or kind. Traditional farmer workgroups of mutual assistance, called *nafir* and *kaben* (Tb), still exist. However, their importance is declining as a result of the commercialization of relations. The average labour demands of main cultivation activities in the Border Area are given in table 5.5. Still other and highly variable demands on labour time are made, which are not presented in the table. These include travel between the residence and place of cultivation. The distances recorded in the L&E Survey are between 1-20 km. These are covered either on foot, or by donkey, camel, or truck. Land users sometimes build temporary seasonal camps in the farming zone to minimize this travel. Travel is usually also required at the place of cultivation, for example to collect brushwood for application in SWC techniques. The recorded maximum time spent travelling for this purpose was five hours daily.

---

7 Manure is collected in a few villages. It is not used in local crop production, but sold in Kassala town. Here, it is applied in the building industry to increase the durability of mud plasters. To a lesser extent, it is also used as fertilizer in horticulture in the area around Kassala town.

Table 5.5 **Average labour demands (in man-days per ha) for the main cultivation activities in the Border Area, eastern Sudan**

	MAN-DAYS PER HA
LAND PREPARATION	
– rainfed	2-21
– teras [A]	3-18
– wildflooding	0-5
SOWING	2-4
WEEDING 1st round	10-12
WEEDING 2nd round	7-10
WEEDING 3rd round [B]	5-7
HARVESTING	38
TOTAL LABOUR DEMAND	
– rainfed	64-92
– teras	65-89
– wildflooding	62-76

Source: Adapted from Bokkers & Dabloub (1986,17-20), Van Dam & Houtkamp (1992,71), Van Dijk & Ahmed (1993,11). Note: [A] labour demands of individual teras activities are given in section 3.5. The range presented in the table is from the minimum of catchment cleaning only (3 man-days/ha), to the maximum of catchment cleaning and an average bund repair combined (18 man-days/ha), [B] optional.

Crop-livestock systems interactions
Usually, a part of the stalks and stover locally produced is stored as forage for livestock. The remainder is used in the local building enterprise for the construction of roofings and sheds. Camels and donkeys are used as pack and riding animals. They provide the common means of transport for people, agricultural equipment and commodities in the Border Area. Livestock not on seasonal trekking graze on the stubble after the harvest. Pack and riding animals and all livestock kept near the cultivated lands add nutrients to the soils in the form of urine and manure. No commercial contracts of any form for stubble-grazing and provision of manure were encountered in the Border Area, however. Also, no use of draft animals was made in land preparation in this area in the 1980s and 1990s.

5.2.2 Livestock raising

Livestock husbandry in the Border Area includes the raising of camels, cattle, goats and sheep. OXFAM's 1989 livestock survey of the Red Sea province provides the most recent and detailed overview of breeds and their characteristics. The following information is mainly taken from their report (ERGO 1990). Camels chiefly perform as riding or pack animals. Some 18 different breeds are mentioned. The main types are of the indigenous *bishari* or Beja riding camel, and the exotic *rashaidi*, or pack camel. Camel milk is an important constituent of the rural Beja diets. Camels and the ritual exchange of their milk play an important role in Beja reciprocity and networking codes. Cattle are predominantly kept for their milk and economic exchange value. An important by-product is clarified butter, called *samin*. The OXFAM survey mentions 7 different cattle breeds. The main species include the Beja breeds of *erashai* which is a variety of the short-horned zebu. Cross-breeds are common in the Gash delta and around Kassala as a result of imports of foreign stock from Eritrea. Sheep are kept for their economic and social exchange value and for meat. Local butchers also collect the fat from the stomach area to produce *wadak*. This is used in cooking and is also a cosmetic used for hair-styling. Sheep milk is either consumed by the household, or processed to make clarified butter. A number of 6 breeds is discerned in the survey, mainly of the desert-sheep type. An equal number of 6 breeds of goats is mentioned. The main species are the indigenous *hadalat* and *baladi*. Goats are kept for their social and economic exchange value. When the rural Beja eat meat, usually on the occasion of social ceremonies, this is often meat of the goat. Its milk is important in the Beja diet and for the production of clarified butter. Goat hair is used to twist ropes and to weave blankets. The skins are tanned locally and processed in artisanal work.

Table 5.6 **Livestock productivity, Red Sea province, 1989**

	CAMEL N=19	CATTLE N=30	GOAT N=30	SHEEP N=30
Average annual birth (%)	38.6	59.6	59.9	63.9
Age at first birth (months)	65.7	43.4	15.7	19.0
Birth interval (months)	31.1	20.1	20.0	18.8
Twinning percentage	na	na	15.0	13.0
Percentage milking	100.0	93.3	64.3	73.3
Average daily milk gift (ltr)	2.5	4.1	0.7	0.7
Average age of sample (year)	5.5	9.3	4.8	4.8

Source: Integrated Livestock Surveys of Red Sea Province 1989 (ERGO 1990: 65,72,78,83). Note: data have been collected in Sep.-Oct. 1989, na means not applicable.

The main livestock productivity parameters are presented in table 5.6. These are for the Red Sea province. The Border Area was found in the L&E Survey to provide comparable outcomes, except for two. The daily milk gift of cows is lower (1-3.5 ltr, against a reported 4.1 ltr in the Red Sea province), while the daily milk gift of goats and sheep is higher (1.5-3.5 ltr, against 0.7 ltr).

Other animals are kept in and around the house. These include donkeys for travel and portage and chicken, pigeons and occasionally rabbits for home consumption. This latter group is not included in the computations of the L&E Survey of household livestock wealth. Livestock migrates seasonally between the Border Area and pastures elsewhere in eastern Sudan, Eritrea and Ethiopia. However, milch cows and baggage camels remain in the village area. Sheep and goats also migrate, but cover shorter distances. The foreign regions were partly under control of the liberation movements EPLF and TPLF until mid 1991. In principle, these regions have always been open to Beja nomads. At the same time, they also remained notorious for looting armed bandits, however. These areas were therefore avoided as much as possible. If the area was traversed, local militiamen were hired for protection. Pastoral cross-border moves are safer at present. These have even become institutionalized to the extent that Sudanese and Eritrean veterinary officers regularly exchange information on livestock health, pasture conditions and seasonal movements of livestock.

Herds on long-distance trekking usually consist of either camels, sheep, cattle, or sheep and cattle combined. Herds, having been greatly reduced in size after the 1984-1985 drought, are usually pooled among households. The herd is looked after either by one of the participating households, or by a contract-herdsman. The contract-herdsman is normally selected from the own tribe and lineage. He receives compensation in cash or kind. The latter is provided in the form of free consumption of milk and disposal of offspring. There usually also is a standard field allowance called *mazareef*. This includes rations of coffee, tea, sugar, ginger, tobacco and sometimes clothes and shoes.

The husbandry of livestock which remain behind in the village is taken care of by the other household members, including women. According to Beja custom, women are not allowed to milk. Livestock in the village, especially lactating animals, receive supplementary feeding. This includes grass cut in the village surroundings, shells and crushed nuts of the branch palm, stalks and stover, fodder grain, and agro-industrial by-products purchased in Kassala. The latter are cakes of sesame and cotton seed called *ombaz*, mixtures of sorghum, wheat bran and grinded groundnut shells called *komfood*, and mixtures of only sorghum and wheat bran called *ruda*.

5.2.3 Labour migration

Household participation in labour migration is quite complex in the Border Area. Usually, more than one household member is engaged, who may combine different destinations in the same year, and frequently does so during the same migration journey. It is for this reason that in the presentations in chapter 6 only the main types and destinations are discussed. These main types and destinations are defined as those contributing most to the household income financially. However, in all phases of data processing, computation and evaluation the full range of migration activities is considered. Nomadic transhumance, forced movement as a result of war and drought and permanent resettlement in urban areas are not considered as labour migration (*cf.* section 2.3.1).

Daily labour migration in the Border Area is mainly directed towards Kassala town. Employment is usually found here as casual labourer. Particularly important in income-generating terms are the jobs found at the Kassala livestock market. The official government policy since the 1980s is to discourage rural-urban migration because of the limited urban capacities (KADA 1988).

The most important destinations of temporary migration are the agricultural schemes in eastern Sudan. These include the Gash delta, New Halfa, and to a lesser extent the irrigation schemes in Rahad and Tokar delta. These destinations also include the mechanized-farming schemes around Gedaref. The greatest demand for migrant labour is during harvest time. In the Border Area and Gash delta, the harvest seasons are more or less one after the other. There is usually a small overlap for the Gash delta and the New Halfa irrigation scheme, where the cotton-pick starts from December onwards. The other more distant schemes are of less overall importance to the households in the Border Area. Their peaks in labour demand are also more competative seasonally.[8] Labour migration to the Gash delta has always been actively promoted by the government (*cf.* section 4.1.3). Migration to New Halfa largely developed out of the first contacts made here by households employed in the construction of New Halfa town and its irrigation scheme in the 1960s. Labour migration opportunities still largely depend on private networks of contacts. In years of exceptional shortage, labour is also actively recruited by government representatives of the agricultural schemes, who operate in the town of Kassala and in the villages of the Border Area.

Long-term migration in the Border Area during the 1980s was mainly directed to Saudi Arabia, Egypt and Kuwait. This migration has become more difficult in recent years as a result of stricter exit regulations, and the increasingly disputed position Sudan takes in the international political forum. Long-term migration inside Sudan was almost exclusively directed to the larger urban centres Khartoum, Gedaref, New Halfa and Port Sudan.

5.2.4 Local off-farm employment

The same complex participation mentioned for labour migration also exists with respect to household engagement in off-farm employment. Households usually have more than one member working in these activities. The individual members usually also participate in more than one activity in the same year. Again, only the main activities (contributing most to the household income financially) are discussed in chapter 6. However, the full range of activities is considered in data processing, computation and evaluation.

We discern four categories of off-farm employment in this study: (i) cultivation activities. These mainly consist of contract-farming inside the village domain; (ii) livestock husbandry. These mainly consist of contract-herding of individual and pooled herds inside the village domain; (iii) collection and production activities. This group includes the cutting and collection of fuelwood, charcoal-making, stone-cutting, gravel collection, artisanal activities (mainly the processing of leaves of the branch palm, which leaves are called *saaf*), water collection and selling, cutting of grass and finally the collection of animal dung. A number of these activities is carried out in the private domain, which allows also women to participate. The most important income-generating activity of women in the Border Area is *saaf*-processing. A variety of products is made of *saaf*, including baskets, mats, fans, brushes and ropes. Other women's activities include the fabrication of ropes made of worn-out cloth, the weaving of goat hair, and the selling of eggs, butter, vegetables and processed food. Beja women finally also engage in paid jobs as tailor, hairdresser, midwife, henna-painter and nurse. Women accumulate their own

---

[8] Blue Nile irrigation water allows nearly year-round high-intensity cropping patterns of cotton, groundnut and sorghum in the Rahad irrigation scheme. The Gedaref area is located at lower lattitudes and its main harvest time of rainfed sesame and sorghum is in the months August and September-October. The agricultural season in the Tokar irrigation scheme is comparable to the Gash delta. However, its poor accessibility makes it less attractive for Border Area households to seasonally migrate to.

income in the form of livestock and jewellery, and keep their own household budget. However, they still could not tell the different denominations of currency in a survey made in the mid 1980s by Ausenda (1987). This indicates a low level of market orientation. The last employment category discerned in this study is (iv) local services. This includes public services, such as in education and health care, religious functions in mosques and Koran schools, and private enterprise mainly in the form of shopkeeping and trade. A number of activities reported in the L&E Survey clearly is illegal. These include cross-border trade of commodities and livestock, cutting of trees and particularly the trade in trunks of the branch palm, charcoal-burning and game hunting. These activities will not be specified further at the village level for reasons of privacy.

5.2.5 Networking

The households of the Border Area have no access to official credit facilities. The main lending condition of the Sudan Agricultural Bank is to have land registered in property, which condition they do not meet. The Sudanese traditional system called *sheil*, by which advance money is lent by private persons at the beginning of the season against the value of a future crop, is not common in the area (Ahmed, pers. comm. Kassala, May 1989). Its use also has not been reported in the L&E Survey. Informal networking as defined in this study (*cf.* section 2.3.1) involves the transfer of wealth between households, and between village leaders and households. This obviously is not an employment category in itself. However, it can be understood as a specific form of social conduct to secure income opportunities which are controlled by others.[9] The mechanisms of informal transfers of wealth find their origin in Islam and tribal custom. Traditionally, alms-giving called *zaka* contributes to a fund applied to by village leaders for the relief of destitutes. The rules for collection and distribution of alms are given in the Koran.

The tribal element in networking is reciprocity. This is a central value in the life of the Beja. The mechanisms which sustain reciprocity include (*cf.* section 3.3) (i) *silif* (Tb). This is the expectation of reciprocity which governs land use, stock loans and the sharing of food; (ii) *gwadab* (Tb). This is the customary tribute in grain or livestock received from other lineages for their seasonal use of natural resources, and finally; (iii) *hamot,* also called *hamoysin, jihamod, karaj*, and *minhet* (Tb). This is the temporary allocation of livestock to poor households. They may use the milk and hair of the livestock, and sometimes may keep their offspring. There are also other mechanisms by which wealth is transferred and which are less strictly attached to Beja cultural codes. These include gifts received at all important *rites de passage* and extensive interest-free credits provided by shopkeepers (*cf.* Morton 1989). Finally, also the participation in, and use made of, voluntary workgroups can be regarded as containing an element of transferring labour income. Women actively engage in networking. They have their own saving-system called *sanduq*. The accumulated entrance fees of the participants are paid as a lump sum to one

[9] No strict rules can be given for this. An obvious form of what would be called "networking behaviour" is to comply with religious directives and tribal rules and regulations.

of them at the time. In the L&E Survey, these different types of transfers have not always been specified by the respondents. They were usually referred to as just "gifts". The most important categories in the research villages of the Border Area are receipts associated with *zaka*, *silif*, *gwadab*, *hamot* and *rites de passage*.

5.2.6 Government and NGO interventions

Government intervention in the Border Area in the 20th century can be divided into measures of (i) pacification and taxation; (ii) development and provision of public services; and (iii) relief aid. The peak in each was discussed in chapter 4 to have taken place in the 1930s, 1960s and 1980s respectively. The most important interventions which still affected the household and village economies of the 1980s and early 1990s are discussed next.

Pacification and taxation

The Border Area was under administrative control of the Anglo-Egyptian Condominium between 1902-1956. All institutions for pacification and taxation were introduced in these years (Marshall 1949).[10] According to Eissa (1990), most institutions continued to operate unaltered after Sudan gained Independence in 1956. Pacification measures included, in the Border Area, the establishment of two police stations in Mamman and Shellalob. The precise dates of establishment remain unknown. The first already existed in 1916. The second did not exist before 1949, according to archival sources. The tasks of the police included to control tribal conflict and to force back the smuggling of grain, sesame, oil, rice and hides to Eritrea (Kassala Province Annual Report 1948, KD/57.A.1). The leaders of the Hadendowa and Beni Amer were officially appointed as *Nazir* under Indirect Rule of the mid 1920s to mid 1930s (*cf.* section 4.3.3). They were to administer their own area and people. Firstly, a range of ordinances provided them with the necessary executive powers. Secondly, the demarcation of the districts Kassala and Beja (*cf.* section 3.2) provided them with defined territorial bases for this purpose. This local empowerment and "territorialism" was a dual strategy to achieve peace, improve tax collection and increase administrative efficiency locally (Marshall 1949, Khalaf 1965).

Since Indirect Rule, taxes have been levied in the Border Area on crops, livestock and private income. However, these have not been collected on a regular basis from the 1970s onwards. The abolition of native rule, subsequent dry years and common hardship, and limited means of transport to collect taxes in the rural area were causes for this (DHV & IES 1989). A new institution named the Department of Alms Giving was established in 1988. This Department of *Zakat* is to revive and manage the system of religious taxation based on traditional *zaka* virtue and guidelines. Formally, the department taxes crop

---

[10] The 1929 Civil Justice Ordinance was based on a first code promulgated in 1900 which was intended to preserve the rights of individuals under local custom and religious laws. The main taxation laws include the 1901 Tribute Ordinance, the 1924 Taxation of Rainlands Act and the 1925 Taxation of Animals Act. A new type of tribute was introduced in 1936. This was meant as a simplification of the earlier Acts on herd and crop tax collection. A poll tax on huts was introduced 1936/1937. The purpose was to include those groups which still remained largely untaxed under the crop and herd taxation Acts, particularly migrant West-Africans (NRO Arber 57/A.9, Marshall 1949).

production when the total production per household exceeds 7 sacks. Taxation is at a rate of 20 % for rainfed lands which include lands under SWC, and 10 % for irrigated land. Its rules for the taxation of herds prescribe rates of 1 sheep or 1 goat for every 40 sheep or goats, 1 goat for every 5 camels, and 1 head of cattle for every 30 head of cattle. Personal income is taxed at a rate of 2.5 % when this annually exceeds £s 50,000. Still, no households in the Border Area interviewed in the L&E Survey had been taxed under these new rules. The limited data available in Kassala indicate that for 1989/1990, the villages researched only received grain from the department. The populations in respectively Ilat Ayot were given 15 sacks of sorghum, in Telkook 105 sacks, in Um Safaree 145 sacks and in Hafarat 47 sacks (files Kassala Department of Alms Giving and interview, Jan. 1992).

Public services and development

The provision of public services and rural development started as a comprehensive programme in the Border Area in the 1960s. Before these years, only incidental interventions were made, particularly during the first active period of the Water Organization 1947-1954 (*cf.* section 4.2.1). During these years, a new *hafir* was constructed in Um Safaree and the existing one in Hafarat was rehabilitated. Ilat Ayot and Telkook did not exist as villages then. The populations in these areas remained dependent on shallow wells. Rural development interventions greatly expanded with the formulation of international programmes after the 1984-1985 drought. Besides in agriculture, the government operated during these years in the fields of infrastructure, domestic water supply, education, and health care. All main villages in the Border Area now provide primary school education and basic health care services. Telkook, in addition, has a hospital and Ilat Ayot a Primary Health Care unit (KADA 1986).

Relief aid

The first documented relief measures in the Border Area date from 1949. Reports of the Governor's Office in Kassala and correspondence between the Governor and his District Commissioners (BD/19.B.4, Weekly Report No.6) refer to the Beni Amer in the Kassala District. In June and July 1949, 14,000 persons received a total of 120 tons of grain. This was to be rationed at 10 *rottle*, per person, per month (1 *rottle* equals about 0.5 kg). Unmentioned quantities of dried milk were also delivered to women and children. In the Togan region (present-day Ilat Ayot and Telkook area), a total of 172 tons of grain was delivered and distributed at 10.5 *rottle*, per person, per month for the same two months in 1949. The commodities were supplied by the government and private secular and religious groups.[11]

---

[11] 36 tons of grain for relief aid in the Beja District were made available by AbdelRahman El Mahdi Pasha (NRO Sudan Government KP/19.B.2). A donation of £s 2,000 for the same district came from the religious Omdurman Khatmia Committee in Khartoum. Another £s 280 came from an unspecified group in the Tokar area of the Red Sea province (NRO Weekly Report No.8, 1949).

The total amount of relief goods delivered during the 1984-1985 drought, and again in the early 1990s, is unknown. Aid agencies operated in a highly uncoordinated fashion in these years (Abdel Ati 1991). Households interviewed in the L&E Survey usually also could not distinguish between all the types of relief provided to them. There was incidental government relief aid coordinated by the Rural Councils. There were distributions in the 1990s made from the revenues of *zaka* coordinated by the Department of Alms Giving. Finally, international organizations implemented their own programmes coordinated by the Sudanese Relief and Rehabilitation Commission (RRC). The first two government programmes mainly distributed sorghum at irregular intervals, usually at rates of 1/8 to 1 *ruba* per household (1 *ruba* equals some 5.5 kg of sorghum). The international rations followed the World Food Programme standards of 400 gr cereals, 50 gr vegetable oil, 30 gr pulses and 30 gr milk powder, per person, per day. These were delivered at regular intervals, but over undocumented periods of time. We estimate, based on interviews, that in the research villages of the Border Area the total household ration per month received from various agencies was about 0.5 sack of sorghum or wheat (some 45 kg), 1 kg milk powder, 1 *rottle* oil, 1 *rottle* dates, and 10 *rottle* sugar for the period from mid 1984 to mid 1986. The available official data for the Border Area are presented in tables 5.7 and 5.8. These lists exclude distributions which are known to have been made in the area by the NGOs CARE and Saudi Red Crescent.

Table 5.7 **Registered relief aid distributions in the research villages, total deliveries over 1984-1986 and 1989-1991**

	WHEAT [A] (mt)	SUGAR [A] (mt)	COOKING [A] OIL (ltr)	MILK [A] POWDER (mt)	WHEAT [B] (mt)
Ilat Ayot	342				90
Telkook	342				128
Um Safaree	4,050	300	95,466	500	423
Hafarat	2,754	200	136,380	350	315

Source: IARA Kassala, RRC Khartoum. Note: [A] period 1984-1986, [B] period 1989-1991. 1 metric ton (mt) equals 1,000 kg. No data exist for known deliveries which have been made in the area by CARE and the Saudi Red Crescent.

Table 5.8 **Registered relief aid distributions of wheat to the Kassala Rural Council and Kassala Border Council including Hameskoreib in the Red Sea province, 1991-1992**

	WHEAT (mt)	VILLAGES (no.)	BENEFICIARIES (no.)
January-March 1991	444		
July-August 1991	382		
September-November 1991	2,000	286	110,500
September-January 1991, 1992	750	310	110,500
March 1992	2,500		

Source: RRC Khartoum. Note: 53 mt cooking oil was also delivered to these Rural Council areas in November 1991.

5.3 SWC interventions and land use

In this study, we focus on the SWC interventions implemented by the Department of Soil Conservation in the villages of Telkook and Hafarat. In the first, a standard earth dam of the spreader type was built in 1985. In second, the Hedadeib pilot scheme was implemented in 1987. These are both SWC techniques of floodwater-harvesting. The households who participate in these projects normally also engage in several other cultivation techniques on the local and non-local lands.

In the next chapter, the cropping pattern, cropping intensity, and level of grain and stalks production is discussed for each of the four research villages separately. This will be done for the household level, by farming zone, and by technique. The main reference years are the normal-to-dry year 1983 and the relatively wet year 1988. A few additional data also cover the situation for the normal-to-dry year 1989 and the very dry year 1990. Production figures are presented in sacks per feddan, and kg per hectare according to standard conversion factors in MANR (1991) (*cf.* Local measures, p.vi).

The cultivation in specific farming zones, the application of specific techniques, and the labour demands involved on the landholdings in all these situations are to an important extent interrelated in the Border Area. It can be assumed that if no constraints in land use would exist, households would allocate most labour time to landholdings (*viz.* farming zones, crop production techniques) they consider as generally performing best. Such an unconstrained situation is not likely to arise, however. A number of potentially limiting factors to crop production in technical and organizational areas was discussed in sections 1.2 and 1.3. Therefore, a certain degree of deviation from the optimal pattern of preferred land use is to be expected. However, we argue that when precisely opposite outcomes are found in the L&E Survey data, this is likely to indicate risk-spreading behaviour. This, accordingly, is the situation where households usually allocate most of their labour time to those landholdings which are not their best. Such "sub-optimal" landholdings are defined for the purpose of this study by (i) subjective criteria suggested by the land user himself; and (ii) an objective criterium of the average level of grain production (1983, 1988-1990, sorghum yields in kg per ha).

In addition to details on SWC interventions in Telkook and Hafarat, the village presentations in the next chapter finally conclude with a discussion of the scores of the set of 18 selected research variables.

6

The village economies

The findings relating to the Border Area villages presented in this chapter are based on the L&E Survey outcomes, unless otherwise indicated. They are presented at the level of households and landholdings. The calculations are sometimes also performed for sub-groups of households. The composition of these sub-groups is discussed in the introductory section first.

6.1 Introduction

Sub-groups include households who use indigenous SWC on at least one of their landholdings. They are called in this study indigenous SWC households (iSWC households). Sub-groups of households who participate in government SWC projects are called project SWC households (pSWC households). Those who do not, are referred to as nopSWC households. Finally, households who neither engage in indigenous, nor in project SWC, are called noSWC households. Sub-groups of iSWC and pSWC households partially overlap in the manner shown in figure 6.1. pSWC households only exist in the villages with government SWC intervention which are Telkook and Hafarat. The village distribution of households and sub-groups is presented in table 6.1. The 1983 and 1988 totals are not identical. Several households sampled in the latter year do not comply with the definition of a household for the situation in 1983 (*cf.* section 2.3.1 *Definitions*).

The main entries made in the presentations per village discussed in sections 6.2-6.5 include (i) the village economy. This is typified by the distribution of household engagements in different livelihood categories, the associated incomes and the role of indigenous and project SWC in these in particular; (ii) Government and NGO

Figure 6.1 **Main groups and sub-sets of households in the L&E Survey. Note: the figure is not a true presentation of the actual numbers of households involved**

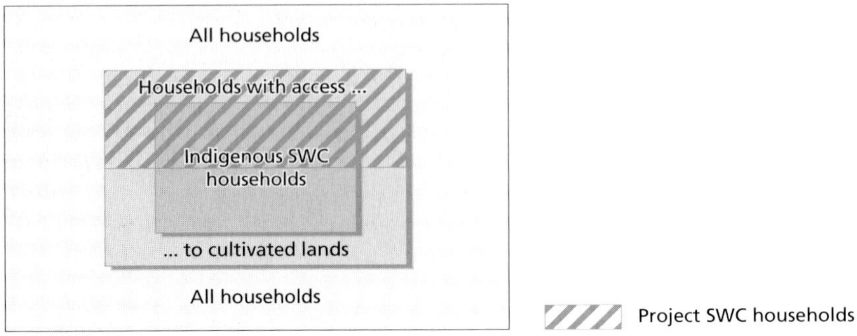

interventions. Detail is given of interventions in land use, and for Telkook and Hafarat, of SWC projects; (iii) SWC land use, land use dynamics and risk-spreading behaviour. The cropping patterns and levels of production are presented by technique. Finally, we discuss (iv) scores on the selected research variables which will be used in the data analysis. The main purpose of comparing 1983 and 1988 is to obtain a more balanced picture of the Border Area livelihoods, using data of a normal-to-dry year and wet year respectively. The findings of two years only, obviously, cannot indicate a trend over time. In the last section of this chapter, the outcomes of four villages will be compared and the first tentative conclusions drawn.

Table 6.1 **Numbers of households, indigenous SWC (iSWC) households, project SWC (pSWC) households, non-project SWC (nopSWC) households and non-SWC (noSWC) households in the L&E Survey, 1983 and 1988 (in numbers and % of all households)**

	Ilat Ayot 1983-88		Telkook 1983-88		Um Safaree 1983-88		Hafarat 1983-88		Border Area 1983-88	
All households (hh.)	54	58	57	61	60	64	57	61	228	244
iSWC hh.	26	30	5	13	49	49	42	41	122	133
pSWC hh.	na	na	na	24	na	na	na	26	na	50
iSWC and pSWC hh. [A]	na	na	na	5	na	na	na	16	na	21
nopSWC hh.	54	58	57	37	60	64	57	35	228	194
noSWC hh.	28	28	52	29	11	15	15	10	106	82
% iSWC hh.	48	52	9	21	82	77	74	67	54	55
% pSWC hh.	na	na	na	39	na	na	na	43	na	21
% iSWC and pSWC hh. [A]	na	na	na	8	na	na	na	26	na	34
% nopSWC hh.	100	100	100	61	100	100	100	57	100	80
% noSWC hh.	52	48	91	48	18	23	26	16	47	34

Note: na means not applicable, [A] households using both indigenous SWC and project SWC (cf. figure 6.1).

6.2 Ilat Ayot

The village of Ilat Ayot is located at latitude 16° 55' North and longitude 36° 30' East. This is in the 250-300 mm rainfall zone of the Border Area. The settlement is built on the high grounds of a gravelly interfluve between khor Youdrut and khor Kokoraeib. The village was established some time between 1952 and 1954 at the instigation of the Ali Betai movement. The village plan reflects the functional separation of public and private domains according to strict Hadendowa tribal codes. The public domain consists of the market area. This originated between 1966 and 1979, according to aerial photographs. Some 9 groceries, 2 bakeries, 1 tailor, 24 storehouses for grain, and 2 guesthouses all built of mud were counted here in 1990. There is also a primary school and a PHC unit and dressing station built of red brick. In this area, the men gather for their daily shopping and exchange of information. The private domain consists of a vast residential area. This is located spatially separate from the public market domain. The residential area exclusively consists of tents of woven branch palm fibre pitched up in parallel lines. Traditionally, these tents change bearing according to the dominant winds. Tents are now increasingly being walled in by thornbushes, however, and also annexes of mud and red brick are being erected. This indicates an inclination of these households towards a more durable settlement in the village. A total of 307 household dwellings were counted in Ilat Ayot in 1990 (Van Dijk 1991). The village area is the tribal land of the Hadendowa section of Jamilab. Eight different sub-sections and lineages have been counted as resident in the village. The most important are Hawan and Atawajab (together some 70 % of the interviewed households in the L&E Survey). Hubach & Marouf (1985) estimated the village population at 6,000 in 1984. Our surveys using aerial photographs and ground checks arrived at a number of 1,800-2,000 in 1989. These figures need not be contradictory. Ilat Ayot functioned as a regional distribution centre for relief goods supplied by the international programmes during the 1984-1985 drought. This temporarily attracted many destitute Hadendowa from the surrounding areas. Local commentators confirm that after the drought peak, most newcomers again left the village.

Nearly all interviewed household heads (93 %) were born in the village area. In all other cases, the households settled in Ilat Ayot in the years between 1965 and 1986. There are several groups who have a more nomadic lifestyle. They continue to live scattered in the village surroundings for at least a part of the year. The average household size in the village is 5.9. An average 2.1 members in the household are economically active (situation in 1988). Their economic orientation is mainly on Kassala town located at some 70 km distance south of the village. The nearest track to Kassala runs 12 km outside the village area. The tracks are rough and not passable for periods of several days, up to one week, after heavy rainfall. The journey to Kassala takes about 1.5 hour in a four-wheel drive vehicle. It takes at least 5 hours by truck in the local system of public transport. Until recently, the only departure point was in the neighbouring village of Timikreef. This is at some 3 hours' walk from Ilat Ayot. The public transport system was improved in 1990. Trucks now collect travellers in Ilat Ayot and the surrounding settlements in the late afternoon, and proceed to Kassala after a nightstop in Timikreef. The arrival time is due next day before 9 am. Only in the dry season, when the Gash riverbed can be crossed, is the town of Aroma in closer reach than Kassala. Aroma had an estimated population of 10,000 in 1990. The town is located in the Gash delta at some 40 km due west of Ilat Ayot. It lodges the seat of the Gash Delta Agricultural Corporation. The town has a varied market and service apparatus, but there is no regular transport schedule with the Border Area. Ilat Ayot is considered in the matrix of L&E research villages (*cf.* section 2.1.2) as located at a relatively great distance to Kassala. There have been no government SWC interventions made in Ilat Ayot.

6.2.1 Village economy

Crop production

The households have access to local and non-local lands for crop production. The latter fall under the GDAC irrigation scheme, and are also situated on the eastern Gash margin. The share of households with access to land increased between 1983 and 1988. This growth mainly developed under sharecrop agreements at the non-local lands (table 6.2). On the other hand, the number of households with non-local titles was found to decrease from 9 (17 %) in 1983, to 7 (12 %) in 1988. The average land title size also slightly decreased when 1983 and 1988 are compared. An exception, however, is non-local sharecropping. The decrease in land title size on local lands was caused by the abandonment of marginal land in the farming zones of Hag Adam and parts of Tebiai (*cf.* table 6.3) and by succession.

Table 6.2 **Entitlement to arable land (number of households) and average household title (fd), Ilat Ayot, (1983 and 1988)**

	1983			1988		
	No.of hh.	Fd	%	No.of hh.	Fd	%
No titles	3	na	6			
Local lands only	18	8.5	33	18	7.9	31
Local and non-local lands	33	12.3	61	40	12.3	69
– of which: non-local sharecrop	24	4.6		33	5.2	
– of which: non-local entitlement	9	5.7		7	5.3	
Total	54	11.3	100	58	10.7	100

Source: L&E Survey. Note: na means not applicable.

The local lands of Ilat Ayot cover some 1,125 ha and are listed by farming zone in table 6.3. The Hag Adam area was abandoned after the 1984-1985 drought. This was mainly for reasons of irrigation water shortage, and its relatively distant location from the settlement. Parts of the Tebiai area were equally abandoned due to shortage of water, but also because of excessive deposition of sand and the frequent occurrence of pests of stemborer. Kokoraeib is of a relatively recent date. The area was developed for cultivation after a new branch in khor Youdrut had been formed. According to aerial photographs, this took place between 1966-1979. All local lands in Ilat Ayot are held under customary right of possession and utilization. Local land was found to be sharecropped under terms of equal sharing of produce on one landholding only. New GDAC land titles were allocated to households in Ilat Ayot in 1990. A total of 900 fd was distributed by village leaders among the poorest families, at shares of 2-5 fd per household. Cultivation on these new lands only started in 1991 which year is not covered by the L&E Survey.

Table 6.3 **Characteristics of the farming zones in the Ilat Ayot village domain (situation 1988)**

Khor name	Size khor catchment (km²)	% Rock outcrop	Farming zone	Size farming zone (ha)	Dominant technique
Kokoraeib	43.69	4.3	Kokoraeib	265	w
Youdrut	43.69	4.3	Ilat Ayot	170	t
Youdrut	ibid.	ibid.	Youdrut	80	t, w
Youdrut	ibid.	ibid.	Tebiai	75	t
Tendelai	19.19	98.6	Tendelai	} 325	t
Tendelai	ibid.	ibid.	Eragot		t, w
Tendelai	ibid.	ibid.	Hemiai		t
Tendelai	ibid.	ibid.	Hag Adam	210	t, (a)

Source: L&E Survey, 1979 and 1986 aerial photographs. Note: rock outcrop as % of the khor catchment, t=teras, w=wildflooding, a=abandoned, areas are based on 1979 aerial photographs.

The average total household crop production over the period 1983-1990 of local and non-local lands combined is presented in table 6.4. Millet clearly is of minor importance in the area. The total annual production of sorghum is variable and ranges from 6.5 to 9 sacks (600 to 829 kg). The annual production of stalks varies in the range from 410 to 670 bundles (1,230 to 2,010 kg). The production in the drought years 1984-1985 is higher than in several other years. However, this is mainly the effect of relatively stable contributions to production from the lands in the GDAC irrigation scheme and on the Gash margin (*cf.* appendix 6.1). The production in the relatively wet year 1988 is high, but lower than would be expected on the grounds of comparative rainfall conditions alone. The reason is that excessive rainfall and flooding in Ilat Ayot was followed by a high incidence of malaria and dengee fever. This significantly reduced the availability of labour at critical periods during harvest time. For all years listed in table 6.4, the average sorghum production falls under the household food-security level. This can be defined for an average household in the Border Area as an annual requirement of 10-13 sacks.[1] An overview of the average household production figures of sorghum, millet and stalks, by cultivation technique, for 1983 and 1988-1990, is presented in appendix 6.1.

---

[1] The caloric intake of the Hadendowa in the Gash delta area is given by Ausenda (1987,531-534,586-588: "bush-dwelling Hadendowa") as 43 Kcal per day, per kg body-weight, under normal conditions. However, the actual intake was regularly found to drop temporarily to some 32 Kcal. At an average household composition of 2 male adults, 1 female adult and 3 children with average weights of 52 kg, 39 kg and 25 kg, and the equivalent of 1 sack of sorghum set at 260,000 Kcal (*Ibid.* 1987,586), the household grain requirements range from 13 sacks (normal), to 10 sacks (minimum) per year.

Table 6.4 **Total household crop production. Averages of sorghum (in sacks), millet (in sacks) and stalks (in bundles), Ilat Ayot (1983-1990)**

	Sorghum	N	Millet	N	Stalks	N
1983	6.6	42	1	1	422	24
1984	8.7	34	---		604	30
1985	8.2	36	---		478	35
1986	7.6	30	0.1	2	532	30
1987	7.8	30	---		669	25
1988	7.1	53	0.5	1	508	30
1989	8.4	28	---		410	23
1990	8.7	30	---		536	27
Av. 1983-1990	7.9		0.5		520	

Source: L&E Survey. Note: 1 sack sorghum is 92.1 kg, 1 sack millet is 92.9 kg, 1 bundle stalks is 3 kg. -- means not cultivated.

Crop production in Ilat Ayot is still mainly for subsistence needs. However, the marketing of produce is becoming more important. 11 % of the households sold a part of the sorghum harvest in 1983. In 1988, this was 15 %. Respectively 32 % and 35 % of the households sold a part of their stalks harvest in these years. These figures are the highest recorded in the research villages. It is mainly the production from non-local lands which is brought to the markets by the households of Ilat Ayot.

Livestock raising

Livestock numbers drastically declined in Ilat Ayot in the 1980s (table 6.5). The village herd, in terms of the total of all households in the L&E Survey sample, was 920 TLU in 1983. This was only 305 TLU in 1988. The share of households owning at least one animal remained at a high 93 %. However, the wealth per household in head of stock decreased for all species. This average household wealth was 18.0 TLU in 1983 (range 2.0-64.8) and 5.4 TLU in 1988 (range 0.2-24.6). This represents a 70 % reduction. Livestock losses as a direct consequence of the 1984-1985 drought reportedly were suffered in 11 out of 58 surveyed households (19 %). Forced sales as a result of these same conditions occurred among 29 households (50 %). Of the remaining 18 households, 4 showed an increase in livestock wealth, 6 showed no change. For 8 households, no data were available for either 1983 or 1988 to make the comparison. The average loss due to drought per household was 25.1 TLU (range 4.0-58.8) and mainly consisted of cattle and sheep. When this figure is compared with the average livestock wealth per household, it is shown that usually larger herds have been affected by the drought. Table 6.5 shows that goats and donkeys are most common in Ilat Ayot, while cattle, camels and sheep are next in importance.

Cattle and sheep are the species most vulnerable to drought. However, usually only cattle and camels are sent on seasonal trekking in Ilat Ayot. The two herds are not mixed, but do share the main destinations of seasonal migration. These are situated in the Border Area, Gash delta and Gash Dai. In the normal-to-dry year 1983, however, the cattle were also sent south to the areas of Gedaref and the Rahad river. This is at a distance of 400 km from Ilat Ayot. The absence from the village of the herdsmen was found in the majority of cases to cover a period of over 200 days per year. The main income-generating activities in Ilat Ayot related to the livestock sector include, besides contract-herding jobs, mainly the processing of milk.

Table 6.5 **Livestock ownership (number of households and %) and average household livestock wealth (number of head), Ilat Ayot (1983 N=54 and 1988 N=58)**

	1983			1988		
	No.of hh.	%	No.of head	No.of hh.	%	No.of head
Chickens	11	20	5.0	11	19	3.9
Goats	43	80	10.9	46	79	10.7
Sheep	31	57	28.2	27	47	9.0
Cattle	42	78	11.2	27	47	3.2
Camels	35	65	6.3	33	57	2.4
Donkeys	45	83	1.5	38	66	1.2
Total livestock	50	93		54	93	

Source: L&E Survey. Note: total livestock means any of all species.

Labour migration

The participation in labour migration in Ilat Ayot was in 1988 some 10 % higher than in 1983. Also the average numbers of migrants and destinations per household increased when these years are compared (table 6.6). There is no daily labour migration, because of the village's relatively distant location from the urban centres and its poor transport facilities. The prevailing main destinations of temporary migration (*cf.* section 5.2.3 for definitions) in 1988 were in the Gash delta (17 out of 24 migrating households, or 71 %), and in New Halfa (15 %). These were the first destinations in this year. The second destinations were in Kassala. Employment in the Gash delta and in New Halfa was predominantly found in contract-herding (49 %) and farm labour (32 %). Employment in Kassala was mainly found in the urban industry and services sectors. The period of continued absence from the village due to labour migration was usually between 30-100 days (50 % of the migrating households) for the first destinations. Generally, shorter durations of stay of less than 30 days were recorded for the secondary destinations. Only one household reported to have a member on long-term labour migration. This was to Saudi Arabia, where employment had been found in livestock husbandry. The pattern of migration destinations, types of employment, and duration of stay in 1983 was largely identical to that of 1988.

Local off-farm employment

The number and share of households in Ilat Ayot engaged in off-farm activities doubled when the situations in 1983 and 1988 are compared. The average number of activities per household, however, remained the same. The increase was in all employment categories, but less frequently in local herding. This is supposedly due to the decline of livestock. The main increase in terms of numbers of households involved is found in collection and production activities. In Ilat Ayot, these mainly comprise of the cutting and collection of fuelwood, which is sold in the village.

Table 6.6 **Labour migration engagement by main type (number of households), average number of migrants (number of household members) and average number of destinations, Ilat Ayot (1983 N=54 and 1988 N=58)**

	1983				1988			
	No.of hh.	%	Av.no. migrants	Av.no. destin.	No.of hh.	%	Av.no. migrants	Av.no. destin.
Daily labour migration	na	na	na	na	na	na	na	na
Temp. labour migration	16	30	1.31	1.44	24	41	1.42	1.71
Longt. labour migration	na	na	na	na	1	2	1.00	1.00
Total migration	16	30			24	41		

Source: L&E Survey. Note: total migration means any of all categories, na means not applicable.

Networking

In Ilat Ayot, informal transfers of wealth are mainly made in cash. In 1988, these contributed to a greater number of household incomes than in 1983. Also incidental relief distributions were made in the area in early 1988.[2] These incomes, however, could not always be distinguished from traditional networking gifts by the respondents. They are included in the 1988 figures as networking income.

Table 6.7 **Off-farm employment by main type (number of households) and average number of activities per household, Ilat Ayot (1983 N=54 and 1988 N=58)**

	1983			1988		
	No.of hh.	Av.no.activ.	%	No.of hh.	Av.no.activ.	%
Contract-farming	0		0	3		5
Contract-herding	1		2	2		3
Collection and production	12		22	24		41
Local services	5		9	9		16
Total local off-farm employment	18	1.28	33	38	1.29	66

Source: L&E Survey. Note: total local off-farm employment means any of all categories.

The average amounts of cash and value of goods involved in individual transfers in 1983 was the equivalent of £s 693. This was £s 888 in 1988 (in current £s). The use of informal workgroups *kaben* (Tb) was reported by 16 out of 21 households (76 %) for the situation in 1983, and by 23 out of 27 households (85 %) for 1988.

[2] Ilat Ayot was visited by a representative of an unknown Arab charity organization. £s 100 banknotes were being distributed on this occasion to the population gathered.

Table 6.8 **Networking engagement (number of households), Ilat Ayot (1983 N=54 and 1988 N=58)**

	1983		1988	
	No.of hh.	%	No.of hh.	%
Livestock transfers	0	0	2	3
Grain transfers	3	6	3	5
Cash transfers	4	7	7	12
Total networking	6	11	11	19

Source: L&E Survey. Note: total networking means any of all categories. Note: 1988 data include NGO gifts in cash.

Household income

The composition of household livelihoods in Ilat Ayot in terms of monetary income and livestock wealth is shown in table 6.9. The decline in livestock between 1983-1988 of some 70 % was already mentioned. The household income is presented, firstly, as the average income per livelihood sector. These figures can not be summed over households, because not all households engage in the same sectors. The household income is presented, secondly, as the sectoral average per household, which figures can be summed.[3] The 1988 budgets are deflated with 160 % to obtain constant 1983 prices. In terms of averages per sector, in 1983 crop production ranks third on this list, after off-farm employment and livestock-related incomes. It ranks second in 1988, after off-farm employment. These incomes increased between 1983 and 1988 for crop production and labour migration. The livestock-related incomes, off-farm employment incomes and networking incomes all decreased, but most notably the first. In terms of average incomes per household, the main changes consist of greater income shares generated by off-farm employment and smaller shares by livestock-related activities, when 1988 is compared with 1983. Some additional data on land use and livestock wealth are also available for 1989 and 1990 (figure 6.2). The decline in livestock wealth between 1983 and 1988 is apparent. After 1988, livestock wealth tends to increase again. The average entitlements to local lands for cultivation remain stable. The entitlements to non-local lands increase slightly over these years.

6.2.2 Interventions by the government and NGOs

Local interventions in the area made by various government institutions, national and international NGOs, are presented chronologically in table 6.10. The common administrative and taxation interventions (*cf.* section 5.2.6) are not included due to a lack of accurate data. After early relief operations in the area in 1948-1949, there is a gradual change in the accent of interventions from labour recruitment campaigns, to provision of public services, emergency relief, and finally to agricultural research and development. However, emergency relief measures had to be taken again in 1990 and 1991. Interventions have clearly intensified in Ilat Ayot after the 1984-1985 drought.

---

[3] In other words: the average income per sector is based on row totals and the average income per household is based on column totals, where rows and columns respectively represent livelihood categories and households.

Table 6.9 **Household income and livestock wealth. Income is presented as household averages per sector (rounded figures in constant 1983 £s) and sectoral averages per household (in %), livestock wealth (in TLU), Ilat Ayot (1983 and 1988)**

	1983			1988		
	Hh. av. per sector	% sectoral av. per hh.	N	Hh. av. per sector	% sectoral av. per hh.	N
Crop production	1,495	50	41	1,650	52	55
Livestock-related income	1,590	20	21	690	6	20
Labour migration	915	11	16	980	13	26
Off-farm employment	2,675	14	15	2,090	25	38
Networking	690	5	6	555	4	11
Total household income	3,200	100	48	3,780	100	57
Livestock wealth	18.0 TLU		50	5.3 TLU		54

Source: L&E Survey. Note: based on partly modelled data. The 1988 figures are deflated to a 1983 base by 160 %.

Figure 6.2 **Average household land entitlement (in fd), average household livestock wealth (in TLU), average total household income (constant 1983 £s x 1,000), Ilat Ayot, 1983-1990**

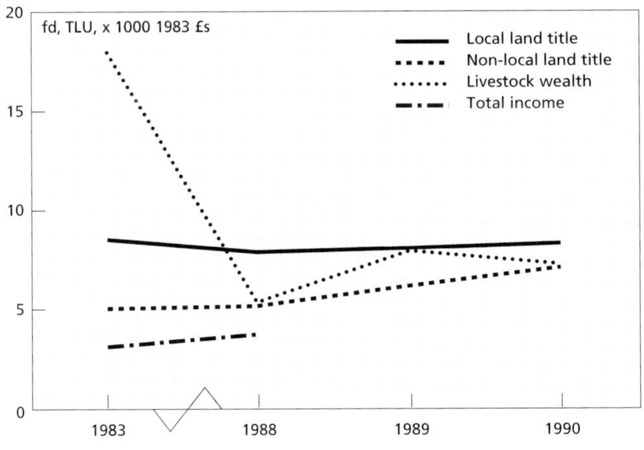

The early drought relief campaigns in the area were discussed in section 5.2.6 *Relief aid.* Labour recruitment in the 1960s was for the cotton harvest in the GDAC irrigation scheme. This recruitment has not been reported for any of the later years. The basic public services of education, health care and domestic water supply were being provided from the 1970s onwards. Relief aid in the mid 1980s and in 1990-1991 was given by the Sudanese government, and the NGOs CARE and the Saudi Red Crescent. These provided food, clothes and also stimulated the organization of health care facilities locally. The food rations included sorghum, wheat, milk powder, sugar, cooking oil and dates. The interventions in crop production started in the Ilat Ayot area in 1988 with pilot surveys carried out by the Horticulture Department and NCDRWR. This was in the framework of the international combined MFEP/KADA and WADS programmes (TNO & NCDRWR 1990). Horticulture development based on pumped groundwater was proposed. However, the quantities in which the resource proved available were not large enough to justify its development. The Department of Soil Conservation commenced in 1986/1987 with a programme of seed distribution of drought-resistant and short-maturing varieties of sorghum *Tajaroob Halfa*, *Gadam el Hamam* and hybrid Pioneer USA. It also distributed seeds of millet, okra, rosella and watermelon.

Table 6.10 **Government and NGO interventions in the Ilat Ayot area**

1948-1949	Drought relief food distributions.
1960s	Labour recruitment for cotton-pick and packing in the Gash delta irrigation scheme.
1970	Building of primary school.
1976	Building of dressing station.
1976-1977	Labour recruitment for cotton-pick in the New Halfa and Rahad irrigation schemes.
1984-1986	International emergency relief programmes. Ilat Ayot was one of the relief distribution points for the Border Area (+).
1987	Construction diversion dike in khor Youdrut to refill well fields (+).
1987-1988	International emergency relief programmes (+).
1986	Contruction 15 m lined well.
1988	Electro-magnetic and Vertical Electrical Sounding groundwater surveys (+).
1988	Research horticultural crops (+).
1988	Seed distribution programme (+).
1988	Construction 15 m cement-lined well in Ilat Ayot khor (+).
1988-1989	Building of new dressing station (+).
1990	Rebuilding of a diversion dike in khor Youdrut to refill well fields (+).
1990-1991	International emergency relief programmes (+).
1991	Upgrading and extension of primary school.

Source: NRO Carlisle BD/19.B.4, Weekly Report No.6,8, KADA (1986), TNO & NCDRWR (1990), files of Kassala NCDRWR and SCLUWP, L&E Survey interviews. Note: (+) means direct or indirect support received from international programmes.

6.2.3 SWC details

The natural environment of Ilat Ayot is highly dynamic. The new branch of khor Kokoreib, which development could be traced from aerial photographs to cover only a period of 13 years, at present locally reaches bed widths of 100 m. At an entirely different scale, Niemeijer (1993) concluded that semi-detailed soil survey and mapping is not possible in this area because of these dynamics which result in extreme lithological variabilities. It is also at this latitude that the first micro-sand dunes appear in the Border Area. Brushwood cores are being used in the building of *teras* bunds to trap the windblown sand. No SWC interventions have been made by the government in the village domain itself. The only dam built by the Department of Soil Conservation in 1987 serves the purpose of improving domestic water supply in a well-field along khor Youdrut. The wells have regained their water table, but part of the farming zone of Kokoraeib now receives less floodwater as a consequence. This negatively affects some 60 households who are cultivating in this zone under wildflooding techniques.

Production performances

About half the number of Ilat Ayot households has access to land under iSWC techniques of either *teras*, or brushwood panels. This share was slightly higher in 1988 (52 %), than in 1983 (48 %) (table 6.11). The most important technique is the *teras* (table 6.12). Average sizes of the land titles by technique are shown in the same table. These decrease when 1983 and 1988 are compared for *teras* and wildflooding. They increase for land under brushwood panels, and for non-local lands in the GDAC scheme and notably on the eastern Gash margin. Table 6.13 presents the cropping patterns at the landholding level. The dominant pattern under all cultivation techniques is a single crop of sorghum of the Aklamoy variety. This is less pronounced under *teras* (52-56 % of the cases) because of more frequent use of combinations of different sorghum varieties (Feterita and Hagartai, not shown in this table). No second crops, or intercrops, were grown in Ilat Ayot in 1983 and 1988. Small patches of okra and rosella were being tried in a number of *terus* close to the village area in 1989 and 1990, however.

Table 6.11 **Engagement in iSWC for crop production (number of households), Ilat Ayot (1983 and 1988, all households)**

	1983		1988	
	No.of hh.	%	No.of hh.	%
Teras	24	44	26	45
Brushwood	5	9	6	10
– of which: teras and brushwood	3	15	2	9
Total indigenous iSWC (teras and/or brushwood)	26	48	30	52
Total households	54	100	58	100

Source: L&E Survey.

Production figures for grain and stalks and cropping intensities are presented in appendix 6.1 for 1983 and 1988-1990 by technique. The production is variable over these years. Millet is of limited importance. The highest average production of sorghum on local lands was reached under brushwood panels in 1989. This was an average 8 sacks per fd (1,754 kg/ha). The lowest average production was under *teras* in 1983. This was only 0.81 sack per fd (178 kg/ha). The production of sorghum at non-local lands usually exceeds the production at the local lands of Ilat Ayot. The average non-local production peaked in 1990 with 7.2 sacks per fd (1,572 kg/ha) in the GDAC scheme. The lowest average non-local sorghum production was still 3.3 sacks per fd (723 kg/ha) under wildflooding on the Gash margin in 1989 and 1990.

Millet was cultivated in Ilat Ayot in 1983 and 1988 on local lands. The yields were extremely poor. The highest average was under *teras* in 1988 (0.8 sack per fd, or 177 kg/ha). The lowest in the same year under wildflooding (0.18 sack per feddan, or 40 kg/ha). The production figures for stalks include stover and are presented for sorghum and millet together. The highest average production on local lands was under brushwood panels in 1989 (400 bundles per fd, some 2,856 kg/ha). The lowest under the same technique in 1990 (55 bundles per fd, 392 kg/ha). The highest and lowest production on non-local lands was 350 and 133 bundles per fd (2,499 and 950 kg/ha) in the GDAC scheme, and on the Gash margin respectively. In both cases, this applies to the situation in 1990. Generally, the differences in production level between local and non-local lands are less pronounced for stalks than for grain (*cf.* appendix 6.1). This is because stalks are not collected when transport costs to the village are unfavourable. This practice is known to have resulted also in under-reporting on the actual stalk production levels reached during the L&E Survey.

Table 6.12 **Entitlement to arable land (in fd) by technique, Ilat Ayot (1983 and 1988, landholdings of all households)**

	1983		1988	
	Fd	N	Fd	N
Teras	6.2	24	6.0	26
Brushwood	4.4	5	6.0	6
Wildflooding	6.1	40	5.6	45
Gash delta scheme	5.0	22	5.5	18
Gash margin	4.4	11	4.9	22

Source: L&E Survey. Note: data are at the level of landholdings.

The cropping intensities in Ilat Ayot tend to decrease under all techniques over the years between 1983 and 1990. On local lands, they generally varied between 60-80 %. The lowest average was found under brushwood panels (20 % in 1989). The highest average was found under *teras* to the maximum of 100 % in 1989. The cropping intensities at the non-local lands were in a similar range from 60 % to 90 %. The lowest average occurred in the GDAC scheme in 1989 (35 %), and the highest on the Gash margins under wildflooding techniques in 1983 (100 %) (*cf.* appendix 6.1).

Table 6.13 **Dominant cropping pattern by cultivation technique and percentage of occurrence, Ilat Ayot (1983 and 1988, landholdings of all households)**

	Cropping pattern 1983	%	N	Cropping pattern 1988	%	N
Local lands						
– Teras	sorghum Aklamoy	56	16	sorghum Aklamoy	52	21
– Brushwood	sorghum Aklamoy	80	5	sorghum Aklamoy	83	6
– Wildflooding	sorghum Aklamoy	67	33	sorghum Aklamoy	78	40
Non-local lands						
– Gash delta scheme	sorghum Aklamoy	100	22	sorghum Aklamoy	88	17
– Gash margin	sorghum Aklamoy	100	8	sorghum Aklamoy	92	13

Source: L&E Survey. Note: data are at the level of landholdings. Note: dominant means highest occurrence per technique.

Land use patterns

Households in Ilat Ayot have access to eight different farming zones. They also cultivate lands in the Gash delta and on the Gash margins. The total number of landholdings per household ranges from 1 to 4, with an average number of 2.1. Land use is variable, by

Figure 6.3 **Number of households in Ilat Ayot, by combinations of number of cultivation techniques applied, and number of landholdings entitled to (local and non-local lands combined), 1988**

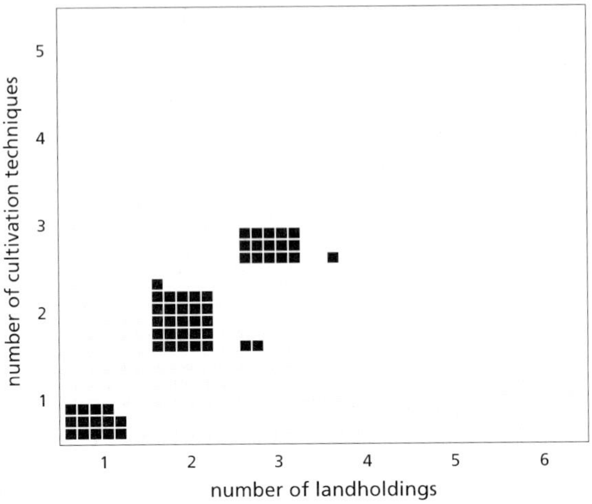

location, and by year. Some 40 out of 55 respondents (73 %) in Ilat Ayot stated that they regularly do not cultivate all their landholdings in any one year. The main reasons given for this behaviour were "shortage of irrigation water" (18 out of these 40, or 45 %) and "shortage of farm labour" (9 out of 40, 23 %). Land users frequently also use different techniques for cultivation, either by choice, or by circumstance. When for the purpose of comparison only the group of households with access to more than one landholding is considered, 26 out of 44 households (59 %) use two different techniques on two landholdings, and 15 (34 %) even use three different techniques on three landholdings (figure 6.3). Risk-spreading behaviour (*cf.* section 5.3 for definitions) can be found among 14 out of 55 households (26 %), when sub-optimal landholdings are defined by the land users according to criteria selected by themselves, the "free criteria". The most frequently used criteria were "great distance to the village" (6 out of these 14, or 43 %), and "unfavourable water supply" (5 out of these 14, 36 %). Risk-spreading behaviour is adopted by 10 out of these 55 households (18 %), when sub-optimal landholdings are defined from L&E Survey data according to their average level of sorghum production per feddan.

6.2.4 Nutrient harvesting in *teras* SWC[4]

Niemeijer (1993), together with staff of the Soil Conservation Department, made a case study of the nutrient status of soils under *teras* in Ilat Ayot. The studied *teras* is situated in the Tebiai farming zone. It was selected from aerial photographs on the grounds of its

[4] The following data are mainly taken from Niemeijer (1993,163-191, appendices c1, c2).

clearly distinguishable sediment harvesting performances in the past. Although we can not consider the outcomes as representative for the Border Area as a whole, these findings do provide good insights into some of the favourable properties of this technique. The land use history and physical conditions pertaining to this location in Tebiai are summarized in table 6.14. The soil chemical characteristics at two sampling points inside the bunded area, and two more at control locations outside the bunded area, are presented in table 6.15. The soil nutrient status of cultivated land under wildflooding in the farming zone of Bernoop in Hafarat (*cf.* section 6.5.1), in addition, was considered as a second "outside" control location. No land use history is known for this Bernoop location. The farming zone, however, is likely to have been cultivated since at least 1941. In this year, the Melassee dam was built in this area (*cf.* sections 6.4.2, 6.5.2). A single crop of sorghum is likely to have been the main type of land use in Bernoop since.

Table 6.14 **Land use history and physical conditions of a selected teras landholding in the Tebiai farming zone of Ilat Ayot**

Age of the teras (est.)	20-35 years
Maximum sedimentation rate (est.)	1.0-3.0 cm per year
Main cropping pattern (est.)	sorghum monoculture
Recent production (est.)	
– grain	1987: absent, 1988: 450 kg/ha, 1989: 169 kg/ha
– stalks	1987: 250 kg/ha, 1988: not collected, 1989: no data

Source: Niemeijer (1993). Note: the teras age was derived from aerial photographs available for 1963 onwards, interviews with farmers and field observation. The sedimentation rates were determined by various methods (Ibid.1993,164-176) and averages over different teras sections are presented here. Information on cropping patterns and production levels was collected through interviews with farmers.

The samples taken from the surface sediment (0-1 cm and 2.5-7.5 cm) represent the soil nutrient status under conditions of recent sediment supply. The samples taken from the topsoil (10-15 cm and 15-20 cm) represent conditions under prolonged land use. The figures in table 6.15 indicate that organic carbon, total nitrogen, total phosphorus, and Olsen's procedure for the estimation of plant available phosphorus P-Olsen are generally higher within the bunded area of the *teras* landholding than outside. These figures are also higher than under wildflooding in Bernoop. In addition, the pH of the soils is also more favourably lower inside the *teras* than at the outside locations and under wildflooding.[5] The pH of the top soil under wildflooding, however, is the same as inside the *teras*. A so-called "primitive nutrient balance" was calculated by Niemeijer. The outcomes indicate that nitrogen, as one of the main limiting macro-nutrients to plant growth, is not likely to constrain crop production under *teras* in the Tebiai area. This holds for the situation at present and the near future, which period was not specified (*Ibid.* 1993,187-189).

---

[5] The optimum range of pH for planth growth is 6-7.5 (ILACO 1981).

Table 6.15 **Soil chemical characteristics in a selected teras landholding (Tebiai farming zone, Ilat Ayot) and wildflooding landholding (Bernoop farming zone, Hafarat)**

	INSIDE TERAS [A]	OUTSIDE TERAS [A]	WILDFLOODING [B]
Soil reaction (pH H_2O)			
– surface sediment	7.84-7.98	8.31	8.09
– topsoil	8.04-8.06	8.41-8.55	8.04
Organic carbon (%)			
– surface sediment	1.23-1.94	0.60	0.63
– topsoil	0.95-0.97	0.39-0.51	0.38
Total nitrogen (ppm)			
– surface sediment	757-1,551	507	551
– topsoil	567-679	172-183	223
Total phosphorus (ppm)			
– surface sediment	847-852	708	814
– topsoil	830-871	527-664	858
P-Olsen (ppm)			
– surface sediment	11.1-11.3	3.1	n.d.
– topsoil	5.7-6.5	1.4-2.3	n.d.

Source: Niemeijer (1993,181-184). Note: [A] samples of surface sediment are taken from 0-1 cm, of topsoil from 10-15 cm. [B] samples of surface sediment are taken from 2.5-7.5 cm, of topsoil from 15-20 cm, n.d. means not determined.

6.2.5 Scores on selected variables

The scores on the set of 18 research variables are discussed next. The presentations are based on all households in the L&E Survey, except when indicated otherwise.

Table 6.16 **The importance of iSWC in crop production. Average household crop production income by technique as % of total household crop production income, Ilat Ayot (1983 and 1988, iSWC households)**

	1983		1988	
	%	N	%	N
Teras	45	12	49	21
Brushwood	54	3	43	6
Teras and brushwood	35	1	100	2
Total indigenous iSWC (teras and/or brushwood)	50	14	52	25

Source: L&E Survey.

The dimension Income

The share of households in Ilat Ayot engaged in iSWC was shown in table 6.11 to have remained at comparable levels of 26 out of 54 households (48 %) in 1983, and 30 out of 58 households (52 %) in 1988. The proportion of households who actually had a production, however, differs more importantly. Table 6.16 shows that this is the case for 14 out of these 26 households (54 %) in 1983, and 25 out of these 30 (83 %) in 1988. The iSWC incomes are shown in the same table to have contributed for this sub-group of iSWC households an average 50-52 % to the total household crop production income in these years. The incomes of the crop production sector as a whole contributed an average 54-59 % to the grand total income of the households in Ilat Ayot (table 6.17). When the iSWC contributions, in turn, are expressed by the percentage of their contribution to this grand total, these figures are 20 % in 1983, and 27 % in 1988 (iSWC households only). *Teras* and brushwood panels were approximately of equal importance in this respect.

Table 6.17 **The importance of crop production in livelihood. Average household crop production income by technique as % of total household income, Ilat Ayot (1983 and 1988, iSWC households and noSWC households)**

	1983		1988	
	%	N	%	N
Total crop production (iSWC and noSWC)	59	41	54	55
Total indigenous iSWC crop production	20	14	27	25
– of which: teras	19	12	24	21
– of which: brushwood	18	3	29	6
– of which: teras and brushwood	7	1	57	2

Source: L&E Survey.

The dimension Time

SWC time

The total household labour time in Ilat Ayot allocated to selected cultivation practices during the months of the growing season (SWC-time *S*) was an average 72 man-hours (range 4-407) in 1983, and 64 man-hours (range 7-306) in 1988 (iSWC households only). These scores of SWC-time *S* are expressed in table 6.18 as a percentage of the total household labour time allocation to crop production in 1983 and 1988. The same data for 1989 and 1990 are presented in appendix 6.2. The figures for "total iSWC" tend to decline when the years 1983 and 1988-1990 are considered. These percentages were respectively 89 %, 92 %, 64 % and 58 %.

Potential crop production time

The average score in Ilat Ayot over the selected growing season months of the measure Potential crop production time *C* (*cf.* section 5.4 for definitions) was 2.20 in 1983, and 2.09 in 1988. The average household size in these respective years was 4.9 and 5.9. The average number of economically active members in the household was respectively 2.2 and 2.1. The score of *C* divided by the latter number of economically active household members is used as the research variable in this study (table 6.19).[6]

---

[6] The scores are used in a comparison of research villages. No importance should be attached to their absolute values because these are based on relatively arbitrary weighing procedures (*cf.* section 2.3.2).

Table 6.18 **The importance of iSWC in crop production. Average household scores of SWC-time S by technique as % of total household labour-time allocations to crop production, Ilat Ayot (1983 and 1988, iSWC households)**

	1983		1988	
	%	N	%	N
Teras	75	20	82	26
Brushwood	76	5	82	4
Teras and brushwood	97	2	95	1
Total indigenous iSWC (teras and/or brushwood)	89	23	92	29

Source: modelled data based on L&E Survey. Note: labour time spent on maintenance, gap-filling, weeding and thinning for the growing season months June to and including October.

Table 6.19 **The importance of crop production in livelihood. Average household scores of Potential crop production time C divided by the number of economically active household members, Ilat Ayot (1983 and 1988, all households)**

	1983		1988	
	Score	N	Score	N
C/no. economically active household members	0.76	53	0.77	58

Source: modelled data based on L&E Survey. Note: calculations over the growing season months June to and including October.

The dimension Land

The area of cultivated lands allocated to iSWC is calculated as a percentage of the total household acreage of respectively local lands, and local and non-local lands combined. The shares of "total iSWC" in table 6.20 do not differ greatly between 1983 and 1988 (iSWC households only). They cover respectively 74 % and 75 % of all local lands, and respectively 56 % and 58 % of the local and non-local lands combined. The data for 1989 and 1990 are presented in appendix 6.2. These indicate largely identical levels for 1989, but higher iSWC land shares in Ilat Ayot for 1990.

The dimension Perception

The perception of land users with respect to the relative importance of iSWC in crop production is presented by average rank scores in table 6.21. The ranking in Ilat Ayot is over five techniques. These include, besides the two iSWC techniques (*teras*, brushwood panels) also wildflooding, flood-recession and gravity irrigation. The calculations are made for the sub-group of iSWC households only. The average rank of iSWC techniques was 1.87 for the situation in 1983 and 1.69 for 1988. This indicates a relatively high importance in the range from 1 to 5, but still lower scores (indicating higher importance given to iSWC) are found in the other research villages.

Table 6.20 **The importance of iSWC in crop production. Average household acreage by technique and farming zone in % of total land entitlement, Ilat Ayot (1983 and 1988, iSWC households)**

	1983 % Local land	1983 % Local and non-local land	1983 N	1988 % Local land	1988 % Local and non-local land	1988 N
Teras	66	51	24	69	54	26
Brushwood	66	47	5	80	54	6
Teras and brushwood	100	74	3	100	100	2
Total indigenous iSWC – (teras/brushwood)	74	56	26	75	58	30

Source: L&E Survey.

Table 6.21 **The importance of iSWC in crop production. Average household rank scores of perception by technique, Ilat Ayot (1983 and 1988, iSWC households)**

	1983 Score	1983 N	1988 Score	1988 N
Teras	2.00	23	1.67	24
Brushwood	1.50	5	1.71	5
Total indigenous iSWC (teras/brushwood)	1.87	26	1.69	29

Source: L&E Survey. Note: range is 1-5, where 1 represents highest importance.

The rank number given by households to crop production in a similar range of five livelihood categories was on average 1.68 for the situation in 1983 and 1.47 for 1988. In terms of perception, crop production was considered as second-best category in 1983, after livestock husbandry. However, it was considered of first importance in 1988. Crop production is the only sector whose importance in livelihood increased when the situations in 1983 and 1988 are compared.[7]

Table 6.22 **The importance of crop production in livelihood. Average household rank scores of perception by livelihood category, Ilat Ayot, (1983 and 1988, all households)**

	1983 Score	1983 N	1988 Score	1988 N
Crop production	1.68	50	1.47	55
Livestock husbandry	1.58	53	1.77	52
Labour migration	2.67	12	2.76	21
Local off-farm employment	2.74	19	3.09	34
Networking	3.40	15	3.77	22

Source: L&E Survey. Note: range is 1-5, where 1 represents highest importance.

[7] When these perception scores are compared with the ranked household income scores of table 6.9 (average incomes per sector in constant 1983 £s), it appears that the perceived importance of crop production (position 2 in 1983 and position 1 in 1988) is higher than the importance assessed in monetary terms (positions 3 in 1983 and position 2 in 1988).

6.3 Telkook

Telkook is located at latitude 16° 05' North and longitude 36° 40' East. This is in the 250-300 mm rainfall zone of the Border Area. The settlement was built in 1952 on the slopes of Jebel Onob overlooking khor Telkook. This place is indicated as a well-field on the early topographical map of the area (1:250,000 Maman NE 27 M, surveyed 1902 and revised 1940). The village itself is only first indicated on the more recent 1:100,000 topographical map (338 Jabal Sheqat surveyed in 1982). The population was given in the 1983 census as 17,562. L&E Survey assessments based on official food ration books indicate a settled population of only 916 households and a total of 4,333 persons in November 1991.[8] The same division of the built-up area into public and private domains as in Ilat Ayot is also found in Telkook. Some three-quarter of the settlement consists of private residential areas of lined up tents. The remaining one quarter is the public domain of the market area. A total of 34 shops, bakeries, tailors, butchers and workshops of leather workers was counted in the market area in 1992. Telkook, in addition, also has 4 guesthouses, 7 grain storehouses, 1 Koran school and 1 mosque. The public service buildings include 1 primary school, 1 hospital, 1 veterinary station and 1 guesthouse of the Department of Soil Conservation. Telkook services a wide area in the region, especially in the fields of health care and religious training. Telkook is, after Hameskoreib, the largest religious centre of the Ali Betai movement in the Border Area. As in Ilat Ayot, tents are increasingly being walled in. The construction of annexes is becoming more common. This indicates an inclination of the households to more settled lifestyles.

Telkook is the tribal land of the Jamilab section of Hadendowa. A total of 22 sub-sections and lineages were counted as resident. Lalai and Idris are the most important (together 25 %). Other resident non-Beja tribes include Eritrean Baria, who have come to Telkook to study the Koran. Some 82 % of the household heads interviewed were born in the village. The remaining household heads were born elsewhere in eastern Sudan and had settled in Telkook between 1950-1988. Small groups, usually lineages of less than 10 families together, still have nomadic lifestyles. They live in the village surroundings, and rely on Telkook for their basic commodities. According to the ration books, this group consisted of 257 families or 2,010 persons in 1991. These are not part of the 916 households mentioned above. The average household size in Telkook in 1988 was 5.4 and an average 1.5 member in the household was economically active.

The economic orientation of the households in Telkook is mainly on Kassala town located 115 km southeast of the village. Two main tracks, connecting Kassala with Tokar and Suakin at the Red Sea coast, call at the village. The journey to Kassala takes about 3 hours by four-wheel drive vehicle. It takes, however, nearly 6 hours in the public transport system by truck. After heavy rainfall in the wet season, the tracks are impassable for a

---

[8] This figure is based on the *Kurut el Tamwun* issued by the Kassala Rural Border Council to ration food supplies in the rural areas. No access into the residential area of Telkook was allowed to check this in the field, however (*cf.* section 2.3.1 *Research methods*). The population figure based on the ration books was acknowledged by local leaders as more reliable than the census figures.

period of up to one week. Travel generally took considerably more time in the years before 1991. The Border Area, at that time, was frequently being patrolled by the Ethiopian air force. Its MIGs, searching for cross-border movements of trucks with goods for the EPLF, and sometimes mistaking, could only fly during daytime. This compelled travellers to make the safer, but also more demanding and longer journey at night. Telkook is considered in the matrix of research villages as located at a relatively great distance to Kassala. The government has made SWC interventions in the village area.

6.3.1 Village economy

Crop production

Telkook has the greatest share of households without access to cultivated land of all the villages researched in the Border Area. This was 68 % in 1983, but declined significantly to 39 % in 1988 (table 6.23). The greater importance of cultivation in 1988 resulted mainly from allocations on local lands. Only 10-11 % of the households in Telkook have access to a combination of local and non-local lands. This is significantly less than the 60-70 % found in Ilat Ayot.

Table 6.23 **Entitlement to arable land (number of households) and average household title (fd), Telkook, (1983 and 1988)**

	1983			1988		
	No.of hh.	Fd	%	No.of hh.	Fd	%
No titles	39	na	68	24	na	39
Local lands only	12	4.5	21	31	5.3	51
Local and non-local lands	6	8.8	11	6	9.2	10
– of which: non-local sharecrop	3	3.3		2	2.9	
– of which: non-local entitlement	3	11.7		4	10.8	
Total	57	6.1	100	61	6.3	100

Source: L&E Survey. Note: na means not applicable.

The households in Telkook have access to seven different farming zones currently in use. These extend over a total area of 830 ha, excluding Gadamai which is not covered by aerial photographs available, and Hamandow which was factually abandoned in the 1920s (table 6.24). The latter area used to be cultivated under wildflooding. A smaller branch of khor Telkook which used to supply this area with floodwater, however, changed its course, and left the area high and dry. The entitlement to local lands is in all but three cases under customary rights of possession and utilization. The exceptions concern land use under sharecropping in the system of equal sharing of *sahib el nus*. A total of 2,000 fd of tenancy titles in the GDAC scheme was allocated to the Telkook households in 1990-1991. These lands are not included in the L&E Survey, however (*cf.* section 5.2.1 *Land titles*).

Table 6.24 **Characteristics of the farming zones in the Telkook village domain (situation 1988)**

Khor name	Size khor catchment (km²)	% Rock outcrop	Farming zone zone (ha)	Size farming technique	Dominant
No khor	na	na	Kadabot	335	t
No khor	na	na	Gadaweet	45	t
No khor	na	na	Gadamai	--	t
Telkook	123.56	0.0	Telkook	290	t
Telkook	ibid.	ibid.	Hamandow	??	(a)
Mindoweet	51.13	9.3	Mindoweet	50	t
Ungwateit	51.44	0.0	Huagas	100	w
Ungwateit	ibid.	ibid.	Ungwateit	355	e,w

Source: L&E Survey, 1979 aerial photographs. Note: t=teras, w=wildflooding, a= abandoned, e=earth dam, areas based on 1979 aerial photographs, -- means no coverage by aerial photographs.

Table 6.25 shows the averages of total household crop production for the years 1983-1990 on local and non-local lands combined. The cultivation of millet is of minor importance. The production of sorghum is variable. It ranges from 3.7 to 8.6 sacks of grain (341-792 kg), and 105 to 731 bundles of stalks (315-2,193 kg) per year. There is a low total production in the first drought year 1984, as could be expected. The production in the wet year 1988, however, is not higher than in most other years on the list. It is even lower than in the second drought year 1985. Like in Ilat Ayot, crop production activities in Telkook were importantly curtailed in 1988 by the widespread occurrence of malaria. The household food-security level of 10-13 sacks of sorghum was not reached in any of the years 1983-1990. An overview of the production figures by cultivation technique is given in appendix 6.1. Crop production in the GDAC scheme significantly contributes to the livelihood of households who hold such titles. Usually, the level of production is twice that reached at the local lands. However, the number of households with such titles was still limited at the time of the survey. Crop production in Telkook is almost entirely for subsistence. No stalks and grain were marketed by any of the interviewed households in 1983. Only 1 out of 13 producing households (8 %) reported to have sold less than half of its production of stalks and sorghum on the local market in 1988.

Livestock raising

The total herd in Telkook of the households in the L&E Survey sample was 722 TLU in 1983 and 223 TLU in 1988. The share of households owning at least one animal remained, with some 60 %, at about equal levels in these two years (table 6.26). The wealth per household in head of livestock decreased for all species, except for goats and donkeys. The average household wealth was 18.0 TLU in 1983 (range 1.0-145.0), and 4.0 TLU in 1988 (range 0.2-40.0). This represents a reduction of 78 %. Losses of livestock in 1984 and 1985 due to direct drought effects were only reported by 2 out of 61 households (3 %). Forced sales were reported by 28 households (46 %). Of the remaining 31 households, 3 mentioned an increase in livestock wealth, 4 mentioned no change, and in 24 cases no data were available to compare 1983 and 1988. The average reported drought-loss per household was 45.8 TLU of a combined herd of cattle and sheep (2 households only). Goats were being raised by about half the number of households in Telkook. The other species, however, were of lesser importance, particularly when this is compared with the other research villages. Donkeys were kept by less than 30 % of the households. No chickens were being raised because these allegedly attract snakes and wildcats.

Table 6.25 **Total household crop production. Averages of sorghum (in sacks), millet (in sacks) and stalks (in bundles), Telkook (1983-1990)**

	Sorghum	N	Millet	N	Stalks	N
1983	8.6	9	1.0	1	517	4
1984	4.9	4	---		250	2
1985	7.8	15	---		275	13
1986	3.7	15	---		221	14
1987	8.4	15	---		105	14
1988	7.0	23	0.5	1	442	16
1989	4.9	16	---		550	17
1990	6.7	22	---		731	25
Av. 1983-1990	6.5		0.8		386	

Source: L&E Survey. Note: 1 sack sorghum is 92.1 kg, 1 sack millet is 92.9 kg, 1 bundle stalks is 3 kg. -- means not cultivated.

Table 6.26 **Livestock ownership (number of households and %) and average household livestock wealth (number of head), Telkook (1983 N=57 and 1988 N=61)**

	1983			1988		
	No.of hh.	%	No.of head	No.of hh.	%	No.of head
Chickens	0	0	0.0	0	0	0.0
Goats	33	58	16.4	28	46	16.6
Sheep	20	35	64.2	16	26	27.8
Cattle	18	32	12.2	8	13	4.1
Camels	20	35	2.4	12	20	1.3
Donkeys	16	28	1.7	16	26	1.8
Total livestock	36	63	37	61		

Source: L&E Survey. Note: total livestock means any of all species.

The destinations of seasonal trekking are mainly located in the Border Area, Gash delta and Gash Dai. Camels and cattle share these destinations, but the herds are not mixed. The absence from the village of camel herdsmen was between 100-200 days per year in both 1983 and 1988. Cattle herdsmen reported this absence was more than 200 days in the normal-to-dry year 1983, and under 100 days in the wet year 1988. The main income-generating activities in the livestock sector are the processing of milk and contract-herding.

Labour migration
Table 6.27 lists the number of households in Telkook engaged in the main types of migration (*cf.* section 5.2.3). Migration in 1988 was, with 46 % of the households participating, more important than in 1983 when this was still 18 %.

Table 6.27 **Labour migration engagement by main type (number of households), average number of migrants (number of household members), and average number of destinations, Telkook (1983 N=57 and 1988 N=61)**

	1983				1988			
	No.of hh.	%	Av.no. migrants	Av.no. destin.	No.of hh.	%	Av.no. migrants	Av.no. destin.
Daily labour migration	na	na	na	na	na	na	na	na
Temp. labour migration	10	18	1.00	1.20	28	46	1.00	1.48
Longt. labour migration	na	na	na	na	1	2	1.00	1.00
Total migration	10	18			28	46		

Source: L&E Survey. Note: total migration means any of all categories, na means not applicable.

There is no daily labour migration from Telkook because of its relatively distant location from the urban centres. The prevailing main destinations of temporary migration for 1988 were the Gash delta (11 out of 28 migrating households, or 39 %), and Kassala town (32 %). The secondary migration destinations were also in these regions. Employment was mainly found in casual agricultural labour (41 %) and contract-herding (35 %). The prevailing secondary types of employment again include casual labour in agriculture, and in the Kassala urban sector. The absence from the village due to temporary labour migration most frequently was between 30-100 days per year (48 %). This is shorter for the secondary migration destinations, which all took less than 30 days per journey. There were two cases of long-term labour migration reported in Telkook, respectively to Saudi Arabia and the town of Gedaref in eastern Sudan. Employment had been found in the agricultural sector in both instances. The migration characteristics for 1983 do not substantially differ from the situation in 1988. Only contract-herding in the Gash delta was, with 50 % of the cases in 1983, more important than it was in 1988.

Local off-farm employment
The participation in off-farm activities almost doubled when the situations in 1983 and 1988 are compared. The average number of off-farm activities per household was also slightly higher in 1988 (1.58), than in 1983 (1.50) (table 6.28). The increase took place in all activities, except in contract-herding. This supposedly is related to the losses of livestock. The main increase was found in the categories local services, particularly for functions related with religious activities, and in collection and production. The latter activities mainly included the collection of grass *Aristida adscensionis* (coll. Eilab), cutting of fuelwood, charcoal-burning and local artisanal works.

Table 6.28 **Off-farm employment by main type (number of households) and average number of activities per household, Telkook (1983 N=57 and 1988 N=61)**

	1983			1988		
	No.of hh.	Av.no.activ.	%	No.of hh.	Av.no.activ.	%
Contract-farming	0		0	1		2
Contract-herding	3		5	2		3
Collection and production	15		26	27		44
Local services	3		5	8		13
Total local off-farm employment	21	1.50	37	38	1.58	62

Source: L&E Survey. Note: total local off-farm employment means any of all categories.

Networking

The number of households in Telkook who received networking incomes was approximately the same for 1983 and 1988. The informal transfers between households mainly consisted of sorghum. The transfers from village leaders to households exclusively consisted of cash. The average gift received per household in 1983 was the equivalent of £s 640. This was £s 979 in 1988 (current £s). The use of informal workgroups of *kaben* (Tb) was reported by 17 out of 30 households (57 %) in 1983, and by 9 out of 17 households (53 %) in 1988.

Table 6.29 **Networking engagement (number of households), Telkook (1983 N=57 and 1988 N=61)**

	1983		1988	
	No.of hh.	%	No.of hh.	%
Livestock transfers	1	2	0	0
Grain transfers	4	7	8	13
Cash transfers	2	4	1	2
Total networking	7	12	9	15

Source: L&E Survey. Note: total networking means any of all categories.

Household income

Table 6.30 shows the composition of household livelihoods in Telkook in terms of monetary income and livestock wealth. The decrease in livestock wealth per household of nearly 80 % was already mentioned. The household average incomes per livelihood category show that crop production ranked third in 1983, after the incomes earned in off-farm employment and labour migration. The same crop production incomes ranked a low fourth in 1988, after all other livelihood categories except networking. The incomes earned in livestock-related activities, labour migration and off-farm employment all increased when the situations in 1983 and 1988 are compared. Especially this increase in the first group of activities is noteworthy, given the important losses of livestock. A plausible explanation is that the smaller herds of 1988 were more frequently kept in the village area, which facilitated the processing and marketing of their produce. The average total household income substantially increased from £s 2,700 in 1983, to £s 4,170 in 1988. The sectoral average incomes per household indicate that the shares of crop production, labour migration and off-farm employment were all greater in 1988 than in 1983. The incomes obtained from livestock-related activities and networking were smaller in 1988 than in 1983. The combined data available for the Telkook households on income, land use and livestock wealth for 1983 and 1988-1990 are shown in figure 6.4. Livestock tends to continue to decline, also after 1988. The average entitlements to local lands and non-local lands tend to slightly increase after 1988.

Table 6.30 **Household income and livestock wealth. Income is presented as household averages per sector (rounded figures in constant 1983 £s) and sectoral averages per household (in %), livestock wealth (in TLU), Telkook (1983 and 1988)**

	1983			1988		
	Hh. av. per sector	% sectoral av. per hh.	N	Hh. av. per sector	% sectoral av. per hh.	N
Crop production	1,540	12	8	1,060	20	26
Livestock-related income	1,020	23	10	1,510	6	11
Labour migration	2,045	15	8	2,955	23	25
Off-farm employment	2,520	41	16	2,685	49	35
Networking	640	9	7	612	2	9
Total household income	2,700	100	31	4,170	100	51
Livestock wealth	18.0 TLU		36	4.0 TLU		37

Source: L&E Survey. Note: based on partly modelled data. The 1988 figures are deflated to a 1983 base by 160 %.

Figure 6.4 **Average household land entitlement (in fd), average household livestock wealth (in TLU), average total household income (constant 1983 £s x 1,000), Telkook, 1983-1990**

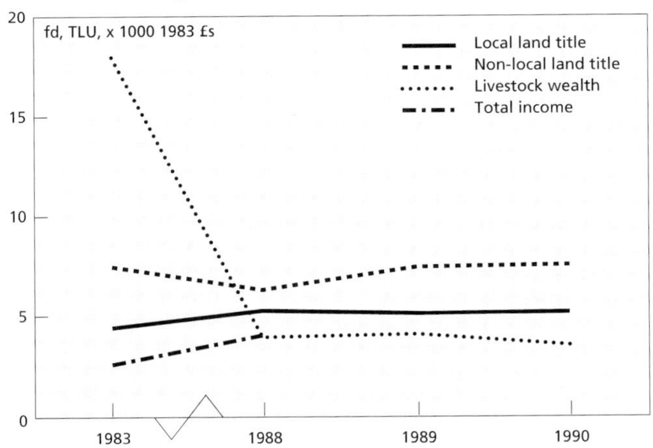

6.3.2 Interventions by the government and NGOs

The pattern of government interventions in Telkook, excluding those in the fields of taxation and common administration for which data is lacking (*cf.* section 5.2.6), is comparable to the one described for Ilat Ayot (table 6.31). Besides early reconnaissance for SWC mentioned in the archival sources, the first intervention recalled by the households is labour recruitment in the late 1930s. The Anglo-Egyptian Administration assigned local tribesmen for espionage in the then hostile Italian Eritrea. The Hadendowa were favoured for these activities because practical use could be made of their traditional rivalry with the Beni Amer who at that time collaborated with the Italians.

The first relief interventions and rural development initiatives were made in Telkook at a relatively early stage. Reports on its favourable potential for floodwater harvesting date from 1931 (DSA Barter 646/2). The first formulations for SWC research to improve the rangelands in the area of khor Mendoweet were made in 1964. This project was part of the national Anti-Hunger Campaign which was never implemented due to logistic difficulties (*cf.* section 4.2.1).

Table 6.31 **Government and NGO interventions in the Telkook area**

1931	Field treks by civil servants to assess the potential of the Telkook area for floodwater harvesting.
1937-1940	Recruitment campaigns by the British Army.
1948-1949	Drought relief food distributions.
1962	Relief distribution after heavy rains and flooding.
1964	Anti-Hunger Campaign and SWC research in the khor Mindoweet area.
1965	Building of primary school.
198?	SCLUWP survey and mapping of khor Mindoweet.
1982	Construction of 8 m lined well.
1982	Building of hospital.
1984-1985	International emergency relief programmes (+).
1985	Construction SWC earth dam in khor Ungwateit.
1985	Construction of two 9 m and 2 m lined wells.
1985	Extension of hospital (+).
1986	Extension SWC earth dam in khor Ungwateit.
1987	Building of SCLUWP guesthouse and store.
1987	Lining of shallow well (+).
1988	Electro-magnetic and Vertical Electrical Sounding groundwater surveys (+).
1988	Provision pesticides after locust plague.
1988	Research Horticultural Department (+).
1988	Construction of two wells for horticulture pilot project (+).
1988	Development 10 fd horticulture pilot project (+).
1989-1990	Rebuilding primary school.
1989	200 m flood protection works constructed by SCLUWP (+).
1989	10 fd agro-forestry project: vegetables and woodlot, construction of 2 wells for irrigation (+).
1989	International emergency relief programmes (+).
1990	Building of veterinary station and training of local nurse (+).
1990	Building of dressing station (+).
1990	Development second 10 fd horticulture project (+).
1991	International emergency relief programmes (+).

Source: NRO Carlisle BD/19.B.4, DSA Barter 646/2, TNO & NCDRWR (1990), files of Kassala Rural Water Corporation, NFC, Department of Horticulture, SCLUWP. Note: (+) means direct or indirect support received from international programmes.

Educational facilities were available in Telkook from the mid 1960s onwards. Other public services were introduced some two decades later. The level of government activities declined in the 1970s as a result of the general deadlock in planning (*cf.* section 4.2.1), and the upsurge in military activity. The Eritrean EPLF, fighting a guerilla-war against the Ethiopian army, was forced to retreat into Sudan near Telkook in 1978. The tense security situation was, in the mid 1980s, still a reason for the Department of Soil Conservation in Kassala to suspend its activities in certain parts of the northern Border Area (Ahmed 1987). A decree of the Sudan military in 1985 temporarily commanded the suspension of all government operations within a distance of 60 km from the international border.

Households in Telkook received emergency relief in the mid 1980s, in 1989 and in 1991. However, only limited data exist on the rations supplied locally (*cf.* section 5.2.6 *Relief aid*). The government started in 1985 with the construction of an earth dam. These activities are discussed separately below. From the mid 1980s onwards, the level of government interventions steadily increased in the fields of health care, education and agricultural development supported by the international MFEP/KADA and WADS programmes. The FAO, in addition, supported the Fuelwood Development Eastern Sudan (FDES) programme between 1989-1991. This was implemented by the Kassala section of the Forestry National Corporation (FNC). The main interventions of the international programmes in local land use include the following. The NCDRWR executed a geo-hydrological survey in the area of Telkook in 1988. A first 10 fd pilot area was developed for horticultural use on the southern bank of khor Telkook. The supply of groundwater proved insufficient. A second area of the same size was developed in 1990 at a location some 200 m upstream along the same khor. An area of 1 fd was cultivated in this year with tomato, okra and onion. Groundwater was delivered by motorpump and conveyed to the fields by means of portable tubes. FDES, in the meantime, had developed a total area of 10 fd for agro-forestry on the northern margin of this khor in 1989. In this year, 3.5 fd was cultivated with onion and rosella. 1.5 fd was reserved as a village woodlot of mainly *Conocarpus lancifolius*. FDES also initiated a programme through which tree-seedlings of *Acacia seyal, A. tortilis, Balanites aegyptiaca, Ziziphus spina-christi* and *Azadirachta indica* were distributed locally. The impact of these recent programmes is still uncertain. The early 1990s were relatively dry, producing little discharge in khor Telkook for a continuous supply of groundwater.

6.3.3 Government SWC interventions

An earth dam was constructed by the Department of Soil Conservation in khor Ungwateit in 1985. Only the background to this SWC intervention is discussed in this section. The intervention effects will be turned to in section 7.1.

Area and people

The project is situated some 15 km northeast of the Telkook village in one of the branches of khor Togan. The area is referred to as Tanqwtitibilli on the 1:100,000 map. The people of Telkook call it khor Ungwateit. The upstream and downstream sections of this khor are known as Tebili and Shashob respectively. The dam was built in the area of Shashob. This is a relatively flat valley-bottom where floodwater naturally spreads over the cultivated lands. Several nomadic communities live scattered in this area. However, the greater part of land users in Ungwateit come from Telkook and other villages of Togan (estimated population of 3,000 in 1991), and Belastaf (estimated population of 500 in 1991). The farming zone is the tribal domain of the Hadendowa section called Jamilab. The 1979 aerial photographs of Ungwateit before the dam had been constructed show one *teras*. The remainder of this 334 ha is cultivated under wildflooding techniques. According to interviews, there was already extensive gully erosion in the area before the construction of the dam. Such erosion is not yet shown on the 1979 aerial photographs, however. From the beginning of the activities in Ungwateit, confusion has existed among the Telkook households as to exactly who was considered by the department as the main beneficiary group of this intervention. The cultivated lands commanded by the dam are largely used by the people of Telkook. The position of the dam, according to informants in Telkook, however, was largely based on instigation of land users from Togan. The latter also wanted their lands to benefit from the diverted floodwater. The number of Telkook households in the L&E Survey with land under command of the dam is 24. The average size of their landholdings is 3.8 fd (range 0.5-10 fd, N=22) in 1988 (table 6.34, 6.35).

Table 6.32 **Selected characteristics of the Ungwateit earth dam project (1985-1991) and indigenous land use in the area in 1979**

	1979	1985	1986	1991
Total dam length (m) [A]	na	634	869	200
Commanded area (ha)	na	nd	nd	42 [B]
Cultivated area (ha)	35	nd	nd	0
Dam breach (%)	na	nd	nd	77

Source: L&E Survey, files SCLUWP Kassala. Note: na means no applicable, nd means no data available, [A] lengths at the start of the season, [B] estimated.

Institutions and achievements

The dam was constructed as part of the Border Area Earth Dam programme of the Department of Soil Conservation. It was also part of the larger regional development programme called "Togan-Mamman Waterspreading". This may explain part of the local confusion about who was supposed to benefit from this intervention. Unlike most other dams, the work in Ungwateit, for being a regional programme, was co-financed by the Soil Conservation Administration in Khartoum. The standard working procedures for construction (*cf.* section 4.2.2 *Earth dams*) were followed in Telkook. The Department of Soil Conservation received assistance from the Department of Irrigation in Kassala for level surveys in the field. A winged-type of earth dam was built by loader in 1985. The total dam length at the date of construction was 634 m. An extension of 235 m was added in the following year. The total length of the dam reached in 1986 therefore was 869 m. During our first visit to Telkook in November 1991, the dam was found to be almost completely out of function. Already in the second year of service 1987, several breaches had occurred. In the years that followed, about 670 m of dam length had been washed away. The length of the structure still in tact for floodwater harvesting in 1991 was only 200 m. This remaining section, in addition, was severely affected by erosion. The estimated area that could be irrigated in 1991 was 100 ha, or 42 ha. No efforts had been made, neither by local users, nor by the Soil Conservation Department, to maintain or repair the dam after it had been completed. The main cause of this damage is likely to lie in the same design and construction flaws reported on earlier for other dams in the Border Area (*cf.* section 4.2.2 *Earth dams*). However, the poor definition of the target group is also likely to have contributed to a situation where locally no preventive action had been taken to avoid the deterioration of performance of the dam over the years.

Table 6.33 **Government institutions involved in the Ungwateit earth dam project and their main fields of activities**

GOVERNMENT INSTITUTIONS	MAIN ACTIVITIES IN PROJECT AREA
Ministry of Agriculture and Natural Resources	
– Dept. of Irrigation	Level-survey, dam design
– Dept. of Soil Conservation	Socio-economic survey, construction

Source: interviews SCLUWP Kassala.

6.3.4 SWC details

Telkook is the most northern of our four research villages in the Border Area. Aridity strongly influences land use locally. Erosion by wind increases in intensity at this lattitude. Permanent structures readily accumulate wind depositions of sand. An indication of the magnitude of this latter process is provided in one mudhouse in the village. This house is strongly exposed to these winds. Its outside entrance level was found to be about one metre higher than the original ground level inside. This situation took three and a half years to develop. The same depositions also cause regular changes in the courses of smaller khors. The prevailing method of local *teras* building in the area of Telkook is by using brushwood cores to trap the windblown sand. The sandy soils of the cultivated lands favourably influence the infiltration characteristics. They unfavourably reduce the capacity for soil moisture storage. Sandy soils are also less suitable for the construction of earth works, and *teras* bunds.

Production performances
The share of households in Telkook with land under iSWC is, with 13 households (21 %) in 1988, greater than 5 households (9 %) in 1983. These are the lowest figures recorded in the research villages. *Teras* is the main indigenous technique applied. The application of brushwood panels alone for water harvesting purposes is almost absent. The Ungwateit earth dam theoretically supplies floodwater to the lands of 24 households in our survey (39 %). However, the number of households who actually received water from the dam in 1988 was only 10. In addition, land use in Ungwateit was in this year factually practised

under wildflooding techniques. It is for this reason that earth dam land is not indicated as a separate category in appendix 6.1, but is referred to as cultivated under wildflooding.

Table 6.34 **Engagement in iSWC for crop production (number of households), Telkook (1983 and 1988, all households)**

	1983		1988	
	No.of hh.	%	No.of hh.	%
Teras	4	7	10	16
Brushwood	1	2	3	5
– of which: teras and brushwood	na		na	
Total indigenous iSWC (teras and/or brushwood)	5	9	13	21
Project pSWC (earth dam)	na		24	39
Total households	57	100	61	100

Source: L&E Survey. Note: na means not applicable.

The average size of land titles in Telkook, by technique, is presented in table 6.35. These data are at the landholding level. They show that the entitlements have increased under all techniques when the situations in 1983 and 1988 are compared, except for rainfed cultivation and GDAC irrigation. The most conspicious increase, however, concerned the area under *teras*. This was from an average 2.7 fd in 1983 (N=5), to an average 6.9 fd (N=11) in 1988. Table 6.36 illustrates that the dominant cropping pattern for all techniques, except *teras*, is a single crop of sorghum Aklamoy. In the *teras*, different sorghum varieties mainly short-maturing Hagartai and Feterita were grown, and intercropping was practised with vegetables (2 households with 2 landholdings in 1988; not shown in this table). Vegetables were also intercropped in 1988 under wildflooding techniques in parts of Ungwateit other than those controlled by the earth dam (1 household with 1 landholding in 1988; not shown in table 3.36).

Table 6.35 **Entitlement to arable land (in fd) by technique, Telkook (1983 and 1988, landholdings of all households)**

	1983		1988	
	Fd	N	Fd	N
Teras	2.7	5	6.9	11
Brushwood	2.0	1	3.3	3
Wildflooding	5.8	6	6.7	3
Rainfed	3.0	3	3.0	3
Gash delta scheme	7.5	6	6.4	9
Earth dam	na	3.8	22	

Source: L&E Survey. Note: data are at the level of landholdings, na means not applicable.

The production of grain and stalks in Telkook varied considerably over the years 1983 and 1988-1990 (*cf.* appendix 6.1). The highest average production of sorghum on local lands was obtained under *teras* in 1983 (2.82 sacks per fd, 618 kg/ha). The lowest average production was in 1989 and 1990 under brushwood panels with an average 0.2 sack per fd (44 kg/ha). However, the number of cultivating households was in all cases small, especially in 1983 (table 6.34). Millet was only cultivated in 1983 and 1988. It produced an average 0.10 and 0.17 sack per fd (less than 40 kg/ha) in these years respectively. The production on non-local lands in the GDAC scheme is less variable. There was an important dip in 1989, however. The highest average sorghum production in the scheme was 4.93 sacks per fd in 1988 (1,081 kg/ha). The lowest was 2.08 sacks per fd (456 kg/ha) in 1989.

Table 6.36 **Dominant cropping pattern by cultivation technique and percentage of occurrence, Telkook (1983 and 1988, landholdings of all households)**

	Cropping pattern 1983	%	N	Cropping pattern 1988	%	N
Local lands						
– Teras	mixed varieties sorghum	67	3	mixed varieties sorghum	50	10
– Brushwood	na			sorghum Aklamoy	100	2
– Wildflooding	sorghum Aklamoy	67	3	sorghum Aklamoy	86	3
– Earth dam	na			sorghum Aklamoy	63	15
Non-local lands						
– Gash delta scheme	sorghum Aklamoy	100	6	sorghum Aklamoy	100	9

Source: L&E Survey. Note: data are at the level of landholdings, na means not applicable. Dominant means highest occurrence per technique.

The average production of stalks of sorghum and millet combined on the local lands was between 667 bundles per fd (4,762 kg/ha) under *teras*, and 40 bundles per fd (286 kg/ha) under brushwood panels. In both cases, this applies to the year 1990. The highest and lowest average productions in the GDAC scheme were at comparable levels. This was 583 bundles per fd (4,163 kg/ha) in 1983, and 54 bundles per fd (386 kg/ha) in 1990. However, the uneconomic returns from transporting stalks to the village are also known to have resulted in underreporting on the level of stalk production on non-local lands among households in Telkook (*cf.* section 6.2.3 *SWC details*).

The cropping intensities on local lands in Telkook vary between some 70-100 %. The highest average intensities were under brushwood panels in 1988, 1989 and 1990. The lowest was found under wildflooding techniques in 1990 (73 %). The highest and lowest cropping intensities recorded for the non-local GDAC scheme lands were 89 % in 1988, and 77 % in 1983 and 1989.

Land use patterns

The households in Telkook have access to seven different farming zones and they also cultivate in the GDAC irrigation scheme. The total number of landholdings per household ranges from 1 to 3, with an average of 1.3. Households with access to land in 1988 (N=37) stated that they regularly do not cultivate all their landholdings mainly because of "shortage of irrigation water" (18 out of these 37 households, or 49 %), and "physical incapability or illness" (8 out of 37 households, or 22 %). Only 7 of these 37 households (19 %) had access to more than one landholding. 4 out of these 7, in turn, used two different techniques on two landholdings (figure 6.5). Risk-spreading behaviour was adopted in Telkook by 3 out of the total of 37 landed households (8 %), when sub-optimal landholdings would be defined according to the free criteria chosen by the land users themselves. The criterium used in all these three cases was "unreliable water supply". Risk-spreading behaviour was found among 4 out of 37 households (11 %), when sub-optimal landholdings were defined according to their average level of sorghum production per feddan (*cf.* section 5.3 for definitions).

Figure 6.5 **Number of households in Telkook, by combinations of number of cultivation techniques applied, and number of landholdings entitled to (local and non-local lands combined), 1988**

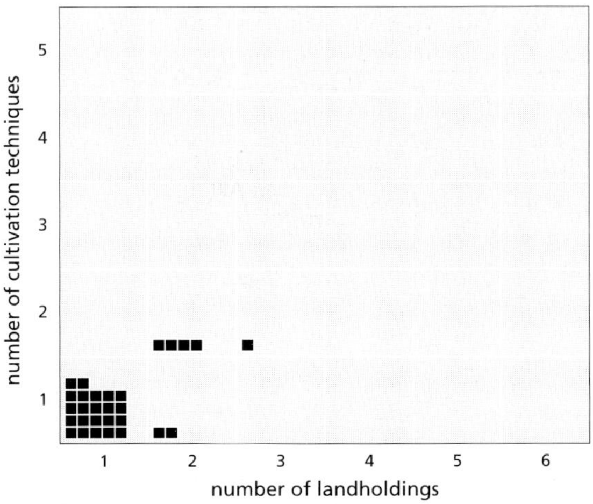

6.3.5 Scores on selected variables

The scores on the set of 18 research variables are discussed next. The presentations are based on all households in the L&E Survey, except when indicated otherwise.

The dimension Income
The number of households in Telkook who used iSWC increased from 5 (9 %) in 1983, to 13 (21 %) in 1988. A total of 24 households (39 %) participated in 1988 in the SWC project in Ungwateit (table 6.34). Table 6.37 shows that among these 5 iSWC households in 1983, only 2 had production on the cultivated lands which were under iSWC. This was the case for 10 out of the 13 households in 1988. The iSWC incomes contributed in both years the maximum 100 % to the total household crop production income. In this particular sub-group of iSWC households, accordingly, no other cultivation techniques had supported the household livelihoods in these years. The land under project SWC, factually cultivated under wildflooding conditions, contributed 83 % to the total household crop production income in 1988 (no earth dam was in use in Telkook in 1983).

Table 6.37 **The importance of iSWC in crop production. Average household crop production income by technique as % of total household crop production income, Telkook (1983 and 1988, iSWC households and pSWC households)**

	1983		1988	
	%	N	%	N
Teras	100	2	100	8
Brushwood	---	100	2	
Teras and brushwood	---	---		
Total iSWC (teras and/or brushwood)	100	2	100	10
Project pSWC (earth dam)	na	83	10	

Source: L&E Survey. Note: na means not applicable, --- means technique not used.

The income of the crop production sector as a whole contributed an average 48 % to the grand total income of all households in Telkook in 1983. This same figure was 39 % for the situation in 1988 (table 6.38). When the iSWC contributions, in turn, are expressed as the percentage of their contribution to the grand total income (sub-group iSWC households), this figure for these years was between 30-31 %. This was also the contribution of pSWC to the grand total household income in 1988 (sub-group pSWC households).

The dimension Time

SWC time
The total household labour time allocated in Telkook to selected cultivation practices during five growing season months (SWC-time *S*) was an average 34 man-hours (range 5-81) in 1983, and 38 man-hours (range 1-154) in 1988. The scores of SWC-time *S* are expressed in table 6.39 as a percentage of the total household labour time allocated to crop production. The scores were both in 1983 and 1988 100 % for iSWC. Approximately the same scores were recorded in 1989 and 1990 (*cf.* appendix 6.3). This means that with respect to selected cultivation techniques and growing season months, no labour had been allocated in the sub-group of iSWC households to any techniques other than iSWC for crop production on local lands. Also in the sub-group of pSWC households, all labour had been allocated to the earth dam lands in 1988. However, these were factually being cultivated under wildflooding techniques.

Potential crop production time
The average score in Telkook over the selected growing season months of the measure Potential crop production time *C* (*cf.* section 5.4 for definitions) was 1.55 for the situation in 1983, and 1.48 for 1988. The average household size in these years was 4.3 and 5.4. The average number of economically active members in the household was respectively 1.6 and 1.5. The scores of *C* divided by the latter number of economically active household members is used as the research variable in this study (table 6.40).[9]

[9] The scores are used in a comparison of research villages. No importance should be attached to their absolute values because these are based on relatively arbitrary weighing procedures (*cf.* section 2.3.2).

Table 6.38 **The importance of crop production in livelihood. Average household crop production income by technique as % of total household income, Telkook (1983 and 1988, iSWC households, pSWC households and noSWC households)**

	1983		1988	
	%	N	%	N
Total crop production (iSWC, pSWC, noSWC)	48	8	39	26
Total indigenous iSWC crop production	31	2	30	10
– of which: teras	31	2	23	8
– of which: brushwood	--		57	2
– of which: teras and brushwood	--		--	
Project pSWC (earth dam)	na		31	10

Source: L&E Survey. Note: na means not applicable, -- means technique not used.

Table 6.39 **The importance of iSWC in crop production. Average household scores of SWC-time S by technique as % of total household labour-time allocations to crop production, Telkook (1983 and 1988, iSWC households and pSWC households)**

	1983		1988	
	%	N	%	N
Teras	100	3	100	9
Brushwood	---		100	2
Teras and brushwood	---		---	
Total indigenous iSWC (teras and/or brushwood)	100	3	100	11
Project pSWC (earth dam)	na		100	9

Source: modelled data based on L&E Survey. Note: labour time spent on maintenance, gap-filling, weeding and thinning for the growing season months June to and including October. na means not applicable, --- means technique not used.

Table 6.40 **The importance of crop production in livelihood. Average household scores of Potential crop production time C divided by the number of economically active household members, Telkook (1983 and 1988, all households)**

	1983		1988	
	Score	N	Score	N
C/no. economically active household members	0.75	45	0.76	57

Source: modelled data based on L&E Survey. Note: calculations over the growing season months June to and including October.

The dimension Land

The area of cultivated lands allocated to iSWC is calculated as a percentage of the total household acreage of respectively local lands, and local and non-local lands combined. Almost the entire range of cultivated lands (99-100 %) was allocated to iSWC in 1983 and 1988 (sub-group iSWC households). It makes little difference whether this is expressed as a share in either two totals, because the non-local lands are relatively insignificant in Telkook. The same figures for 1989 and 1990 indicate equal high land shares under iSWC (*cf.* appendix 6.1). The number of Telkook households involved in iSWC, however, is in all years relatively small. Lands under the earth dam equally covered the entire household acreage of local lands (sub-group of pSWC households) in 1988. However, this represented for the same year 84 % of their total of local and non-local lands combined. The pSWC households in Telkook, accordingly, cultivated more than others also at non-local lands in the Gash delta.

The dimension Perception

The perception of land users of the relative importance of iSWC in crop production is presented by average rank scores in table 6.42. The ranking in Telkook is over five techniques. These include for the sub-group of iSWC households[10] besides the two iSWC techniques, also rainfed cultivation, wildflooding and gravity irrigation in the GDAC scheme. The average rank of iSWC techniques was 1.11 for the situation in 1983, and 1.17 for 1988. For comparison (not shown in the table), an average score of 1.27 (N=9) was given to pSWC in 1988, but ranking in this case was over a range of 1-6 techniques.

Table 6.41 **The importance of iSWC in crop production. Average household acreage by technique and farming zone in % of total land entitlement, Telkook (1983 and 1988, iSWC households and pSWC households)**

	1983			1988		
	% Local land	% Local and non-local land	N	% Local land	% Local and non-local land	N
Teras	100	100	4	100	98	10
Brushwood	100	100	1	100	100	3
Teras and brushwood	---	---		---	---	
Total indigenous iSWC – (teras/brushwood)	100	100	5	100	99	13
Project pSWC – (earth dam)	na	na		100	84	22

Source: L&E Survey. Note: na means not applicable, --- means technique not used.

Table 6.42 **The importance of iSWC in crop production. Average household rank scores of perception by technique, Telkook (1983 and 1988, iSWC households)**

	1983		1988	
	Score	N	Score	N
Teras	1.14	3	1.10	10
Brushwood	1.00	1	1.50	2
Total indigenous iSWC (teras/brushwood)	1.11	3	1.17	11

Source: L&E Survey. Note: range is 1-5, where 1 represents highest importance.

[10] This particular group did not include households using "earth dam" techniques, albeit five iSWC households in Telkook also use pSWC (*cf.* section 6.1, table 6.1).

Table 6.43 **The importance of crop production in livelihood. Average household rank scores of perception by livelihood category, Telkook, (1983 and 1988, all households)**

	1983		1988	
	Score	N	Score	N
Crop production	2.10	20	1.69	36
Livestock husbandry	1.30	30	2.00	27
Labour migration	1.75	8	1.71	24
Local off-farm employment	1.71	14	1.94	33
Networking	2.69	16	3.17	30

Source: L&E Survey. Note: range is 1-5, where 1 represents highest importance.

The average rank number given to crop production in a similar range of five livelihood categories was 2.10 in 1983, and 1.69 in 1988. In terms of average scores over these categories, the households of Telkook considered crop production the most important sector in 1988. This, however, came only fourth after livestock husbandry, off-farm employment and labour migration in 1983. The only categories for which the perceived importance was higher (lower scores) in 1988 than in 1983, were crop production and labour migration (table 6.43).[11]

6.4 Um Safaree

Um Safaree is located at latitude 15° 29' North and longitude 36° 30' East. This is in the 300-350 mm rainfall zone of the Border Area. The village is a narrow long strip of some 4 km length. It is situated on the high ridge between khor Um Safaree and the wide floodplain of khor Aderjawab. Um Safaree means "the mother of travelling". The village was established by Beni Amer nomads who seasonally migrated with their herds to this area, originating from Tokar at the Red Sea coast. The village was established in the 1920s. The village is not indicated on the early 1:250,000 topographical map (Kassala 56 A surveyed 1900-1902, revised 1940). It only first appears on the 1:100,000 map (Khor Abu Allaga 396) surveyed in 1984. The village plan clearly differs from the Hadendowa settlements in the Border Area. The separation of public and private domains is less strict. Um Safaree has no defined market place, and its 7 shops with a standard grocery stock are situated among the residences. At regular intervals, there are large open spaces where the livestock are penned in at night. The dwellings are mainly circular houses with conical straw roofs. They are built of mud and straw and the doors and window covers are made of corrugated iron. Usually, sheds of straw are attached to the main house. The living-

[11] When the perception scores are compared with the ranked household income scores of table 6.30 (average incomes per sector in constant 1983 £s), it is shown that the perceived importance of crop production (position 4 in 1983 and position 1 in 1988) differs from the importance assessed in monetary terms (positions 3 and 4 in 1983 and 1988 respectively).

quarters are fenched in by thornbushes. Only a few households in Um Safaree are living in the traditional tent of palm-fibre. This local architecture illustrates the generally longer sedentary tradition of the tribe when compared with the Hadendowa of Ilat Ayot and Telkook. Shortage of domestic water is a major problem in Um Safaree. Recent surveys indicated that there are no opportunities to improve this situation through local groundwater development (TNO & NCDRWR 1990). For this reason, some 40 households (12 % of the 1988 Um Safaree population of 328 households) reside most of the year in Kassala town. They only seasonally return to the village for cultivation.

A total of 19 tribal sections was counted as resident in Um Safaree. The most important are Beet Malah and Ay Tamah (together 50 %). This relatively great pluriformity can be understood, firstly, from the specific configuration of the Beni Amer tribal lands (*cf.* section 3.2 *Organization and gender*) and, secondly, from greater mobility before settlement as was mentioned above. Hubach & Marouf (1985) estimated the village population at 2,050 in 1984. Surveys based on aerial photographs indicated this to be approximately 2,500 in 1989. Some 85 % of the heads of household interviewed were born in the village, 11 % were born in the Tokar area, and 4 % in other parts of the Border Area and in Eritrea. Their dates of settlement in the village was found to be well distributed over the years between 1927-1976. The relatively small share of Eritrean households in Um Safaree is mainly explained from the local shortage of domestic water. The greater part of the Eritrean refugees who settled in the rural areas in the 1960s (*cf.* section 4.3.4), did so in the villages of the Border Area south of Um Safaree. In 1988, the average household size in Um Safaree was 7.7, and the average number of economically active members in the household 2.8. Their economic orientation is strongly directed to Kassala town, located at a distance of 15 km southeast of the village. Good tracks over smooth and level terrain allow travelling times to Kassala of only 30-45 minutes. This is by truck in the local system of public transport. A small bridge constructed over khor Amerei in 1991 now provides access to Kassala town during all seasons. Um Safaree is considered in the matrix of research villages as located at a relatively small distance to Kassala. There have been government SWC interventions in the past.[12] However, these have left no tangeable structures which are used, or even can be can be observed, today.

6.4.1 Village economy

Crop production

The households of Um Safaree predominantly use only local lands for cultivation. The average title on these lands, 18.4 fd, remained the same in 1983 and 1988 (table 6.44). This is the highest figure recorded in the research villages in the Border Area. The entitlements to non-local lands only concern two households. These titles were held in the GDAC irrigation scheme. The total household entitlement to local and non-local lands

---

[12] In the years before 1941, an earth dam for floodwater harvesting was built in the Aderjawab area. This dam no longer exists (*cf.* section 6.4.2). In 1987, in addition, preparations for a SWC pilot scheme were made which were aborted in the same year (*cf.* section 4.2.3 *Implementation in Hedadeib*).

combined, accordingly, also remained almost constant when the situations in 1983 and 1988 are compared.

Table 6.44 **Entitlement to arable land (number of households) and average household title (fd), Um Safaree, (1983 and 1988)**

	1983			1988		
	No.of hh.	Fd	%	No.of hh.	Fd	%
No titles	1	na	2			
Local lands only	57	18.4	95	61	18.4	95
Local and non-local lands	1	25.8	1	3	23.7	5
– of which: non-local sharecrop	1	1.0				
– of which: non-local entitlement	1	7.5		2	4.3	
No data	1	na	2			
Total	60	18.5	100	64	18.6	100

Source: L&E Survey. Note: na means not applicable.

The households of Um Safaree have access to 12 different farming zones. This again is the highest number of all research villages. The total area covered is about 3,435 ha. Access to these lands, however, is shared with the neighbouring villages including Hafarat and Awad (table 6.45). Parts of the farming zones of Hubor, Silkiai and Bernoop have been abandoned since 1975 because of deteriorating rainfall conditions. Land is held under customary rights of possession and utilization. There is one household in Um Safaree who uses land under the sharecropping conditions of *sahib el nus*.

The average annual total household production of grain and stalks between 1983-1988 on local and non-local lands combined is given in table 6.46. The cultivation of millet is of greater importance than in Ilat Ayot and Telkook. This expectedly is mainly a matter of taste preference. The soils do not significantly differ between the villages and precisely an opposite outcome would be expected, with more millet in the dryer northern part of the Border Area, were rainfall conditions to be considered an important factor in this respect.

The production of sorghum varied in Um Safaree between a total of 2.3-7.1 sacks (212-654 kg). For millet, this was between 0.8-3.6 sacks (74-334 kg). The production of stalks of millet and sorghum combined was between 200-811 bundles (600-2,433 kg) in these years. Production levels of sorghum, millet and stalks reached their lowest point in the first drought year 1984. In the second drought year 1985, the sorghum levels recovered, those of millet remained low, while the production of stalks even reached the highest level on this list of eight years. This probably mainly resulted from a greater collection effort in this year of stress. The highest sorghum production was registered in Um Safaree in the wet year 1988 (table 6.46).

The 1983-1990 average of the total household sorghum production in Um Safaree is, with 5.1 sacks, lower than in Ilat Ayot (7.9 sacks) and Telkook (6.5 sacks). This is mainly the result of the limited contribution of the GDAC irrigation scheme and Gash margin lands to the household economies in Um Safaree. The local grain production in most of the years listed in table 6.46 is under the food-security level of 10-13 sacks of sorghum, per household, per year. The average household production figures of sorghum, millet and stalks by cultivation technique for the years 1983 and 1988-1990 are given in appendix 6.1. Crop production in Um Safaree is still mainly for subsistence. Only 8 % of the households reported to have sold part of their grain harvest on the market in 1983. This same figure was 15 % in 1988. 13 % and 28 % of the households reported to have sold a part of their stalk harvest in these respective years.

Table 6.45 **Characteristics of the farming zones in the Um Safaree village domain (situation 1988)**

Khor name	Size khor catchment (km²)	% Rock outcrop	Farming zone zone (ha)	Size farming technique	Dominant
no khor	na	na	Wad Kaduj	230	t
no khor	na	na	Ja' ahbir	25	t
Amerai	205.00	16.5	Hubor	80	w
Kawateib	12.75	96.2	Kawateib	40	t
Kawateib	ibid.	ibid.	Sherifei	35	w,t
Idi Kideb	4.09	38.9.	Idi Kideb	90	t, w,b
Idi Kideb	ibid.	ibid.	Haykol	120	w,b
Aderjawab	205.00	16.5.	Silkiai	}	t
Aderjawab	ibid.	ibid.	Fil Natar	}	t
Aderjawab	ibid.	ibid.	Bernoop	} 2,740	w
Aderjawab	ibid.	ibid.	Aderjawab	}	w
Aderjawab	ibid.	ibid.	Taham	75	w,b

Source: L&E Survey, 1979 and 1986 aerial photographs. Note: t=teras, b=brushwood panels, w=wildflooding. Areas are based on 1979 aerial photographs.

Table 6.46 **Total household crop production. Averages of sorghum (in sacks), millet (in sacks) and stalks (in bundles), Um Safaree (1983-1990)**

	Sorghum	N	Millet	N	Stalks	N
1983	3.8	39	2.0	20	499	52
1984	2.3	8	0.8	3	200	15
1985	5.4	39	1.5	10	811	42
1986	5.2	41	3.3	10	592	43
1987	5.4	35	3.6	16	580	42
1988	7.1	54	2.2	29	711	57
1989	4.7	19	1.5	2	547	28
1990	6.7	10	3.0	2	424	31
Av. 1983-1990	5.1		2.2		546	

Source: L&E Survey. Note: 1 sack sorghum is 92.1 kg, 1 sack millet is 92.9 kg, 1 bundle stalks is 3 kg. -- means not cultivated.

Livestock raising
Livestock numbers drastically declined in Um Safaree in the 1980s (table 6.47). The village herd in terms of the total of all households in the L&E Survey sample was 1,025 TLU in 1983, but only 350 TLU in 1988. The share of households owning at least one animal slightly decreased from 98 % in 1983, to 94 % in 1988. The livestock wealth per household in head of stock decreased for almost all species, and most importantly so for cattle and sheep. The average household livestock wealth was 17.1 TLU in 1983 (range 0.2-320.0), and 5.3 TLU in 1988 (range 0.2-40.0). This represents a decline of 69 %. Only 2 out of 64 households (3 %) mentioned losses as a direct consequence of the 1984-1985 drought. Forced sales proved more important (43 households, 67 %). Among the remaining 19 households in Um Safaree, an increase of livestock between 1983 and 1988 was mentioned by 10 households (16 %), 2 mentioned no change (3 %), and no data were available for either year for 7 households (11 %). The reported two drought-losses concerned 22.2 TLU and 278.0 TLU of cattle and sheep combined. This indicates that relatively large herds had been affected. In 1988, mainly goats were being raised in Um Safaree, with cattle and sheep coming next in importance. Donkeys were the common means of local transport (table 6.47).

The main destination for seasonal trekking from Um Safaree with camels is Eritrea. More in particular, this concerns the areas along the upstream sections of the rivers Atbara and Gash. The main destinations of cattle differ between 1983 and 1988. In the first, which was a normal-to-dry year, cattle were mainly sent to the Setit and Atbara rivers and their tributaries. These provide the best pastures of eastern Sudan (FAO & Lund 1989). On their way, herdsmen also concluded stubble-grazing agreements in the mechanized-farming schemes around Gedaref. A herd of 100 head could graze 1,000 fd of sorghum stubble at prices of £s 200-300 in 1983, £s 2,000-3,000 in 1988, and £s 6,000 in 1990. In the wet year 1988, cattle were sent for stubble-grazing to Gedaref only. The absence from the village of these herdsmen in 1983 and 1988 was usually between 100-200 days per year. The income-generating activities related to the livestock sector included, in Um Safaree, the processing of milk, the selling of clarified butter and eggs, and contract-herding jobs.

Table 6.47 **Livestock ownership (number of households and %) and average household livestock wealth (number of head), Um Safaree (1983 N=60 and 1988 N=64)**

	1983			1988		
	No.of hh.	%	No.of head	No.of hh.	%	No.of head
Chickens	10	17	6.6	8	13	8.6
Goats	46	77	8.9	52	81	8.8
Sheep	36	60	21.5	32	50	14.4
Cattle	47	78	12.8	33	52	4.9
Camels	14	23	3.0	12	19	1.6
Donkeys	56	93	1.7	56	88	1.5
Total livestock	59	98		60	94	

Source: L&E Survey. Note: total livestock means any of all species.

Labour migration

The share of households in Um Safaree participating in labour migration was 62 % in 1983, and 84 % in 1988. All types of daily, temporal and long-term migration took place. All three types were also more important in 1988 than in 1983. The average number of migrants per household and destinations per household, was usually also higher in 1988 than in 1983. There are two exceptions namely an equal number (1 household member) of migrants on long-term migration, and a smaller average number of temporary migration destinations in 1988 (1.27) than in 1983 (1.39) (table 6.48).

Table 6.48 **Labour migration engagement by main type (number of households), average number of migrants (number of household members), and average number of destinations, Um Safaree (1983 N=60 and 1988 N=64)**

	1983				1988			
	No.of hh.	%	Av.no. migrants	Av.no. destin.	No.of hh.	%	Av.no. migrants	Av.no. destin.
Daily labour migration	25	42	1.04	1.04	34	53	1.15	1.15
Temp. labour migration	22	37	1.18	1.39	37	58	1.22	1.27
Longt. labour migration	1	2	1.00	1.00	2	3	1.00	1.50
Total migration	37	62			54	84		

Source: L&E Survey. Note: total migration means any of all categories, na means not applicable.

Daily labour migration from Um Safaree in 1988 was, without exception, to the town of Kassala. Most commonly, this involved work at the livestock market (20 out of 34 households, or 59 %). Three main types of employment can be found here. The first is an official appointment as licensed middleman called *damin*. This middleman negotiates between buyers and sellers of livestock, and carries out all necessary administrative procedures which are part of the transaction. By way of custom, the middleman and selling-party are usually of the same tribe. This provides the Beni Amer clear advantages in those cases where Eritrean herds are being brought to the market in Kassala. The second job is local trader, called *tadjeer*. The trader participates in livestock transactions, either as buyer or seller. The third job is as (illegal) intermediary, called *sababee*. This intermediary operates on the outskirts of Kassala town. His aim is to intercept selling-parties travelling to the market who are unfamiliar with its procedures and, more importantly for the *sababee*, with its prices. The intermediary later sells this livestock with considerable profit in Kassala himself. Such incomes, which are considered as daily labour migration incomes in this study, are relatively high when compared with those gained in other livelihood activities in the Border Area. The average income earned in livestock trade in Um Safaree was £s 4,160 per household in 1983 and £s 15,124 per household in 1988. These incomes alone already contributed on average 50 % and 70 % to the total household incomes in these respective years.

The prevailing main temporary migration destinations (*cf.* section 5.2.3 for definitions) from Um Safaree in 1988 were the New Halfa irrigation scheme (60 %), and the Gash region (20 %). Employment was found in these places respectively in casual agricultural labour (57 %) and contract-herding (32 %). The absence from the village due to temporary migration was usually between 30-100 days (61 %). The second destinations in the same year were all in Kassala town. The absence from the village in these occasions was less than 30 days. There were three households in Um Safaree who had members on long-term labour migration. These worked in livestock husbandry in the Gedaref area (2 out of 3 households) and in the bakery of New Halfa town (1 household). The pattern of migration destinations, types of employment, and duration of stay is largely identical when 1988 is compared with 1983. Only agricultural labour in the mechanized-farming schemes around Gedaref was an additional secondary destination in 1983, besides employment related to the livestock market in Kassala (2 and 3 out of 5 households in 1983 respectively).

Table 6.49 **Off-farm employment by main type (number of households) and average number of activities per household, Um Safaree (1983 N=60 and 1988 N=64)**

	1983			1988		
	No.of hh.	Av.no.activ.	%	No.of hh.	Av.no.activ.	%
Contract-farming	6		10	4		6
Contract-herding	2		3	9		14
Collection and production	10		17	9		14
Local services	2		3	3		5
Total local off-farm employment	20	1.20	33	25	1.48	39

Source: L&E Survey. Note: total local off-farm employment means any of all categories.

Local off-farm employment

The households of Um Safaree participate less in off-farm employment than was found in other research villages in the Border Area. This probably relates to their better access to financially more attractive job opportunities in Kassala town, such as the livestock market. The main increase in off-farm employment jobs was in local herding (table 6.49), despite the great losses of livestock. This work entailed the herding of sheep and goats which were kept in the village around the house. The average number of off-farm activities per household in Um Safaree was higher in 1988 (1.48) than in 1983 (1.20).

Networking

The share of households who received networking incomes was in 1988 (13 %) slightly smaller than in 1983 (15 %) (table 6.50). There were no gifts of livestock reported in Um Safaree, but exclusively gifts of cash and grain. The average amount of cash and value of grain received was the equivalent of £s 1,232 in 1983. This same amount was £s 912 in 1988 (in current £s). The use of informal workgroups of *nafir* was reported by 42 out of 45 households (93 %) in 1983, and 47 out of 51 households (92 %) in 1988.

Table 6.50 **Networking engagement (number of households), Um Safaree (1983 N=60 and 1988 N=64)**

	1983		1988	
	No.of hh.	%	No.of hh.	%
Livestock transfers	na	na		
Grain transfers	5	8	3	5
Cash transfers	4	7	5	8
Total networking	9	15	8	13

Source: L&E Survey. Note: total networking means any of all categories, na means not applicable.

Table 6.51 **Household income and livestock wealth. Income is presented as household averages per sector (rounded figures in constant 1983 £s) and sectoral averages per household (in %), livestock wealth (in TLU), Um Safaree (1983 and 1988)**

	1983			1988		
	Hh. av. per sector	% sectoral av. per hh.	N	Hh. av. per sector	% sectoral av. per hh.	N
Crop production	1,610	35	53	2,775	34	57
Livestock-related income	7,550	9	23	4,820	9	29
Labour migration	6,120	42	39	10,840	48	52
Off-farm employment	1,120	12	19	1,725	8	21
Networking	1,230	2	9	570	1	8
Total household income	8,980	100	59	14,100	100	64
Livestock wealth	17.1 TLU		59	5.3 TLU		60

Source: L&E Survey. Note: based on partly modelled data. The 1988 figures are deflated to a 1983 base by 160 %.

Household income

The composition of the household livelihoods in Um Safaree in terms of monetary income and livestock wealth is shown in table 6.51. We already mentioned the decline in livestock of some 70 %. In terms of the average incomes per livelihood sector, crop production ranked third in 1983 and 1988, after incomes earned in livestock-related activities and labour migration. The first were mainly based on incomes earned from the selling of milk and clarified butter in Kassala, and the selling of animal produce in the village itself. The second mainly on incomes earned in activities related to the livestock market in Kassala. Although the level of income of livestock-related activities almost halved when the situations for 1983 and 1988 are compared, in 1988 this work still fetched, on average, a higher income than crop production. In terms of the sectoral average incomes per household, the greatest contribution was made by labour migration with 42 % in 1983, and 48 % in 1988. No important changes have taken place in Um Safaree when 1983 and 1988 are compared with respect to these sectoral average incomes per household.

Figure 6.6 **Average household land entitlement (in fd), average household livestock wealth (in TLU), average total household income (constant 1983 £s x 1,000), Um Safaree, 1983-1990**

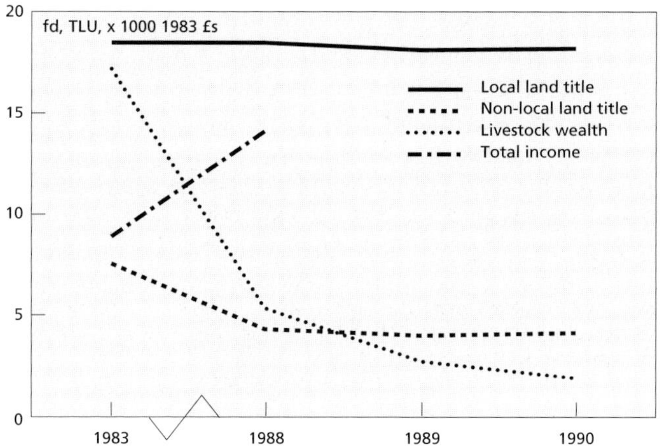

The combined data on household income, land use and livestock wealth over 1983 and 1988-1990 are shown in figure 6.6. Livestock wealth continues to decline, also after 1988. The entitlement to local cultivated lands remains stable. The entitlement to non-local lands decreased between 1983 and 1988, but remains stable after 1988 (two households only).

6.4.2 Interventions by the government and NGOs

The interventions in the area of Um Safaree are chronologically listed in table 6.52. The list excludes the common administrative and taxation interventions due to a lack of data (*cf.* section 5.2.6). The first references to an intervention by the Anglo-Egyptian Administration were found in the archives. In a letter from the Kassala District Commissioner to the Governor written in 1941 (NRO file KD/2 N.1), mention is made of an earth dam called "Melassee". This dam succesfully irrigated the cultivated lands along khor Aderjawab up to Jebel Mokram. This is an area some 5 km south of the present-day village of Um Safaree. No further details are provided and the dam no longer exists. Early relief operations in 1948-1949 were also reported in the archival sources (*cf.* section 5.2.6 *Relief measures*). In the 1950s, a *hafir* was constructed near the village for domestic and livestock water supply. After these interventions, the common cycle of events described for Ilat Ayot and Telkook set in. This included successive phases of labour recruitment campaigns, public services provision, and rural development programmes.

Casual labourers were recruited in Um Safaree for building work in New Halfa in the early 1960s. Public education and health care facilities were introduced in the village in the 1970s. The first land use interventions were made in the Idi Kedib farming zone in 1987. This area had initially been selected for the development of a SWC pilot scheme as part of the MFEP/KADA and BAPP activities in the Border Area. The Department of Soil

Conservation made a survey in Idi Kideb and started the preparation for the building of four earth embankments in this same year. However, this project was aborted (*cf.* section 4.2.3 *Implementation in Hedadeib*). During the mid 1980s drought, relief supplies of food and clothing were delivered to the households in Um Safaree (*cf.* section 5.2.6 *Relief aid*). These operations were locally coordinated by the Islamic African Relief Agency (IARA). Groundwater surveys were made in the village surroundings as part of the WADS programme in 1988. However, these failed to identify any suitable aquifers for development. MFEP/KADA supported the educational facilities in the village. IARA constructed a water-storage reservoir. Domestic water is now delivered by tanker-lorry which first fills up in Kassala town. The same tanker-lorry also supplies individual households on demand at prices of £s 20 per barrel of 200 ltr in 1991. The population of Um Safaree, finally, also benefits from the health-care facilities provided at the nearby refugee reception camp Wad Sherifee.[13] The camp, located 18 km southwest of the village, is run by the Sudanese organizations COR and SRC. International support was received mainly in the 1980s and the early 1990s from the UNHCR, ICRC, SwRC, MSF and The League of Red Cross. An estimated 60 % of the outpatients of the camp hospital in the early 1990s came from surrounding villages in the Border Area including Um Safaree (SwRC Health Coordinator, Interview Kassala, Nov. 1991).

Table 6.52 **Government and NGO interventions in the Um Safaree area**

1941	Reference in government correspondence to the "Melassee" dam in khor Aderjawab (construction date unknown).
1948-1949	Drought relief food distributions.
1950s	Construction hafir in khor Um Safaree.
1960s	Labour recruitment for construction work in New Halfa.
1970s	Building primary school.
1975	Building dressing station.
1984-1985	International emergency relief programmes (+).
1985	Construction attempt 31 m lined well (failed).
1987	Preparation for construction of 4 embankments for floodwater harvesting in khor Idi Kideb. The SWC project was later cancelled (+).
1987	Primary school extended.
1988	Electro-magnetic and Vertical Electrical Sounding groundwater survey (+).
1988	Construction water-storage reservoir and water deliveries made by tanker from Kassala town (+).
1990-1991	International emergency relief programmes (+).
1991	Construction hafir in upstream section of khor Um Safaree.

Source: NRO Carlisle BD/19.B.4, Weekly Report 6, 8, NRO Sandison KD/2 N.1, TNO & NCDRWR (1990), files of Kassala NCDRWR and SCLUWP. L&E Survey interviews. Note: (+) means direct or indirect support received from international programmes.

[13] Some 180,000 refugees were registered in the Wad Sherifei camp at the height of the drought in 1985. This figure was 36,000 in August 1986 (Kuhlman 1990,52). It had risen again to 70,000 in October 1991 (Health Coordinator Swiss Red Cross, Interview Kassala, Nov. 1991).

6.4.3 SWC details

The physical environment at the latitude of Um Safaree is, generally speaking, less hostile to land use than in the northern parts of the Border Area. However, the same dynamics in drainage systems and associated processes described for this northern area also apply to Um Safaree. The change in the course of khor Kawateib in 1985 is a case in point. This left the lands in the farming zone of Sherifei without floodwater. Land users from Um Safaree wanted to dig a channel, but met with fierce opposition from cultivators of these same lands living in Hafarat. The construction work was abandoned to avoid any escalation of this conflict.

Table 6.53 **Engagement in iSWC for crop production (number of households), Um Safaree (1983 and 1988, all households)**

	1983		1988	
	No.of hh.	%	No.of hh.	%
Teras	44	73	44	69
Brushwood	16	27	14	22
– of which: teras and brushwood	11	18	9	14
Total indigenous iSWC (teras and/or brushwood)	49	82	49	77
Total households	60	100	64	100

Source: L&E Survey.

Production performances

The share of iSWC households in Um Safaree is the highest recorded of all research villages in the Border Area. However, this share was smaller in 1988 (49 out of 64 households, or 77 %), than in 1983 (49 out of 60 households, or 82 %) (table 6.53). The main indigenous technique locally applied for SWC is *teras*. The average sizes of household land titles by technique are presented in table 6.54 at the landholding level. When the two research years are compared, it is shown that the average size of landholdings under *teras*, brushwood panels and wildflooding techniques remained stable or increased. However, this size decreased under rainfed cultivation and in the GDAC irrigation scheme. The dominant cropping patterns per technique are listed in table 6.55. These account for only 20-25 % of the landholdings cultivated under a given technique for the situation in 1988, and 20-50 % for the situation in 1983. This means that the variety of cropping patterns was great (*viz.* without being "dominant").

Feterita and Wad Feraj were the preferred varieties of sorghum under brushwood and wildflooding techniques. Single crops of millet were dominant under the *teras*. In the GDAC scheme, exclusively Aklamoy was grown. The households in Um Safaree did not grow vegetables in 1983 and 1988. However, watermelon and okra were succesfully intercropped with different sorghums in 1990 by 6 households using wildflooding techniques and *teras* (not shown in this table).

Table 6.54 **Entitlement to arable land (in fd) by technique, Um Safaree (1983 and 1988, landholdings of all households)**

	1983		1988	
	Fd	N	Fd	N
Teras	7.7	56	8.2	59
Brushwood	7.6	18	8.0	17
Wildflooding	10.5	46	10.5	48
Rainfed	10.0	3	7.3	5
Gash delta scheme	7.5	1	4.3	2

Source: L&E Survey. Note: data are at the level of landholdings.

Table 6.55 **Dominant cropping pattern by cultivation technique and percentage of occurrence, Um Safaree (1983 and 1988, landholdings of all households)**

	Cropping pattern 1983	%	N	Cropping pattern 1988	%	N
Local lands						
- Teras	millet	21	48	millet	20	41
- Brushwood	sorghum Feterita	40	15	sorghum Wad Feraj	23	13
- Wildflooding	sorghum Wad Feraj	48	29	mixed varieties of millet & sorghum	24	29
Non-local lands						
- Gash delta scheme	sorghum Aklamoy	100	1	sorghum Aklamoy	100	2

Source: L&E Survey. Note: data are at the level of landholdings. Dominant means highest occurrence per technique.

Appendix 6.1 shows that the production levels over the years 1983 and 1988-1990 of sorghum, millet and stalks varied considerably in Um Safaree. The highest average production of sorghum on local lands was obtained under wildflooding with 4.0 sacks per fd (877 kg/ha) in 1988. The lowest average was reached in 1989 under *teras* with a production of 0.98 sack per fd (215 kg/ha). The production under *teras* in the wet year 1988 was almost double the level reached in the normal-to-dry year 1983 (2.92 against 1.50 sacks per fd, or 640 kg/ha and 329 kg/ha respectively). This same 1988 level, in addition, was almost three times the level of the normal-to-dry year 1989 of 0.98 sack per fd (215 kg/ha). The highest and lowest average production of sorghum in the GDAC scheme was 2 sacks per fd (438 kg/ha) in 1990, and 1 sack per fd (219 kg/ha) in 1988 and 1989. However, the figures on the use of non-local lands are based on two households only.

The highest average production of millet was in 1983 under wildflooding with 1.37 sack per fd (303 kg/ha). The lowest was under *teras* in 1989 with 0.31 sack per fd (69 kg/ha). The production of stalks of sorghum and millet combined on the local lands was between 279 bundles per fd (1,992 kg/ha) under wildflooding in 1988, and 89 bundles per fd (636 kg/ha) under *teras* in 1990. Stalk production in the GDAC scheme was between 100-200 bundles per fd (714 kg/ha and 1,428 kg/ha) for 2 households in Um Safaree in 1988 and 1990 respectively. The cropping intensities on the local lands were between 50 % and 90 %. The highest average cropping intensity was recorded under brushwood panels in 1990 (91 %). The lowest was recorded under wildflooding techniques in 1988 (55 %).

The cropping intensities in the GDAC scheme were between 20-83 % (two households only) (*cf.* appendix 6.1).

Millet is mainly grown on the sandy soils in the area of Um Safaree. Two households reported to include millet in a rotation with sorghum for the mere purpose of eradicating the root parasite *Striga hermontica*.

Land use patterns

The households of Um Safaree have access to 12 different farming zones and also to the non-local lands in the GDAC scheme. The total number of landholdings per household ranges from 1 to 4, with an average number of 2.1. A number of 51 out of 63 respondents (81 %) stated that they regularly do not cultivate all their landholdings in any one year. The main reasons given for this behaviour were "high operational costs" (29 out of these 51, or 57 %), and "labour shortage" (17 out of 51 households, or 33 %). In the group of households with access to more than one landholding (N=45), a number of 12 cultivated two landholdings with two different techniques (27 %) and 4 households cultivated three landholdings with three different techniques (8 %). Also other combinations of different techniques and landholdings were found in Um Safaree (figure 6.7). These combinations were made either by choice or circumstance. Risk-spreading behaviour (*cf.* section 5.3 for definitions) was found among 12 out of 63 households (19 %), when sub-optimal landholdings would be defined according to the free criteria chosen by the land users themselves. The most frequently used criteria were "irregular water supply" (7 out of these 12, or 58 %), and "bad soils" (4 out of these 12, or 33 %). Risk-spreading behaviour was adopted by 6 out of these 63 households (10 %), when sub-optimal landholdings were defined according to the average level of sorghum production per feddan.

Figure 6.7 **Number of households in Um Safaree, by combinations of number of cultivation techniques applied, and number of landholdings entitled to (local and non-local lands combined), 1988**

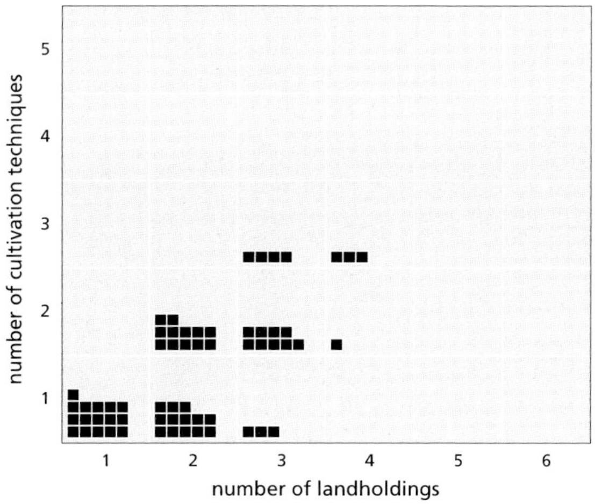

6.4.4 Scores on selected variables

The scores on the set of 18 research variables are discussed next. The presentations are based on all households in the L&E Survey, except when indicated otherwise.

Table 6.56 **The importance of iSWC in crop production. Average household crop production income by technique as % of total household crop production income, Um Safaree (1983 and 1988, iSWC households)**

	1983		1988	
	%	N	%	N
Teras	79	32	81	34
Brushwood	77	12	85	11
Teras and brushwood	82	5	100	4
Total indigenous iSWC (teras and/or brushwood)	88	39	90	41

Source: L&E Survey.

The dimension Income

The share of households in Um Safaree engaged in iSWC was presented in table 6.53 to be sligthly smaller in 1988 (49 out of 64, or 77 %) than in 1983 (49 out of 60, or 82 %). However, the proportion who actually had a production under iSWC was, with 41 out of these 49 (84 %), higher in 1988, than the 39 out of 49 households (80 %) in 1983 (table 6.56). iSWC techniques are shown in the same table to have contributed a high 88-90 % to the total household crop production income in Um Safaree (sub-group iSWC households). The *teras* and brushwood panels made approximately equal contributions in this respect (table 6.56).

The incomes of the crop production sector as a whole contributed an average 38-39 % to the grand total income of the households in Um Safaree (table 6.57). When the iSWC incomes are expressed as the percentage of their contribution to this grand total income, these figures were 28 % for the situation in 1983, and 36 % for 1988 (sub-group iSWC households). In the latter year, this figure approached the contribution of the entire crop production sector, indicating the relatively great importance of iSWC in this sub-group in Um Safaree.

Table 6.57 **The importance of crop production in livelihood. Average household crop production income by technique as % of total household income, Um Safaree (1983 and 1988, iSWC households and noSWC households)**

	1983		1988	
	%	N	%	N
Total crop production (iSWC and noSWC)	39	53	38	57
Total indigenous iSWC crop production	28	39	36	41
– of which: teras	25	32	34	34
– of which: brushwood	24	12	28	11
– of which: teras and brushwood	32	5	29	4

Source: L&E Survey.

The dimension Time

SWC time
The total household labour time allocated in Um Safaree to selected cultivation practices during five growing season months (SWC-time *S*) was an average 44 man-hours (range 5-346) in 1983, and 37 man-hours (range 3-170) in 1988. The scores of SWC-time *S* are expressed in table 6.58 as a percentage of the total household labour-time allocation to crop production for the years 1983 and 1988. The same figures for 1989 and 1990 are presented in appendix 6.2. The scores of *S* are shown to steadily decline over the years 1983 and 1988-1990 with values of 85 %, 83 %, 66 % and 56 % respectively.

Potential crop production time
The average score in Um Safaree over the selected months of the growing season of Potential crop production time *C* (*cf.* section 5.4 for definitions) was 2.70 in 1983, and 2.72 in 1988. The average household size in these years was 6.8 and 7.7. The number of economically active members in the household was respectively 2.6 and 2.8. The score of *C* divided by the latter number of economically active household members is used as a research variable in this study (table 6.59).[14]

Table 6.58 **The importance of iSWC in crop production. Average household scores of SWC-time S by technique as % of total household labour-time allocations to crop production, Um Safaree (1983 and 1988, iSWC households)**

	1983		1988	
	%	N	%	N
Teras	74	20	75	23
Brushwood	77	5	87	6
Teras and brushwood	100	3	100	2
Total indigenous iSWC (teras and/or brushwood)	85	22	83	27

Source: modelled data based on L&E Survey. Note: labour time spent on maintenance, gap-filling, weeding and thinning for the growing season months June to and including October.

Table 6.59 **The importance of crop production in livelihood. Average household scores of Potential crop production time C divided by the number of economically active household members, Um Safaree (1983 and 1988, all households)**

	1983		1988	
	Score	N	Score	N
C/no. economically active household members	0.73	60	0.74	64

Source: modelled data based on L&E Survey. Note: calculations over the growing season months June to and including October.

[14] The scores are used in a comparison of research villages. No importance should be attached to their absolute values because these are based on relatively arbitrary weighing procedures (*cf.* section 2.3.2).

The dimension Land

Table 6.60 shows the areas of cultivated land allocated to iSWC as a percentage of the total household acreage of respectively local lands, and local and non-local lands combined. In Um Safaree, these two figures are identical because access to non-local lands is absent in the sub-group of iSWC households for these years. There was also relatively little change when the years 1983 and 1988, and also 1989 and 1990 in appendix 6.2, are compared. However, when the individual iSWC techniques are considered, the *teras* shares appear to have continuously increased from 61 % in 1983, to 65 % in 1988 and 68 % in 1989 and 1990. The land shares under brushwood panels increased from 61 % in 1983, to 65 % in 1988, but these declined after 1988 to 57 % in 1989 and 55 % in 1990.

The dimension Perception

The perception of land users in Um Safaree of the relative importance of iSWC in crop production is presented by average rank numbers in table 6.61. The ranking is over five techniques. These include besides two iSWC techniques also rainfed cultivation, wildflooding and gravity irrigation in the GDAC scheme. The calculations are made for the sub-group of iSWC households only. The average rank of iSWC techniques was 1.29 for the situation in 1983 and 1.23 for 1988.

Table 6.60 **The importance of iSWC in crop production. Average household acreage by technique and farming zone in % of total land entitlement, Um Safaree (1983 and 1988, iSWC households)**

	1983			1988		
	% Local land	% Local and non-local land	N	% Local land	% Local and non-local land	N
Teras	61	61	44	65	65	44
Brushwood	58	58	16	60	60	14
Teras and brushwood	93	93	11	92	92	9
Total indigenous iSWC – (teras/brushwood)	74	74	49	75	75	49

Source: L&E Survey.

Table 6.61 **The importance of iSWC in crop production. Average household rank scores of perception by technique, Um Safaree (1983 and 1988, iSWC-households)**

	1983		1988	
	Score	N	Score	N
Teras	1.31	42	1.29	42
Brushwood	1.44	18	1.27	15
Total indigenous iSWC (teras and/or brushwood)	1.29	48	1.23	48

Source: L&E Survey. Note: range is 1-5, where 1 represents highest importance.

The rank number given by households in Um Safaree to crop production in a similar range of five livelihood categories was an average 1.33 for the situation in 1983 and 1.29 for 1988. In both years, the respondents considered crop production as their most important sector in the household economy. Crop production and also labour migration and off-farm employment were ranked lower, indicating a higher importance attached to them, for the situation in 1988 than for the situation in 1983 (table 6.62).[15]

Table 6.62 **The importance of crop production in livelihood. Average household rank scores of perception by livelihood category, Um Safaree (1983 and 1988, all households)**

	1983		1988	
	Score	N	Score	N
Crop production	1.33	58	1.29	62
Livestock husbandry	1.95	58	2.42	60
Labour migration	2.71	35	2.34	50
Local off-farm employment	3.15	20	2.96	26
Networking	3.92	39	4.20	44

Source: L&E Survey. Note: range is 1-5, where 1 represents highest importance.

6.5 Hafarat

Hafarat is located at latitude 15° 30' North and longitude 36° 35' East. This is in the 300-350 mm rainfall zone of the Border Area. The settlement was built on the slopes of Jebel Haura in 1840. Hafarat is already indicated on the early topographical map of the region (1:250,000, Kassala 56A) and is among the oldest settlements in the Border Area. The Hafara and Sabdera are the most important tribal sections of Beni Amer in the village (together 67 % of the households in the L&E Survey). A total of 15 other sections was counted as resident, including Eritrean Beni Amer groups of Sodara and Al Mada, and sections of the non-Beja tribes of Ja'aliyin and Halanga. The village was established by the Hafara section, but is factually situated in the tribal domain of the Sabdera which stretches over both sides of the Eritrean-Sudanese border. This situation has been accepted by the Sabdera for a long time. However, the significance of access to resources on the Sudanese side of the border has considerably changed. Since the 1960s, the atrocities of civil war and droughts hit Eritrea. The Sabdera, in response, have tried to reclaim all their relatively safe Sudanese lands. These developments are also at the base of traditional rivalry in Hafarat. Here, this rivalry even resulted in a self-proclaimed division of the village into two separate areas called Hafarat and Sabderat. In the L&E Survey, we only consider the

[15] When these perception scores are compared with the ranked household income scores of table 6.51 (average incomes per sector in constant 1983 £s), it appears that the perceived importance of crop production (position 1 in 1983 and 1988) is higher than its importance assessed in monetary terms (position 3 in 1983 and 1988).

Hafara community because the SWC interventions implemented all concern this group. Consequently, Hafarat referred to below means the self-proclaimed area of the Hafara community excluding the part of the Sabdera.

The total population in the 1983 census which was presented under the village name of Sabderat is 294 households, or 1,376 persons. It is not known if this includes the Hafarat area. Hubach & Marouf (1985) estimated the population of only Hafarat at 1,000-2,100 in 1984. Surveys of Hafarat based on aerial photographs gave a total of 230 households and a population of 1,400 in 1990. Most heads of household interviewed in the L&E Survey were born in Hafarat (82 %). The remaining 18 % in other parts of the Border Area and Eritrea (respectively 11 and 3 heads of household). The recorded dates of settlement in the village range from 1915 to 1988. An estimated 11 households came to live permanently in the town of Kassala at the onset of the drought in 1984. The average household size in Hafarat in 1988 was 5.3 and an average 2.1 household members were economically active. The built-up area mainly consists of straw-roofed circular houses, usually with walls of straw, and sometimes of mud. Almost all houses have a shed attached which indicates a relatively long sedentary history. The private and public domains in the village are not separated. There are 9 shops in the residential area. These include 1 tailor and 8 small groceries which supply a standard assortment of daily commodities. There are also 2 Koran schools in Hafarat. The economic orientation of the households is mainly on Kassala town, located at some 20 km southwest of the village. The system of public transport has a daily schedule and comprises of four trucks and one bus. These make the journey to Kassala in about one hour, calling in at other villages on the way. Hafarat is considered in the matrix of research villages as located relatively close to Kassala. The government has made SWC interventions in the village area.

6.5.1 Village economy[16]

Crop production

The households in Hafarat used predominantly local lands for crop production. The average title of these lands was 14.3 fd in 1983 and 13.6 fd in 1988 (table 6.63). The small decrease was due to the abandonment of marginal land in the farming zone of Aftila and succession.

The households in Hafarat have access to 13 farming zones which cover a total 2,120 ha. This excludes the areas of Baha Bina, El Kitim and Aftila for which the precise demarcation has remained unknown. The latter two farming zones, in addition, were abandoned in the course of the last two decades (table 6.64). El Kitim is located closest to the built-up area. In 1975, it was left exclusively for grazing after recurrent intrusions by livestock from the village. Aftila used to be cultivated under *teras*, but it was abandoned in the early 1980s due to a lack of rainfall. All lands in the Hafarat area are held under

[16] Most of the following baseline socio-economic data for the Hafarat village have been collected by Van Dam & Houtkamp (1992).

customary rights of possession and utilization. No sharecropping was reported on any of the local lands by the respondents in the survey.

Table 6.63 **Entitlement to arable land (number of households) and average household title (fd), Hafarat, (1983 and 1988)**

	1983			1988		
	No.of hh.	Fd	%	No.of hh.	Fd	%
No titles	2	na	3	1	na	2
Local lands only	50	14.3	88	55	13.6	90
Local and non-local lands	5	16.1	9	5	16.3	8
--of which: non-local sharecrop	1	2.0		1	2.0	
--of which: non-local entitlement	4	3.0		4	3.0	
Total	57	14.2	100	61	13.8	100

Source: L&E Survey. Note: na means not applicable.

Table 6.64 **Characteristics of the farming zones in the Hafarat village domain (situation 1988)**

Khor name	Size khor catchment (km²)	% Rock outcrop	Farming zone	Size farming zone (ha)	Dominant technique
No khor	na	na	Aftila	??	t, (a)
No khor	na	na	Wad Kaduj	230	t
Hedadeib	8.73	81.1	Hedadeib	100	(w, t, b), p
Abu Allaga	81.06	25.8	Abu Allaga	95	w
Garbabna	3.94	99.8	Garbabna	30	w
Idi Kideb	4.09	38.9	Idi Kideb	90	w
Amerai	205.00	16.5	Baha Bina	??	t
Kawataib	12.75	96.2	El Kitim	??	(a)
Kawateib	ibid.	ibid.	Sherifei	35	w, t
Kawateib	ibid.	ibid.	Don Bobia	40	w
Aderjawab	205.00	16.5	Bernoop	} 1,500	w, t
Aderjawab	ibid.	ibid.	Orun		t
Aderjawab	ibid.	ibid.	Aderjawab/Gut		w

Source: L&E Survey, 1979 and 1986 aerial photographs. Note: t=teras, b=brushwood panels, w=wildflooding, p=pilot scheme (established 1987), a=abandoned. Areas are based on 1986 aerial photographs.

The average total household crop production for the years 1983 and 1988-1990 on the local and non-local lands combined is presented in table 6.65. In these years, only four households also cultivated non-local lands in the GDAC scheme. The series in this table is not complete. However, the production levels for the missing years can be expected to be of the same order of magnitude as found in the neighbouring village of Um Safaree (*cf.* table 6.46). The production in the wet year 1988 is the highest on the list despite widespread malaria and dengee fever which, in Hafarat also, severely restricted labour allocation to cultivation (Van Dam & Houtkamp 1992).

Table 6.65 **Total household crop production. Averages of sorghum (in sacks), millet (in sacks) and stalks (in bundles), Hafarat (1983-1990)**

	Sorghum	N	Millet	N	Stalks	N
1983	3.9	26	2.1	24	385	28
1984	nd		nd		nd	
1985	nd		nd		nd	
1986	nd		nd		nd	
1987	nd		nd		nd	
1988	4.6	54	3.6	48	576	54
1989	4.0	18	2.7	12	529	19
1990	4.3	13	2.0	10	402	32
Av. 1983-1990	4.2		2.6		473	

Source: L&E Survey. Note: 1 sack sorghum is 92.1 kg, 1 sack millet is 92.9 kg, 1 bundle stalks is 3 kg, nd means no data available.

The annual production of sorghum was between 3.9 and 4.6 sacks (359 to 424 kg). The production of millet was between 2.0 and 3.6 sacks (186 and 334 kg). The annual yields of stalks of sorghum and millet combined were in the range from 385 to 576 bundles (1,155 to 1,728 kg). The average total production in Hafarat was in all years under the household food-security level of 10-13 sacks of sorghum. The average household production figures of sorghum, millet and stalks, by cultivation technique, for 1983 and 1988-1990 are listed in appendix 6.1.

Cultivation in Hafarat is mainly for subsistence. However, 40 % of the households did sell a part of their grain production on the local market and in Kassala town in 1983. This figure was, with 13 %, much lower in 1988. The share of households who sold a part of their production of stalks was 10 % in 1983 and 28 % in 1988.

Livestock raising

The village herd, in terms of the total livestock of all households in the L&E Survey sample, was 570 TLU in 1983 and 357 TLU 1988. The share of households in Hafarat who own at least one animal slightly decreased from 88 % in 1983, to 85 % in 1988 (table 6.66). Livestock wealth of households decreased for all species, but particularly so for cattle and sheep. The average wealth per household was 11.0 TLU in 1983 (range 0.4-81.0), and 6.4 TLU in 1988 (range 0.2-42.0). This represents a decline of 42 %, which is well under the 60-70 % level of loss found in the other research villages in the Border Area. Losses of livestock due to direct drought effects in 1984-1985 were reported by 1 out of 61 households (2 %). Forced sales by 24 households (39 %). In the group of 36 remaining households, 10 reported an increase in livestock wealth between 1983 and 1988, 11 reported no change, and in 15 cases no data were available to compare the situations in the two years. The household which reported a drought-loss, lost a herd of 79.0 TLU of mixed cattle and sheep. This was a large herd by local standards. The households in Hafarat also keep donkeys for local transport, and mainly the women raise chickens.

Table 6.66 **Livestock ownership (number of households and %) and average household livestock wealth (number of head), Hafarat (1983 N=57 and 1988 N=61)**

	1983			1988		
	No.of hh.	%	No.of head	No.of hh.	%	No.of head
Chickens	13	23	9.0	14	23	8.2
Goats	45	79	18.0	45	74	17.9
Sheep	24	42	15.0	19	31	10.4
Cattle	23	40	12.7	19	31	7.3
Camels	32	56	2.2	30	49	2.0
Donkeys	44	77	1.8	48	79	1.7
Total livestock	50	88		52	85	

Source: L&E Survey. Note: total livestock means any of all species.

The sheep which were raised in Hafarat usually remained in the Border Area for the whole year. They only frequented the pastures on the margin of the Gash river. Cattle were taken to the Atbara river and its upstream sections in Ethiopia in the dry season. They returned in the rainy season to graze the pastures along the main khors in the Border Area. Only in exceptional cases were cattle also sent into Eritrea. Before 1991, however, the risk of losing them to armed bandits was generally considered too high. This risk was deemed smaller for camels. Their meat is not eaten and their milk is allegedly disliked by Eritreans. Unlike cattle, camels have therefore in all years been sent into Eritrea during the dry season. They travel as far as the highlands around Akordat and Asmara at distances of over 400 km from Hafarat. These highlands are favourable destinations because of their climate. Gamachu (1977) describes the Akordat and Asmara areas with altitudes of 1,500-3,000 m above msl as having a long rainy seasons of 5-6 months, high total annual rainfall of 200-600 mm, locally up to 1000 mm per year, and mild temperatures of 20-24 °C in the warmest month. This Eritrean option may partly explain the smaller losses of livestock in Hafarat in the mid 1980s. The income-generating activities related to the livestock sector include the processing of milk, the selling of chickens and eggs in the village and in Kassala town, and contract-herding. Animal dung is collected by youths who sell it to visiting merchants from Kassala. It is not applied in local farming, however.

Labour migration

The share of households in Hafarat participating in labour migration in 1983 and 1988 was, with 25-26 %, about the same (table 6.67). These are the lowest figures recorded for the research villages in the Border Area. The average number of migrants and destinations per household remained the same for daily migration, decreased for temporary migration, and increased for long-term migration, when the two years are compared.

The destinations of daily migration of four households in 1988 were all to the town of Kassala. Employment was found in small industries (3 out of 4) and by one household in tea selling on the street. The prevailing main destination of temporary migration was New Halfa (7 out of 11 households, or 64 %). Casual agricultural labour in the irrigation scheme was the most important type of employment (85 %). The period of absence from the village due to labour migration was usually under 30 days (65 %). This was shorter than was recorded for any of the other research villages. One household in Hafarat had a member on long-term migration in 1988. This was to Gedaref town where permanent employment had been found in the government service. Hafarat is also the only research village which displays a different migration pattern when the situation of 1988 is compared with 1983. Daily labour migration in the latter year was mainly to Kassala town (4 out of 7, or 57 %), where employment was found at the livestock market. Temporary migration in 1983 was mainly to the Gash region (8 out of 10, or 80 %), where employment was found in casual labour in the GDAC scheme and contract-herding. All destinations were visited in 1983 for periods of less than 30 days. There was no long-term labour migration reported by any of the households for the situation in 1983.

Table 6.67 **Labour migration engagement by main type (number of households), average number of migrants (number of household members) and average number of destinations, Hafarat (1983 N=57 and 1988 N=61)**

	1983				1988			
	No.of hh.	%	Av.no. migrants	Av.no. destin.	No.of hh.	%	Av.no. migrants	Av.no. destin.
Daily labour migration	7	12	1.00	1.00	4	7	1.00	1.00
Temp. labour migration	10	18	1.20	1.30	11	18	1.09	1.09
Longt. labour migration	na	na	na	na	1	2	1.00	1.00
Total migration	15	26			15	25		

Source: L&E Survey. Note: total migration means any of all categories, na means not applicable.

Local off-farm employment

The number of households in Hafarat engaged in off-farm employment increased for all categories discerned when the situations in 1983 and 1988 are compared (table 6.68). The main activities were in collection and production. These included stone-cutting and the collection of gravel for the building industry in Kassala, and the collection of fuelwood which is sold in the village and also in Kassala. A relatively recent activity is the cutting of grasses, mainly *Panicum coloratum* (coll. *shoos*). This is used in the construction of houses as roofing material. Households changed to grass collection in the second half of the 1980s. Trees in the surrounding area, where felling is allowed by village rules, have almost all been cut. The average number of off-farm activities per household was 1.74 in 1983 and 1.78 in 1988. These are the highest recorded for the research villages in the Border Area.

Networking

No networking incomes were reported in Hafarat for the situation in 1983. Only five households reported to have received such incomes in 1988 (8 %). Three received grain and three received cash, while one of these households received both in 1988. The average amount of cash and the value of grain received per household was the equivalent of £s 128 (current £s). The use of informal workgroups of *nafir* was reported by 6 out of 8 households (75 %) in 1983, and by 9 out of 13 households (69 %) in 1988.

Table 6.68 **Off-farm employment by main type (number of households) and average number of activities per household, Hafarat (1983 N=57 and 1988 N=61)**

	1983			1988		
	No.of hh.	Av.no.activ.	%	No.of hh.	Av.no.activ.	%
Contract-farming	1		2	2		3
Contract-herding	na			na		
Collection and production	20		35	25		41
Local services	2		4	4		7
Total local off-farm employment	23	1.74	40	31	1.78	51

Source: L&E Survey. Note: total local off-farm employment means any of all categories.

Table 6.69 **Networking engagement (number of households), Hafarat (1983 N=57 and 1988 N=61)**

	1983		1988	
	No.of hh.	%	No.of hh.	%
Livestock transfers	na		na	
Grain transfers	na		3	5
Cash transfers	na		3	5
Total networking	na		5	8

Source: L&E Survey. Note: total networking means any of all categories.

Household income

The composition of household livelihoods in Hafarat in terms of monetary income and livestock wealth is presented in table 6.70. The relatively small decrease in livestock wealth per household of 42 % was already mentioned. With respect to the average incomes per sector, it is shown that crop production incomes ranked third in 1983 and second in 1988, after off-farm employment and livestock-related incomes. When the situations in 1983 and 1988 are compared, these average incomes increased for all sectors in Hafarat except for livestock-related activities. With respect to the sectoral average incomes per household, the main changes, when 1988 is compared with 1983, are greater contributions of crop production, and smaller contributions of labour migration and off-farm employment. This is contrary to what was found in other research villages. At first sight, the greater importance of crop production could be explained by the positive effects of government SWC interventions in the area (*cf.* section 6.5.3).

Figure 6.8 shows the combined data available for Hafarat on household income, land use and livestock wealth for 1983 and 1988-1990. The decline in livestock wealth tends to continue after 1988. The average entitlement to local lands also tends to decline. This entitlement to the non-local lands in the GDAC scheme (four households in Hafarat only) remains stable.

Table 6.70 **Household income and livestock wealth. Income is presented as household averages per sector (rounded figures in constant 1983 £s) and sectoral averages per household (in %), livestock wealth (in TLU), Hafarat (1983 and 1988)**

	1983 Hh. av. per sector	1983 % sectoral av. per hh.	N	1988 Hh. av. per sector	1988 % sectoral av. per hh.	N
Crop production	1,570	55	33	3,050	69	56
Livestock-related income	1,960	5	4	525	3	6
Labour migration	1,230	14	12	2,190	6	12
Off-farm employment	3,470	26	17	3,990	22	25
Networking	na			80	0 [A]	5
Total household income	2,900	100	46	5,090	100	59
Livestock wealth	11.0 TLU		50	6.4 TLU		52

Source: L&E Survey. Note: based on partly modelled data. The 1988 figures are deflated to a 1983 base by 160 %, [A] average household networking contribution in 1988 was 0.1 %.

Figure 6.8 **Average household land entitlement (in fd), average household livestock wealth (in TLU), average total household income (constant 1983 £s x 1,000), Hafarat, 1983-1990**

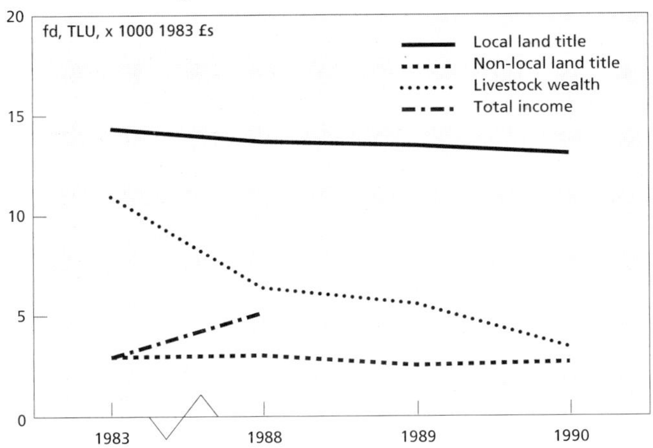

6.5.2 Interventions by the government and NGOs

The interventions in the area of Hafarat are listed chronologically in table 6.71. This list excludes the common administrative and taxation interventions due to a lack of data (*cf.* section 5.2.6). The first intervention recorded in the archival sources was in water resources development. In the late 1930s, surveys were made by the Anglo-Egyptian Administration for the purpose of constructing a dam and a 18,000 m³ water reservoir in Jebel Haura. This project area was located in the mountain range just outside the village of

Hafarat. The project was never implemented, however. The records mention as reason the anticipated high evaporation losses from the reservoir (DSA Smith 498/6, 499/1). The Melassee dam built in khor Aderjawab before 1941 (*cf.* section 6.4.2) probably also supplied floodwater to parts of the cultivated lands of Hafarat. Early relief operations took place in Hafarat in the late 1940s (*cf.* section 5.2.6 *Relief aid*). In the mid 1950s, an indigenous *hafir*, which provides part of the domestic and livestock water supply of the village up to the present-day, was cleaned and deepened. From the 1960s onwards, the same cycle of interventions as found in the other villages was initiated. This included activities in the field of labour recruitment, relief operations, the provision of public services and rural development. Relief aid was given during the 1984-1985 drought, and again in 1989-1991. The rations known to have been distributed were presented in section 5.2.6.

The SWC interventions which were part of the MFEP/KADA programme from 1987 onwards are separately discussed in section 6.5.3 below. Besides this Hedadeib pilot scheme, MFEP/KADA started in 1988 with a Woman in Development (WID) project (MFEP 1989a). Vegetable gardens built next to the house were promoted. The introduced crops irrigated with water from a new well (see below) include jew's mallow, okra, rocket cress and fruit trees of lemon and guava. Training courses in growing vegetables and keeping poultry were given. Local initiatives in forestry were linked with the activities in the Hedadeib pilot scheme. These included the establishment of a shelterbelt and tree plantation in Hedadeib, tree-planting in the village, and the development of a community woodlot along khor Garbabna. The main species promoted locally include *Acacia tortilis, A. mellifera, A. seyal, Ziziphus spina-christi, Balanites aegyptiaca,* and to a lesser extent also *Parkinsonia aculeata, Azadirachta indica, Albizia lebbeck, Tamarindus indica, Hyphaene thebaica*, and *Cassia* spp. The forestry activities were part of the FAO/FDES programme which promoted tree-planting and village woodlot development in the Border Area.

Several other government interventions in the area have also received international support. Groundwater surveys were locally made as part of the WADS programme. The situation proved slightly more favourable than in Um Safaree. A deep-well equipped with a hand pump was installed along khor Garbabna in 1985. The NGO Plan Sudan provided financial support for the extension of the primary school in Hafarat in 1989. The Ford Foundation supported the National Council for Research in the WARK project for SWC field studies.[17] The EEC supported a programme of the Sudanese COR to rehabilitate the same *hafir* earlier maintained in the 1950s.[18]

Hafarat households, like in Um Safaree, finally also benefitted from relief and health care provided in the Wad Sherifei refugee camp (*cf.* section 6.4.2). An unknown number, in addition, is known to have been registered in the camp as refugee. They received food supplies from the camp in the years 1984-1985.

---

[17] As part of the Ford Foundation programme "Field Studies in Sudan's Rainlands".

[18] As part of the EEC programme called "Hafir Rehabilitation and Maintenance".

Table 6.71 **Government and NGO interventions in the Hafarat area**

1937	Field surveys and proposal Jebel Haura Water Storage Project.
1941	Reference in government correspondence to the "Melassee" dam in khor Aderjawab (construction date unknown).
1948-1949	Drought relief food distributions.
mid 1950s	Rehabilitation hafir along khor Idi Kideb.
1960s	Labour recruitment for construction work New Halfa.
1984-1985	International emergency relief programmes (+).
1985	Construction 26 m deep-well (+).
1985	Building PHC dressing station (+).
1987	Construction SWC pilot scheme Hedadeib (+).
1987	2 fd forestry plantation and windbreaks (+).
1988-1991	Research and monitoring SWC pilot scheme (+).
1988	Seed distribution programme (+).
1988	Research horticultural crops (+).
1988	WID programme fruit trees, vegetable gardens, chicken-raising and health care (+).
1988	Tree-planting in school yards (+).
1989	Extension primary school (+).
1989	Village woodlots and tree-planting (+).
1989-1991	International emergency relief programmes (+).
1990	Extension primary school.
1990	Construction local office of Extension Department (+).
1991	Maintenance hafir along khor Idi Kideb (+).

Source: NRO Carlisle BD/19.B.4, Weekly Report 6,8, DSA Smith 498/6, 499/1, TNO & NCDRWR (1990), files of Kassala Rural Water Corporation/NCDRWR and SCLUWP, Note: (+) means direct or indirect support received from international programmes.

6.5.3 Government SWC interventions[19]

Details of the floodwater-harvesting technique which was introduced in the Hedadeib pilot scheme were given in section 4.2.3. The main characteristics of the area, people and institutions, and several of the achievements are discussed in the next sections. The intervention effects of the pilot scheme are turned to in more detail in section 7.1.

Area and people

The pilot scheme is located 5 km north of the village of Hafarat in the farming zone called Hedadeib. The total area developed covers about 70 ha. This includes all areas in use for monitoring and observation and all areas developed under forestry plantations and shelterbelts. The pilot scheme itself is situated near the dividing line which separates the tribal territories in the Border Area of the Beni Amer and Hadendowa.[20] However, access to the cultivated lands is held by both the Beni Amer and Hadendowa. The first live in

[19] The main data used in this section are taken from Kraayenhagen (1987), El Mosbah (1987), Cosijn & Van Dijk (1989) and Van Dijk (1991).

[20] Khor Kawateib running 2 km south of the pilot scheme was part of this boundary dividing the Hadendowa and Beni Amer administrative regions as had been drawn by the Anglo-Egyptian Administration (NRO file KD/20 H.10).

Hafarat and include the two main sections of Hafara and Sabdera. Other Beni Amer live in the area of Jebel Haura. The Hadendowa cultivating in Hedadeib are of the Puscap lineage. They have still mainly nomadic lifestyles. In physical terms, the pilot scheme is situated in the transition zone between the pediment slopes of Jebel Haura and a flat, gently northwest sloping, pediplain. This is an area of high grounds. It is used by travellers, herdsmen and nomadic groups of different tribes to traverse the Border Area in north-south directions during the rainy season. The available aerial photographs of the area dated 1963, 1966, 1979 and 1986 indicate that Hedadeib was not yet used for cultivation in 1963. The first indigenous *terus*, promptly, must have been built in the next years for they are shown on the 1966 aerial photographs. A vast area was brought under cultivation in 1979. The techniques applied included *teras* and brushwood panels and also the non-SWC technique of wildflooding. The total cultivated area in this year was about 83 ha. Khor Hedadeib drastically changed its course in the years between 1979 and 1986. Initially, it had arrived in the area from a southeastern direction, but then changed to a northeastern approach (*cf.* figure 4.10). The aerial photographs indicate a decrease in the total cultivated area as a consequence from 83 ha in 1979, to 47 ha in 1986. Kraayenhagen (1987,15-16) suggests that the drainage dynamics resulted from the heavy silt load of the floodwater. This, in turn, is caused by upstream deforestation in the Jebel Haura area. These events had all taken place before the implementation of the pilot scheme in 1987.

Institutions and achievements

The initial pilot scheme design refers to the development of the entire 100 ha Hedadeib delta. This includes the domains of the Hafarat, Sabdera and Puscab land users (*cf.* figure 4.10). The first 32 embankments were built in 1987. These reached a total length of 31,100 m and the area under command was about 93 ha (table 6.72). Of these, 63 ha were cultivated in the first project year 1987. The construction activities had begun in the downstream area which is largely being cultivated by the Sabdera. When in the course of operations it became clear to them that the greater part of the embankments would be built on lands of the Hafara, the Sabdera began to protest. In this same year also, a number of 17 embankments showed breaches over a total length of 15,200 m. The Sabdera, from then onwards, refused any further co-operation in the intervention. The greater part of the embankments on their lands were erased in 1987. In the next year, the activities in Hedadeib continued in the upstream section. These are the lands of the Hafara. As a result of these adjustments, the scheme lay-out was modified. Factually, the embankments came to cover an area more upstream than planned. A total of 14 embankments with lengths between 400-1,000 m each was built in 1988. A total area of 75 ha came under command of earth embankments in this second project year. Of these, 67 ha were cultivated. The number of households participating in the scheme in 1988 was 32. Of these, 26 are respondents in the L&E Survey.[21] The average landholding size in the pilot scheme was in this year 4 fd (range 0.8-10 fd).

[21] These 26 are all Hafara households. Sabdera households have been interviewed separately as part of the L&E research programme in Feb. 1992 to record also their view on the developments in Hedadeib.

Heavy rainfall resulted in exceptional high flooding in Hedadeib in 1988. This caused damage to 39 % of the total length of 10,350 m of embankments constructed. Because their repair had not been completed before the start of the new season, only 58 ha were cultivated in the next year 1989. New damage, in addition, was caused later in the season over some 9 % of the total remaining embankment length. In this year, the Department of Soil Conservation made the last repair of embankments in the pilot scheme. A small area of 2 ha was cultivated in the pilot scheme in 1990, mainly because of unfavourable rainfall. Still, early floodwater had caused, in this year also, more damage over a length of 77 m embankment.

Table 6.72 **Selected characteristics of the Hedadeib pilot scheme (1987-1991) and indigenous land use in the area in 1986**

	1986	1987	1988	1989	1990	1991
Total embankment length (m) [A]	na	31,100	10,350	6,500	5,934	5,857
Commanded area (ha)	na	93	75	nd	nd	50
Cultivated area (ha)	47	63	67	58	2	0
Embankment breach (%)	na	11.8	38.7	8.7	1.3	0

Source: Van Dijk (1991), files SCLUWP Kassala. Note: [A] lengths at the start of the season, na means no embankments used, nd means no data available.

Finally, a relatively stable area under command of earth embankments of 50 ha was reached at the beginning of the 1991 season. It had been learned from earlier experiences in Hedadeib that repair with permeable brushwood panels, instead of raising new earth works, improved the functioning of the overall system (*cf.* section 4.2.3 *Deviations from initial design*). This can be explained from the fact that such brushwood structures function as spillways. The installation of stone spillways had also been recommended earlier for the improvement of the pilot scheme performances by Mulder (1990). The size of this relatively stable area, in the end, does not significantly differ from that cultivated under indigenous techniques before 1987. This would suggest that, factually, the introduced SWC technique gradually has been adjusting itself to the prevailing physical and socio-economic conditions in the area (table 6.72). The international programmes which supported the government interventions in Hedadeib ended in 1991.[22] In the subsequent years, little rainfall and floodwater were received in the area. The pilot scheme is presently being operated in its "adjusted" form. Repairs are only carried out by the land users themselves using brushwood panels (*cf.* section 4.2.3). No research and monitoring has been carried out in this area by the Department of Soil Conservation after 1991.[23]

---

22 The WARK project was completed in May 1991. The MFEP/KADA programme was terminated in July 1991.

23 The department changed its priorities for field research to the pilot scheme testing stone bunds in Shellalob (*cf.* section 4.2.3).

Table 6.73 **Government institutions involved in the Hedadeib pilot scheme and their main fields of activities**

GOVERNMENT INSTITUTIONS	MAIN ACTIVITIES IN PROJECT AREA
Ministry of Agriculture and Natural Resources	
– Dept. of Soil Conservation	Level survey, construction, monitoring, repair, land preparation, seed-distribution
– Dept. of Range and Pasture Management	Embankment protection, rangeland monitoring,
– Dept. of Unified Agricultural Extension	Extension
Agricultural Research Centre	
– Kassala Agricultural Research Station	Crop and cultivation experiments
Forestry National Corporation	
– Kassala Forestry section	Shelterbelt, tree plantations
National Council for Research	Research and monitoring

Source: KADA (1988).

A large number of government institutions has been working in and around the pilot scheme since 1987 (table 6.73). The Department of Soil Conservation was involved in the main operations and monitoring activities. It also distributed seeds of short-maturing sorghum *Gadam el hamam* and Pioneer USA, and seeds of okra, rosella and watermelon free of charge. The Department of Range and Pasture Management was mainly involved in the development of embankment protection measures using biological means, such as grasses. The Department of Unified Agricultural Extension provided extension support to all government activities in the area. The Kassala Agricultural Research Station (KARS) made trials in the pilot scheme in the period 1988-1990. The performances of different crops and varieties of sorghum, millet, okra, tomato, cucumber and watermelon were compared. Also the effect of different sowing-dates, plant-spacings, seed-rates and seed-dressings were tested (Osman 1989). The Kassala Forestry section of FNC established in 1987 two plantations in the area with *Acacia mellifera, A. seyal, A. tortilis, A. nilotica, Ziziphus spina-christi, Parkinsonia aculeata, Atriplex nummularia, Tamarix nilotica, Tamarindus indica, Balanites aegyptiaca* and *Hyphaene thebaica*. Different planting materials have been tested here for applications in shelterbelts and as erosion protection measures for earth embankments. A shelterbelt of 1,040 m length was established at the southern margin of the pilot scheme in 1988 (4 rows of *Acacia mellifera, Ziziphus spina-christi, Parkinsonia aculeata* and *Balanites aegyptiaca*). The research and monitoring activities in the pilot scheme were supported by the National Council for Research over the years 1989-1991.

6.5.4 SWC details

The environmental conditions in the Hafarat area are comparable to those found in Um Safaree. The regularly changing water courses have, also in Hafarat, an important effect on local land use possibilities.

Table 6.74 **Engagement in iSWC for crop production (number of households), Hafarat (1983 and 1988, all households)**

	1983		1988	
	No.of hh.	%	No.of hh.	%
Teras	34	60	35	57
Brushwood	17	30	14	23
– of which: teras and brushwood	9	16	8	13
Total indigenous iSWC (teras and/or brushwood)	42	74	41	67
Project pSWC (pilot scheme)	na		26	43
Indigenous iSWC and project pSWC	na		16	26
Total households	57	100	61	100

Source: L&E Survey. Note: na means not applicable.

Production performances
The share of households in Hafarat with land under iSWC decreased from 74 % in 1983, to 67 % in 1988 (table 6.74). This is only partly a result of the introduction of pSWC techniques because most households used, besides brushwood panels, wildflooding techniques in this area before 1987. The single most important iSWC technique applied in Hafarat in 1988 is still the *teras* (35 out of 61 households, 57 %). The average sizes of land titles in Hafarat are presented by technique in table 6.75. These are data at the landholding level. Little change is found when the situations in 1983 and 1988 are compared.

The dominant cropping pattern in Hafarat is a single crop of sorghum, millet or a combination of these two. The dominant crop under *teras* is, in about 45 % of the cases, millet (table 6.76). Vegetable growing was reported in Hafarat by 4 households on 5 landholdings in 1983. This was on 8 landholdings of 8 households in 1988. An attempt to grow sesame was made under wildflooding techniques in the farming zone of khor Aderjawab by one household in 1988. The experiment failed for unknown reasons.

The production figures for sorghum, millet and stalks are presented for the years 1983 and 1988-1990 in appendix 6.1. Production in these years was highly variable. The highest average sorghum production on local lands was obtained under wildflooding in 1988 with an average 3.36 sacks per fd (737 kg/ha). The lowest in the pilot scheme in 1990 with an average production of 0.25 sack per fd (55 kg/ha). The figures for sorghum production on non-local lands in the GDAC scheme are for four households only. The highest average production was 6 sacks per fd (1,315 kg/ha) in 1990. The lowest was 1.25 sack per fd (274 kg/ha) in 1983.

Table 6.75 **Entitlement to arable land (in fd) by technique, Hafarat (1983 and 1988, landholdings of all households)**

	1983		1988	
	Fd	N	Fd	N
Teras	5.6	48	5.5	48
Brushwood	4.9	24	4.5	19
Wildflooding	5.6	72	5.6	64
Gash delta scheme	3.0	4	3.0	4
Pilot scheme	na		4.0	26

Source: L&E Survey. Note: data are at the level of landholdings, na means not applicable.

Table 6.76 **Dominant cropping pattern by cultivation technique and percentage of occurrence, Hafarat (1983 and 1988, landholdings of all households)**

	Cropping pattern 1983	%	N	Cropping pattern 1988	%	N
Local lands						
– Teras	millet	44	25	millet	47	34
– Brushwood	mixed varieties millet & sorghum	29	12	sorghum Wad Feraj	39	18
– Wildflooding	millet	38	45	sorghum Feterita	36	45
– Pilot scheme	na			mixed varieties millet & sorghum	58	24
Non-local lands						
– Gash delta scheme	sorghum Aklamoy	100	4	sorghum Aklamoy	100	4

Source: L&E Survey. Note: data are at the level of landholdings, na means not applicable. Note: dominant means highest occurrence per technique.

Millet production on local lands reached the highest average under *teras* in 1988 with 2.53 sacks per fd (559 kg/ha). The lowest average production was obtained in the pilot scheme in 1990 with 0.25 sack per fd (55 kg/ha). The highest average production of stalks of sorghum and millet combined on local lands was in the pilot scheme in 1988. This reached 275 bundles per fd (1,964 kg/ha). The lowest average production of stalks was 95 bundles per fd (678 kg/ha) under wildflooding in 1989. The highest and lowest stalk yield in the GDAC scheme was an average 500 bundles per fd (3,570 kg/ha) in 1990, and 400 bundles per fd (2,856 kg/ha) in 1989 (4 households only). The cropping intensities on local lands in Hafarat range from 75 % to 100 %. The highest average was recorded on landholdings under brushwood panels in 1983 (100 %). The lowest average was recorded under *teras* in 1988 (75 %). The cropping intensities in the GDAC scheme were about 40 % in 1983, 1989 and 1990, and 60 % in 1988 (4 households only).

Land use patterns

The households in Hafarat have access to cultivated lands in 11 different farming zones, while four hold land titles in the GDAC scheme. The total number of landholdings per household ranges from 1 to 6, with an average number of 2.8. This is the highest figure of all research villages in the Border Area. A number of 42 out of 54 households (78 %) in Hafarat stated that they regularly do not cultivate all their landholdings in any one year. The main reasons given for this were "labour shortage" (15 out of 38 households, or 40 %), and "shortage of irrigation water" (10 out of 38 households, 26 %). In the group of households who have access to more than one landholding (N=51), 12 cultivated two landholdings with two different techniques (24 %), 7 cultivated three landholdings with three different techniques (14 %). In two more cases, one household cultivated four landholdings with four techniques and another household even applied five different techniques on five landholdings (figure 6.9). Also other combinations of different techniques and landholdings were made in Hafarat, either by choice or circumstance. Risk-spreading behaviour was adopted by 19 out of 54 households (36 %), when sub-optimal landholdings would be defined according to the free criteria chosen by the land users themselves. The most frequently used in this respect were "unreliable water supply" (8 out of these 19 households, or 42 %) and "poor soils" (5 out of these 19 households, or 26 %). Risk-spreading behaviour was found among 16 of these 54 households (29 %), when sub-optimal landholdings were defined according to their average level of sorghum production per feddan (*cf.* section 5.3 for definitions).

Figure 6.9 **Number of households in Hafarat, by combinations of number of cultivation techniques applied, and number of landholdings entitled to (local and non-local lands combined), 1988**

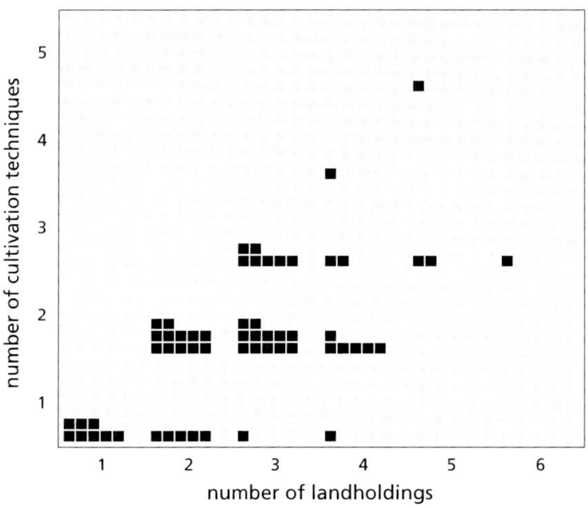

6.5.5 Scores on selected variables

The scores on the set of 18 research variables are discussed next. The presentations are based on all households in the L&E Survey, except when indicated otherwise.

Table 6.77 **The importance of iSWC in crop production. Average household crop production income by technique as % of total household crop production income, Hafarat (1983 and 1988, iSWC households and pSWC households)**

	1983		1988	
	%	N	%	N
Teras	48	14	57	27
Brushwood	62	9	49	14
Teras and brushwood	87	2	78	6
Total iSWC (teras and/or brushwood)	59	21	64	35
Project pSWC (pilot scheme)	na		42	24

Source: L&E Survey. Note: na means not applicable.

The dimension Income

The share of iSWC households in Hafarat declined from 42 out of 57 (74 %) in 1983, to 41 out of 61 (67 %) in 1988 (table 6.74). However, the number of households who had a production under iSWC is shown in table 6.77 to have increased in both absolute and relative terms. This was from 21 out of these 42 (50 %) in 1983, to 35 out of these 41 (85 %) in 1988 (sub-group iSWC households only). The iSWC incomes are shown in the same table to have contributed among iSWC households an average 59-64 % to the total household crop production income in 1983 and 1988 respectively. The pilot scheme contributions of 42 % in 1988 were slightly lower in this respect.

The incomes of the crop production sector as a whole contributed in Hafarat an average 73-77 % to the grand total household income (table 6.78). These are the highest figures recorded for the research villages in the Border Area. When the iSWC incomes, in turn, are expressed as the percentage of their contribution to this grand total, this figure was 46 % both for 1983 and 1988 (sub-group iSWC households). These are also the highest figures recorded in the villages researched. By comparison, the pSWC technique contributed 27 % to the grand total household income in Hafarat in 1988.

Table 6.78 **The importance of crop production in livelihood. Average household crop production income by technique as % of total household income, Hafarat (1983 and 1988, all households)**

	1983		1988	
	%	N	%	N
Total crop production (iSWC, pSWC, noSWC)	77	33	73	56
Total indigenous iSWC crop production	46	21	46	35
– of which: teras	36	14	42	27
– of which: brushwood	51	9	32	14
– of which: teras and brushwood	42	2	45	6
Project pSWC (pilot scheme)	na		27	24

Source: L&E Survey. Note: na means not applicable.

The dimension Time

SWC time

The total household labour time in Hafarat allocated to selected cultivation practices during the five growing season months (SWC-time *S*) was an average 60 man-hours in 1983 (range 5-151), and 56 man-hours in 1988 (range 4-170). The scores of SWC-time *S* are expressed in table 6.79 as a percentage of the total household labour-time allocation to crop production in 1983 and 1988. The same scores for 1989 and 1990 are listed in appendix 6.2. The entire series for "total iSWC" over 1983 and 1988-1990 is 72 %, 79 %, 55 % and 53 %. The score of *S* for pSWC in the Hedadeib pilot scheme was 47 % in 1988, and this declined in the subsequent years 1989 (42 %) and 1990 (31 %).

Potential crop production time

The average score in Hafarat over the selected five months of the growing season of the measure Potential crop production time *C* (*cf.* section 5.4 for definitions) was 1.51 in 1983, and 1.53 in 1988. The average household size in these years was 4.0 and 5.3. The average number of economically active members in the household was in both years 2.1. The scores *C* divided by the latter number of economically active household members is used as the research variable in this study (table 6.80).[24]

Table 6.79 **The importance of iSWC in crop production. Average household scores of SWC-time S by technique as % of total household labour-time allocations to crop production, Hafarat (1983 and 1988, iSWC households and pSWC households)**

	1983		1988	
	%	N	%	N
Teras	61	23	69	28
Brushwood	64	14	66	14
Teras and brushwood	96	5	98	6
Total indigenous iSWC (teras and/or brushwood)	72	32	79	36
Project pSWC (pilot scheme)	na		47	26

Source: modelled data based on L&E Survey. Note: labour time spent on maintenance, gap-filling, weeding and thinning for the growing season months June to and including October, na means not applicable.

The dimension Land

The area of cultivated lands allocated by the households in Hafarat to iSWC and pSWC techniques is calculated as a percentage of the total household acreage of respectively the local, and local and non-local lands combined. The shares for "total iSWC" of just over 60 % (table 6.81) do not greatly differ between 1983 and 1988 (sub-groups iSWC households). Since the use of non-local lands was relatively unimportant in Hafarat, these figures equally do not greatly differ for either local lands, or local and non-local lands

---

24 The scores are used in a comparison of research villages. No importance should be attached to their absolute values because these are based on relatively arbitrary weighing procedures (*cf.* section 2.3.2).

combined. The table also shows that an average 34 % of the total household acreage in 1988 was under project SWC land (sub-group pSWC households). The 1989 and 1990 data are presented in appendix 6.2. These indicate slightly lower iSWC land shares of some 55 % in 1989, but again a similar share of 60 % for 1990. The shares of pSWC land remained approximately the same, at about 30 % in 1989 and 1990.

Table 6.80 **The importance of crop production in livelihood. Average household scores of Potential crop production time C divided by the number of economically active household members, Hafarat (1983 and 1988, all households)**

	1983		1988	
	Score	N	Score	N
C/no. economically household members	0.76	57	0.78	61

Source: modelled data based on L&E Survey. Note: calculations over the growing season months June to and including October.

Table 6.81 **The importance of iSWC in crop production. Average household acreage by technique and farming zone in % of total land entitlement, Hafarat (1983 and 1988, iSWC households and pSWC households)**

	1983			1988		
	% Local land	% Local and non-local land	N	% Local land	% Local and non-local land	N
Teras	54	54	34	53	53	35
Brushwood	44	42	17	49	47	14
Teras and brushwood	85	85	9	83	83	8
Total indigenous iSWC – (teras and/or brushw.)	62	61	42	62	61	41
Project pSWC – (pilot scheme)	na	na		34	34	26

Source: L&E Survey. Note: na means not applicable.

Table 6.82 **The importance of iSWC in crop production. Average household rank scores of perception by technique, Hafarat (1983 and 1988, iSWC households N=4)**

	1983		1988	
	Score	N	Score	N
Teras	1.00	1	1.00	1
Brushwood	1.00	1	1.00	3
Total indigenous iSWC (teras and/or brushwood)	1.00	2	1.00	4

Source: L&E Survey. Note: range is 1-4, where 1 represents highest importance.

The dimension Perception

Data of only four households in Hafarat are available on the land users' perception of the importance of iSWC in crop production (table 6.82). This does not allow us to draw any conclusions from them. The average rank score given by households to crop production in a range of five livelihood categories was 1.38 in 1983, and 1.17 in 1988 (table 6.83). Crop production was considered the most important sector in the household economy in these two years. The perceived importance of crop production, and to a lesser extent also networking, increased (lower scores in the table) when the situations in 1983 and 1988 are compared.[25]

Table 6.83 **The importance of crop production in livelihood. Average household rank scores of perception by livelihood category, Hafarat, (1983 and 1988, all households)**

	1983		1988	
	Score	N	Score	N
Crop production	1.38	16	1.17	23
Livestock husbandry	1.83	12	2.00	13
Labour migration	3.00	1	3.11	6
Local off-farm employment	1.80	5	2.00	6
Networking	3.33	6	3.13	8

Source: L&E Survey. Note: range is 1-5, where 1 represents highest importance.

6.6 Concise presentation of village findings

The two research questions formulated in section 2.1 address (i) the effect of government SWC interventions; and (ii) the importance of indigenous SWC and crop production in local household livelihoods. The SWC interventions and subsistence means have now been examined for the individual research villages in the Border Area. In the last two sections of this chapter, a first presentation of findings is given. The effect of government SWC interventions is turned to in more detail in section 7.1 of the next chapter. The role of distance and the effects of the associated labour-opportunity costs on the importance of indigenous SWC and crop production are discussed in section 7.2 of the next chapter.

6.6.1 Interpretation of the survey outcomes

The outcomes of the L&E Survey indicate that the 1983 and 1988 data together, can be considered to represent an average situation in the Border Area for the 1980s. Still, important changes have taken place in the livelihoods during this decade. These can not be

[25] When the perception scores are compared with the ranked household income scores of table 6.70 (average incomes per sector in constant 1983 £S), it is shown that the perceived importance of crop production (position 1 in 1983 and 1988) is consistently higher than its importance assessed in monetary terms (position 3 in 1983 and position 2 in 1988).

"averaged" out. Firstly, households witnessed substantial losses of livestock. As a consequence, they were forced to seek additional means of living, particularly during the years after the 1984-1985 drought. Secondly, they received international assistance in the form of relief aid. However, the magnitude of these operations in the Border Area remains obscure for lack of documentation. These developments must be considered as an intervening factor from the viewpoint of this study. This will be called "livelihood change". The survey outcomes also indicated that the Border Area is less uniform in socio-economic and environmental terms than initially was assumed. We must therefore identify two more intervening factors, which will be called "tribal-signature" and "aridity-gradient".

Tribal-signature is the collection of characteristics by which the Beni Amer and Hadendowa distinguish themselves, despite belonging to the same Beja group. The first relevant characteristic in this respect is sedentarization history. The Beni Amer settlements of Hafarat and Um Safaree (established at about 1840 and 1920) are importantly older than the Hadendowa settlements of Ilat Ayot and Telkook (established in the 1950s).[26] This is likely to affect, among other things, local patterns of land use. The second relevant characteristic is the difference between Beni Amer and Hadendowa in access to non-local cultivated lands. The Gash region is the tribal domain of the Hadendowa, and they hold preferential access to these lands. The Beni Amer have no customary access to these lands (albeit several households did succeed in acquiring a tenancy title in the GDAC irrigation scheme). Finally, a third characteristic is the difference in the level of women's participation in crop production activities. This is higher for the Beni Amer than for the Hadendowa (albeit at a general low absolute level of participation) (*cf.* section 3.2 *Organization and gender*). Aridity-gradient refers to the south-to-north direction by which annual rainfall totals decrease in the Border Area. This gradient proved steeper than was expected on the basis of older meteorological information (*cf.* section 5.1 *Rainfall and evaporation*). It implies that Telkook and Ilat Ayot are located in a slightly drier zone than Um Safaree and Hafarat. Still, the precise effect of this on indigenous SWC and crop production is difficult to assess for two reasons. Firstly, cultivation in the Border Area partially depends on floodwater of non-local origin. Secondly, areas under rock outcrop which multiply rainfall to produce run-off also become greater in the Border Area in the same south-to-north direction. Finally, the aridity-gradient also has an effect on population distribution in the Border Area and supposedly on the carrying capacity of its natural resources. Although these phenomena can be identified in absolute terms, we cannot evaluate their mutual relationship and their significance in relative terms of "pressure of population on scarce resources" and "critical population densities". This is unfortunate, because these allegedly are important factors in the process of indigenous SWC adoption (*cf.* section 4.3.4).

---

[26] Such differentiation is not found at the household level, however. When all families interviewed in the L&E Survey are divided into two groups according to the duration of village settlement: "residents-since-birth" and "settlers", no important differences occur. Shares of 85-95 % residents-since-birth (and 5-15 % settlers) apply to all villages, irrespective of whether they are Hadendowa or Beni Amer.

This means that we finally have arrived at a situation where the importance of indigenous SWC and crop production is expected to be affected not only by economic distance to the urban centre, but also by (i) livelihood change; (ii) tribal-signature; and (iii) aridity-gradient. The first can reasonably be expected to have influenced the outcomes of the L&E Survey for all households largely in the same manner. Drought and relief operations have affected the entire Border Area in the 1980s. However, tribal-signature and aridity-gradient can not. Moreover, these latter two factors exercise their influence precisely in the same direction as the distance factor does. This means that at a greater distance to Kassala, the research villages are exclusively Hadendowa and the level of aridity is relatively high. Alternatively, at a smaller distance to Kassala, the villages are exclusively Beni Amer and the level of aridity is relatively low. Ways to control the effect of these intervening factors in the analysis of research data are discussed in more detail in section 7.2.

Village typification
In the context of this study, the research villages of the Border Area are meaningfully characterized by considering their date of establishment, prevailing type of household economies, and share of households applying indigenous SWC. For the typification of household economies, we consider the survey data on the sectoral average incomes per household (*cf.* sections 6.2.1, 6.3.1, 6.4.1, 6.5.1 *Household income*) and the definitions used by Wilson (1986) and others (*cf.* section 4.3).

Table 6.84 shows that the older villages Um Safaree and Hafarat have higher shares of iSWC households, but no clear association exists between these two characteristics and the prevailing type of household economy. Figure 6.10 shows the same sectoral average incomes per household. These illustrate the differentiation of the livelihoods in the Border Area, by village, and by research year. Finally, figure 6.11 also presents the household average incomes per sector, by village, and by research year. Typically, incomes earned in local off-farm employment were among the highest of all categories discerned. The situation in Um Safaree differed to an important extent from other villages in this respect. Relatively high incomes have been earned here in labour migration and livestock-related activities.

Table 6.84 **Typification of research villages in the Border Area**

	Ilat Ayot	Telkook	Um Safaree	Hafarat
Date of village establishment	1952-1954	1952	1920s	1840
Prevailing type of hh economy				
– 1983	agro-pastoral	mixed	mixed	agricultural
– 1988	agricultural	mixed	mixed	agricultural
% iSWC households				
– 1983	48 %	9 %	77 %	67 %
– 1988	52 %	21 %	82 %	74 %

Source: L&E Survey. Note: prevailing types of household economies are based on L&E Survey data and definitions given by Wilson (1986) (cf. section 4.3).

Figure 6.10 **Sectoral average incomes per household (% shares in livelihood), Ilat Ayot, Telkook, Um Safaree and Hafarat, 1983 and 1988**

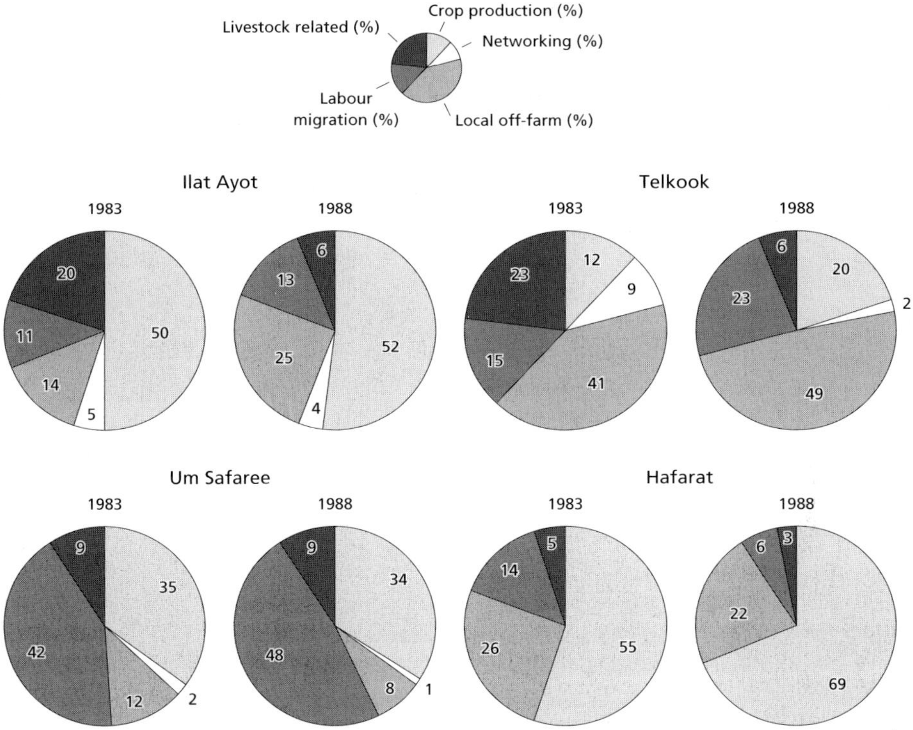

Figure 6.11 **Household average incomes per sector (in 1983 £s), Ilat Ayot, Telkook, Um Safaree and Hafarat, 1983 and 1988**

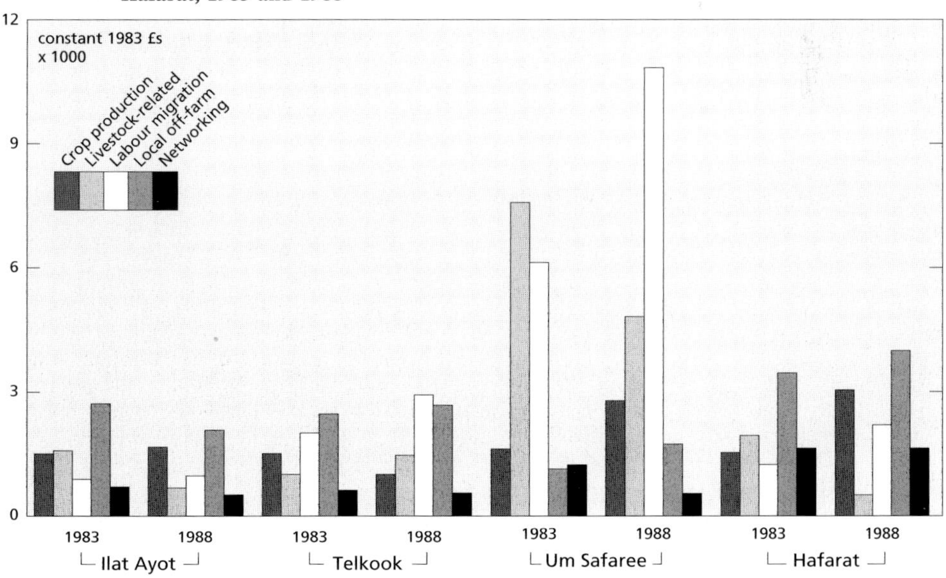

Table 6.85 **Rank scores for the selected set of research variables by village (1983 and 1988, iSWC households and all households)**

	Ilat Ayot	Telkook	Um Safaree	Hafarat
INCOME				
% iSWC in total hh crop production income				
– 1988 SWCCROP8	L	H	mh	ml
– 1983 SWCCROP3	L	H	mh	ml
% Crop production in total hh livelihood income				
– 1988 CROPLIV8	mh	L	ml	H
– 1983 CROPLIV3	mh	L	ml	H
% iSWC in total hh livelihood income				
– 1988 SWCLIVE8	L	ml	mh	H
– 1983 SWCLIVE3	L	mh	ml	H
TIME				
% iSWC-time S in total hh labour time				
– 1988 SWCTIME8	ml	H	mh	L
– 1983 SWCTIME3	ml	H	mh	L
Score C for the pot. hh labour time available [A]				
– 1988 POTTIME8	mh	ml	L	H
– 1983 POTTIME3	H [B]	m	L	H [B]
LAND				
% iSWC land in total hh local land entitlement				
– 1988 SWCBORD8	ml	H	mh	L
– 1983 SWCBORD3	ml	H	mh	L
% iSWC land in total hh land entitlement				
– 1988 SWCTOTA8	L	H	mh	ml
– 1983 SWCTOTA3	L	H	mh	ml
PERCEPTION				
Rank iSWC priority in crop production				
– 1988 SWCRANK8	L	ml	mh	H
– 1983 SWCRANK3	L	ml	mh	H
Rank crop production priority in hh livelihood				
– 1988 CROPRAN8	mh	L	ml	H
– 1983 CROPRAN3	mh	L	H	ml

Source: L&E Survey. Note: hh means household, H means highest score, L means lowest score, mh means medium high score, ml means medium low score, [A] calculations made over the number of economically active members in the household, [B] equal scores. Variable names are described in full in appendix 2.1.

The research variables
Table 6.85 shows the distribution over villages of the highest, lowest and intermediate scores of the selected research variables. This presentation still does not provide any direct insights into the effects of distance, tribal-signature and the aridity-gradient. Two statistical tests will be used in chapter 7 to facilitate the interpretation of these outcomes. The tests themselves are based on a similar comparison of ranked scores. However, by each time delineating different villages and sub-groups of households and landholdings also the effects of the intervening factors can be analyzed. Finally, a statement can be made on whether the relationships found in the Border Area are statistically significant, or merely occur by chance.

6.6.2 Constraints on SWC, crop production and SWC intervention

Irrigation water
Water is an obvious limiting factor to crop production in the drylands. In the Border Area, some 600 mm would be required per year to grow sorghum under optimal conditions. However, rainfall alone supplies only 150-400 mm per year in this area (*cf.* section 5.1). Rainfed cultivation is therefore largely absent. Crops can only be cultivated when floodwater is used, or run-off is harvested. In the latter case, still a minimum amount of rain is required to fall on the catchment, and to fall at the right time. In the drylands, this rainfall variability over place and time is considerable and increases with the level of aridity. The L&E Survey outcomes indicate that shortage of irrigation water was occasionally reason for land users in Ilat Ayot and Telkook to abandon entire farming zones. It was also among the main reasons given to account for uncultivated landholdings. Poor rainfall was among the main reasons given by land users in Um Safaree and Hafarat to explain the abandonment of farming zones in these areas.

Soil nutrients
Depletion of soil nutrients was argued in section 1.2 to frequently impose another important constraint on crop production in dryland SSA. Sometimes, its limiting effect would be more important than the availability of irrigation water. However, the data collected in a case study *teras* in Ilat Ayot seem to indicate otherwise. At this particular location, the dynamic drainage system and processes of sediment deposition continuously replenish the nutrients of soils within the bunded area. Soils within the *teras*, accordingly, were found to have a more favourable nutrient status than the soils of outside uncultivated control areas and soils under wildflooding cultivation. Nitrogen, as major nutrient for plant growth, is not likely to constrain crop production in the studied *teras* in the short term (*cf.* section 6.2.4). Although this finding cannot be generalized, it does indicate that this indigenous technique has important potentials for dryland development in areas where similar conditions exist as in Ilat Ayot.

Labour availability

The average labour demand for manual *teras* construction ranges between 6-16 man-days per hectare. Another 3-18 man-days per hectare, per season, are normally needed for repair and for the cleaning of the catchment. These figures are low when compared with the labour demands for other SWC techniques used world-wide (*cf.* section 1.3). However, the data collected in the L&E Survey indicate that also the production of grain and stalks under *teras* is relatively low. This holds in absolute terms, for the average sorghum production under *teras* is 1.7 sacks per feddan (370 kg/ha). It also holds in relative terms when compared with the other techniques which are applied in the Border Area, especially wildflooding (2.0 sacks per fd, or 438 kg/ha) (*cf.* appendix 6.1). The survey outcomes furthermore indicate that among the households who apply *teras* (and to a lesser extent brushwood panels), the greater part of their labour time is also allocated to this particular technique (and to a lesser extent to brushwood panels): these are the scores of generally around 75 % of SWC-time *S* (for selected cultivation and land preparation activities). Given the low labour demand of *teras*, the equally low level of production it supports, and the relatively high allocation of labour time it receives, the conventional interpretation of the constraining role of labour clearly must be refined. In the next chapter, we consider the cost-benefit relations of different crop production techniques and livelihood activities in the Border Area to further investigate this.

Institutions

Institutional constraints to SWC intervention, given the current Participatory Approach to rural development, may occur at two scale levels (*cf.* section 1.3). At the first and higher level, this concerns the relationship between donor and government. In this situation, constraining conditions may arise from improper working approaches. These approaches are usually formulated by international programmes, and are subsequently to be implemented by government institutions operating in the field. At the second and lower level, this concerns the relationship between government and target group. The constraining conditions in this situation may arise from the limited organizational potential among this group. The outcome is that working methods are usually formulated by the government, and are to be executed by this target group. The survey outcomes indicate that participatory approaches in government SWC interventions in the Border Area have largely been absent. An example of the first set of constraints is the failure to implement a pilot scheme at the appropriate location. Despite existing procedures of the Department of Soil Conservation which prescribe that a "valid request" must precede project implementation (*cf.* section 4.2.2 *Earth dams*), the identification in Hedadeib was mainly made on technical grounds proposed by the international programme. Time pressure, built up within the framework of the same programme, finally made the proposal less non-committal than perhaps was initially envisaged.[27] The second set of constraints was found to apply to the Earth Dam programme. The Department of Soil Conservation assumes, by way of policy,

---

[27] However, this does not mean that in all other situations the procedures of the Department of Soil Conservation resulted in optimal project locations (*cf.* section 6.3.3).

an executive role. However, this implies that land users do not feel responsible to maintain or repair dams. The result is that these fall in disrepair. Still, local organization in the Border Area is relatively well developed. This can be used if, for the purpose of SWC development, also the appropriate techniques are selected. In this context, it would mean techniques which can be managed by land users themselves. The main types of local organization in the area are the *nafir* and *kaben* (Tb) work parties and religious groups. The survey outcomes indicate that a majority of households was involved in work parties in 1983 (generally over 75 % and slightly lower in Telkook), but generally less in 1988. Religious organization is strong in the Border Area and growing. This provides opportunities to place appropriate techniques in a suitable organizational framework for support by government interventions.

7

The effect of SWC interventions and the importance of indigenous SWC and crop production

The two research questions formulated in section 2.1 will be answered in this chapter. Firstly, the effects of government SWC interventions are assessed from secondary sources and data collected in the L&E Survey. The first mainly include reports of two joint evaluation missions of the Sudanese-Netherlands governments, and background documents of the Department of Soil Conservation and MFEP/KADA. Secondly, the importance of indigenous SWC and crop production is exclusively assessed from L&E Survey data, in particular from the selected research variables. These data are available, by village, at the level of households and landholdings. Villages, households and landholdings, in turn, can be grouped according to the occurrence of government SWC interventions and distance to the town of Kassala. The sub-groups of households and landholdings are discerned by the prefixes iSWC, pSWC, nopSWC and noSWC as defined in section 6.1. We briefly discuss the way the L&E Survey data will be used in this chapter.

With respect to the first research question, the impact of government SWC interventions in economic terms is assessed in two ways (section 7.1.2 Impact: enhancing economic positions). Firstly, we compare the crop production performances for the common "with project" and "without project" situations (Gittinger 1984). However, since a tendency exists to overestimate the levels of crop production for the "without project" case[1] we also compare, secondly, the situations "before project" and "after project". The economic returns per unit of land and per unit of labour time allocated, are also studied for project SWC and other cultivation techniques over time for the years 1983, 1988, 1989 and 1990 (section 7.1.4). In the same section, we study the effect of household participation in government SWC projects on their application of iSWC techniques. Finally, we consider the effect of mechanized land preparation in the Hedadeib pilot scheme on the pattern of labour allocation to other cultivation techniques in Hafarat.[2]

With respect to the second research question, indigenous SWC and crop production importance are assessed by comparing the scores of research variables. We take account of the expected intervening factor distance, but also of the unexpected factors tribal-signature and the aridity-gradient. The latter effects are partly controlled by not only considering the distance to Kassala town, but also another distance to the town of Aroma in the Gash delta. This relationship is not affected by effects of aridity-gradient and tribal-signature. However, the grouping of villages, households and landholdings in this case is not optimal, and allows only tentative conclusions to be drawn.

7.1 Evaluation of SWC interventions

The government SWC interventions in Hafarat and Telkook are evaluated in terms of the four criteria (i) accordance with policy; (ii) effectiveness; (iii) efficiency; and (iv) impact and project-sustainability. These are based on the DAC monitoring and evaluation guidelines and their interpretations in SIM (1990, 1991) (*cf.* section 2.1.1). The criteria have been adapted for the situation of SWC interventions in the Border Area. The findings in terms of the first three criteria are mainly based on observations of two evaluation missions to the MFEP/KADA programme in September 1988 (GOS/GON 1988) and October 1990 (GOS/GON 1990). The report of the technical backstopping mission to this programme is also used (Mulder 1990). Finally, frequent reference will be made to reports of the WARK project (Van Dijk 1990, 1991). The fourth criterion of intervention impact is discussed in section 7.1.2. This presentation is entirely based on L&E Survey data collected in Telkook and Hafarat. The evaluation of project-sustainability (section 7.1.3) uses the mentioned secondary sources. Finally, we underline that the DAC criteria provide a general guideline for evaluation for the purpose of this study. However, no attempt is made to be comprehensive in answering all operational questions that can possibly be derived from them.

---

[1] This is because average figures based on all landholdings per household are used. In the "without project" group of households, most land users can seasonally select from different landholdings to cultivate only the best (*cf.* chapter 6). This is not an option in the "with project" group because of the fixed location of the intervention (using other than averages based on all landholdings in the "without project" group would raise the question of which landholdings to include or exclude in these calculations).

[2] No such mechanized land preparation was introduced in the government SWC project in Telkook.

7.1.1 Information from secondary sources

The objectives of interventions in land use in the Border Area are formulated by the government of the Sudan (GOS). At the regional level, these formulations are made by the Ministry of Agriculture and at the local level, in the field of SWC, by the Department of Soil Conservation (*cf.* section 4.2.2). The Border Area Earth Dam programme of this department falls entirely under the responsibility of GOS. The Hedadeib pilot scheme as part of the MFEP/KADA programme is also controlled by policy objectives of the government of The Netherlands (GON). More in particular at the local level, the pilot scheme is controlled by the objectives of MFEP/KADA and BAPP in which latter body the Ministry of Agriculture and the Department of Soil Conservation were represented (*cf.* section 4.2.3).

• Accordance with policy

The following questions are important when the "accordance with policy" criterion is used to evaluate government SWC intervention projects in the Border Area (SIM 1990,6 adapted):

1. *"Do the objectives of the Border Area Earth Dam programme tally with the objectives formulated by GOS; and do the objectives of the MFEP/KADA and BAPP programmes tally with the objectives formulated by GOS and GON ?"*

2. *"Is the delineation of the target group warranted ?"*

The pilot scheme

The GOS/GON evaluation missions concluded that the general objectives of SWC interventions in the Border Area by MFEP/KADA and BAPP were sound and in concert with the main policy lines of the respective governments. The pilot scheme was evaluated in 1988 as a socio-economically viable intervention. Particularly its objectives for training and replicability were positively evaluated (GOS/GON 1988, appendix 9). However, important weaknesses were also reported by the first mission. These concerned the question of how the goals and objectives were to be achieved through the formulated approaches of popular participation (*cf.* section 4.2.3). The mission which visited the project two years later concluded that with respect to this question little progress had been made since (GOS/GON 1990,7,25-34). These observations have later been placed in a broader context (DHV 1991, DGIS 1991), which reports underline that contradictory higher level objectives of the Sudanese and Netherlands governments are partly to blame for this poor coupling with approaches followed at the lower level of the field. According to DGIS (1991), the GOS policy regarding MFEP/KADA was largely ruled by objectives of economic growth and the GON policy by objectives of rural development. The Sudanese growth objectives, eventually, did not tally with the participatory rural development approaches proposed by The Netherlands (*cf.* section 1.1) (*Ibid.* 1991).[3]

---

[3] This is strictly a matter of MFEP/KADA programme emphasis. For other programmes, the DGIS section IOV points to favourable interactions between the respective objectives of economic growth and rural development (IOV 1991,49-60).

One reason why this discrepancy has never been solved in the course of the programme expectedly relates to the dual objective of MFEP/KADA. The developments in the pilot scheme between 1987-1989 provide a good case in point to illustrate this. This dual objective was based on rural development *cum* institution-building (GOS/GON 1988). MFEP/KADA, accordingly, did not communicate directly with its target group (as part of the rural development objective), but was to delegate these responsibilities to government extension services (as part of the institution-building objective). These services had to be reorganized first, which was one of the institutional objectives of the programme. The reorganization started in 1989[4]. By consequence, these developments factually resulted in a situation where for two years, all departments working in Hedadeib remained in their traditional role of project-executor with little target group involvement (GOS/GON 1990,16). The people of Hafarat considered them as mere providers of public services (Van Dijk 1991,69-70) and were relunctant to assume active roles when the new Unified Agricultural Extension Department finally came to convey its message of participatory approaches.

The target group delineations have also been subject to frequent change. The first change was from urban groups, initially supported by KADA from 1978 onwards, to rural groups with an emphasis on the population living in the low-potential areas including the Border Area (Green Mission 1985). The target group of MFEP/KADA was again reconsidered after the evaluation mission of 1988. It recommended to place more emphasis on the high-potential areas in the region. The 1990 mission, finally, concluded that more attention should be given to the region's economic comparative advantages (GOS/GON 1990,34). This would in certain situations imply yet another change of target groups.

The various reports also consider accordance with policy from the perspective of donor commitment to the programme. The Netherlands government DGIS is considered to have set contradictory objectives, but also to have left too little room for manoeuvre. On the one hand, a policy called "process approach" was propagated based on local needs, identification, implementation and follow-up. This requires a long-term commitment to safeguard stability and continuity of the programme. On the other hand, the DGIS policy of regionally concentrated development assistance resulted in high investments being made in a relatively small area. This put considerable spending-pressure on the programme. In the end, this came to conflict with the process approach envisaged (GOS/GON 1988,16,53,57, GOS/GON 1990,21, DHV 1991,6-9, DGIS 1991).

The earth dam

The Telkook project is an entirely Sudanese operation of which no official evaluations have been made. The Department of Soil Conservation has set no specific objectives and only its general objectives for the Border Area apply (*cf.* section 4.2.2). These objectives

---

[4] The Ministry of Agriculture in Kassala initially operated with different extension services in each of its Departments of Soil Conservation, Horticulture, Veterinary Services, Range and Pasture Management. These were combined as part of the MFEP/KADA institution-building objective into one new Department of Unified Agricultural Extension in 1989.

are in accordance with those formulated by the regional Ministry of Agriculture. However, the goal to let earth dams in the Border Area contribute to the growth of the national economy must be considered unrealistic, given the low environmental and economic potential of the region. No target group was defined beforehand and also extension and local participation were absent in the Border Area (Mulder 1990). The Department of Soil Conservation executes projects and the local beneficiaries become a target group on-the-spot (El Hassan, Director of SCLUWP, Interview Kassala, Feb. 1990). This approach must be appreciated against the background of Sudanese rural development tradition, which is one of rendering services (*cf.* section 4.2.2). However, such approach was mentioned in section 6.3.3 to create confusion and to frustrate the operation of projects, because land users feel no responsibility to maintain and repair dams built by the government for them.

• Effectiveness
The following questions are important when the "effectiveness" criterion is used to evaluate government SWC intervention projects in the Border Area (SIM 1990,6 adapted):

1. *"Have the project objectives been realized, and if so, was this within the predetermined period of time ? Were objectives formulated in a realistic way and was it necessary to adjust them ?"*

2. *"What are the results, are there any unintended and unforeseen outcomes, and if so, how do these compare with the objectives ?"*

3. *"In the event objectives have not been realized, what were the reasons for this?"*

4. *"Were certain assumptions taken as premise in project design and formulation, did these assumptions turn out to be correct ?"*

The pilot scheme
In the original MFEP/KADA formulations, a modest package of field studies in Hedadeib was supposed to be followed by replication of the introduced floodwater-harvesting technique in the wider Border Area. However, the first evaluation mission commented that a more detailed research programme would be required to achieve this goal (GOS/GON 1988,25). This programme was set up in 1989, for which purpose MFEP/KADA worked in close co-operation with the WARK project of the National Council for Research. When, firstly, the initial pilot scheme objectives given in Kraayenhagen (1987,8) (*cf.* section 4.2.3) are considered, the discussion on the realization of objectives can already importantly be narrowed down. A system of floodwater harvesting by means of earth embankments was developed and tested in Hedadeib. However, the effectiveness of the system was low because it showed a number of serious imperfections. By consequence, other more elaborate objectives, such as achieving adequate soil moisture contents for crop production, were also not realized in any of the subsequent phases of the project.

The imperfections in the system of floodwater harvesting were discussed in section 4.2.3. The causes include in the technical field, firstly, an inappropriate design in which environmental dynamics were underestimated. Secondly, the causes include imperfect implementation. This was a result of the combined effects of conflict over land use and relocation of the pilot scheme into less favourable areas for floodwater harvesting. The causes in the socio-economic field mainly include an insufficient appreciation of the complexity of local household livelihoods. Land users were presumed to voluntarily participate in construction and maintenance work. This view did not take any account of the socio-economic realities of the households of Hafarat. These imperfections first lead to a stagnation and later even to a decline in the area developed in Hedadeib (*cf.* section 6.5.3). Other activities such as the planting of trees and grasses for the protection of embankments completely failed due to very dry years after 1989. Still, the 1990 evaluation mission concluded that the objectives of increasing crop production levels and developing forestry activities had been achieved in general (GOS/GON 1990,30). The trained staff of the Department of Soil Conservation initiated, after 1989, two new SWC pilot projects in the Border Area without significant involvement of the international programmes[5]. The training component, which was still insufficiently developed in 1988, was therefore more favourably evaluated for the situation in 1990 (GOS/GON 1988,22; 1990,26-30). The 1990 evaluation mission did not assess negative environmental effects of the MFEP/KADA programme (GOS/GON 1990,10). However, in section 7.1.2 "Impact: interaction with environment" we discuss that there were several in the area of the Hedadeib pilot scheme.

The pilot scheme objectives concerning self-reliance and popular participation of land users have not been met. In 1988, the project was still considered by the evaluation mission to be mainly a tool for top-down research and planning (GOS/GON 1988,27). Improvements were made in the later years after the new extension services had become operational.

The objectives of the research programme in the pilot scheme are given in (El Hassan & Van Dijk 1988,3-7). These include the research fields of (i) local household livelihoods; (ii) farming-systems; and (iii) land tenure and water rights. Additional field studies in the pilot scheme were to cover (i) suitable measures to protect earth embankments against erosion; (ii) soil and water relations; and (iii) conditions for successful replication of the experimental technique. This programme was positively evaluated by the mission in 1990 (GOS/GON 1990,26). The WARK research was completed in 1991 and its outcomes partially underlie the discussions in this book.

Several initial objectives of the pilot scheme were not realistic, and others were based on outright false assumptions. The harshness of the environment was underestimated. This holds especially with respect to the amount of floodwater that could be expected in the area and its damaging impact on earth embankments. The amount of rainfall, on the contrary, was overestimated. This had important repercussions, among others, on the possibilities to plant trees and grasses in the pilot scheme area (*cf.* GOS/GON 1990,30). The soils in the upstream section of Hedadeib also proved unsuitable for the building of

---

[5] These are the projects in Kawateib and Shellalob (*cf.* section 4.2.3).

embankments. This problem was aggravated by the failure to development suitable protection materials. The implicit assumption that Hafarat households were predominantly engaged in crop production after their loss of livestock in the mid 1980s was also incorrect (*cf.* section 6.5).

From 1989 onwards, as a result of the research activities, a number of modifications were made in the pilot scheme and a different working approach was followed (WARK 1989, 1990a). Firstly, stone-reinforced spillways were built in the embankments of the upstream section. This followed the recommendations made by the MFEP/KADA technical backstopping mission (Mulder 1990). Their effectiveness could not be gauged, however, since no significant amounts of floodwater reached the area after their installation in 1990 (WARK 1990b). Secondly, interventions in the form of embankment construction, maintenance and repair were carried out, when these had been proposed by the land users themselves. In addition, their participation had to be secured without provision of any compensation. One important exception formed the provision of mechanized land preparation. The impact of this on the participation of Hafarat households in the scheme is separately discussed in section 7.1.4 (*cf. Two exercises*). Finally, preference was given in the project from 1989 onwards to the application of indigenous techniques, such as brushwood panels. In the end, maintenance and repair was taken up by the land users themselves. This only took as long as favourable rains could be expected and when the remainder of the season was judged as not promising for crop production in Hedadeib, other farming zones were visited or other livelihood activities were turned to.

The earth dam[6]

There are no data available, other than aerial photographs, for the situation in Telkook before implementation of the earth dam. Also no monitoring and evaluation was carried out in Telkook after project completion. The objectives of increasing the area of irrigated and cultivated land, and increasing crop yields and farmer incomes can therefore not be evaluated. The dam did not meet the other objectives of contributing to national economic growth, to improve natural rangeland, and to reduce land degradation (*cf.* section 4.2.2 *Earth dams*). Field visits indicated that about one third of the dam length constructed in 1985 had already disappeared in 1991 due to breaching. This observation makes it safe to assume that the effectiveness of this project is low. The main reasons for the poor performance of these types of dams were discussed in section 4.2.2 by Mulder (based on observations elsewhere in the Border Area). These reasons relate to (i) the use of inappropriate techniques; (ii) poor overal design; and (iii) insufficient experience of the SCLUWP staff to control the enormous volumes of floodwater involved (*Ibid.* 1990,12,18-19). Similar conclusions had been drawn in an earlier study on this type of dam in the Border Area by the ILO (ILO 1986,69-70). Since the construction of earth dams is regarded as a public service, no elaborate assumptions other than contained in the workplan of the department (*cf.* section 4.2.2 *Earth dams*) underlie the project in Telkook.

---

[6] The information in this and the next sections on the earth dam in Telkook is taken from files of the Department of Soil Conservation in Kassala, ILO (1986), Ahmed (1987) and Mulder (1990).

One of the latter assumptions that the area would be conflict-free proved not correct, because also the interests of land users from another village Togan were at stake. This probably negatively affected local participation and the urge of households in Telkook to maintain and repair the dam.

• Efficiency

The following questions are important when the third criterion of "efficiency" is used to evaluate government SWC intervention projects in the Border Area (SIM 1990,6-7; adapted):

1. *"Is there a balance between staff size and overhead costs on the one hand, and the set of tasks executed on the other ?"*

2. *"Are results achieved with means earmarked at the outset and is there a reasonable ratio between inputs and results ?"*

3. *"What were the total costs of the activity, what is the cost-effectiveness in terms of costs per unit output, could means have been used in other ways with similar or better results ?"*

The pilot scheme

The efficiency of MFEP/KADA activities in the Hedadeib pilot scheme cannot be assessed directly. Firstly, the accounts of the evaluation missions on efficiency are not broken down into individual projects of MFEP/KADA. Secondly, also in the own account systems of respectively MFEP/KADA, BAPP and the Soil Conservation Department no such breakdown is possible for the individual pilot schemes. A total of four of these was under implementation in the Border Area since 1987.[7] However, the costs per hectare developed can be approximated. With information derived from other sources, also a comparison can be made between the pilot scheme and the earth dam project.

The evaluation missions concluded with respect to the entire MFEP/KADA programme that its overhead costs had been too high. These reached 35 % over the 1980-1986 period. This figure increased to some 40 % in the workplan for 1987 when the first SWC projects were formulated (GOS/GON 1988,13-14, 25-26). The evaluation mission of 1988 concluded that the first results in the pilot scheme had been obtained with the financial means allocated, but that occasionally different physical means had been used to achieve the objectives. Embankment construction and repair was originally planned to be carried out by tractor. However, this was executed in the field in 1988 by road-grader. The use of heavy machinery had initially been justified for the specific situation in Kassala during and after the 1984-1985 drought at a special KADA seminar (KADA 1985,53) and by the first evaluation mission (GOS/GON 1988,1). This approach was later refuted by the

---

[7] These include besides Hedadeib, Kawateib and Shellalob also the Mokram experimental farm (*cf.* section 4.2.3).

second mission because, given the structure of MFEP/KADA and its spending-pressure, "[..] the programme cannot judge whether investments are appropriate and in proportion to the benefits delivered [..]". This would contribute to "[..] equipment-driven approaches and generally low cost-effectiveness [..]" (GOS/GON 1990,11).

Table 7.1 **MFEP/KADA programme expenditures 1987/88-1990/91 and the approved 1991/92 budget (Counter Value Fund CVF in current £s and Foreign Exchange Component FEC in Dfl., rounded)**

	1987/88	1988/89	1989/90	1990/91 [A]	1991/92	Total 1987/91
CVF (1,000 £s)						
Total sectors	4,500	8,964	5,977	6,349	(10,560) [B]	25,790
Agriculture	781	1,075	1,087	1,682		4,625
Waterspreading	314	180	160	207		861
FEC (1,000 Dfl.)						
Total sectors	7,885	10,300	7,226	14,400		39,811
Agriculture	439	2,562	1,838	811		5,650
Waterspreading	388	280	484	182		1,334

Source: KADA (1988, 1989a, 1989b), DHV (1991). Note: [A] MFEP/KADA was terminated on 1 July 1991 with CVF balance £s 1,642,642 for total sectors, £s 373,322 for agriculture and no balance remaining for waterspreading, [B] 1991/92 reserve which has not been allocated.

Table 7.2 **Development Budgets of the Ministry of Agriculture and Natural Resources (MANR) and the Department of Soil Conservation Land Use and Water Programming (SCLUWP) for the Eastern Region/Eastern State, 1987/88-1991/92, (in current £s, rounded)**

(1,000 £s)	1987/88	1988/89	1989/90	1990/91 [A]	1991/92	Total 1987/91
Total Eastern Region/State	20,500	25,800	37,350	81,293	60,377	164,943
– MANR	1,450	3,850	1,800	5,385	6,815	12,485
– SCLUWP	430	550	380	945	1,270	2,305

Source: Development Budget Administration of MFEP Khartoum, files MFEP Kassala.

The financial budget of pilot scheme-related activities in the BAPP project of MFEP/KADA are called "waterspreading". The total 1987/88-1990/91 waterspreading expenditures of four pilot schemes were £s 861,000 and Dfl. 1,334,000 (table 7.1). These expenditures account for 19 % of the Sudanese pounds budget of the sector agriculture (£s 861,000 out of £s 4,625,000). This same figure is 24 % of the Dutch guilder budget (Dfl. 1,334,000 out of Dfl. 5,650,000). Waterspreading, in turn, accounted for only 3 % of the total budget of all sectors of MFEP/KADA of £s 26 million and Dfl. 40 million. The waterspreading expenditures, accordingly, were modest in the context of MFEP/KADA. However, these must be considered as very high in the context of the budget of the regional government. The total amount allocated to waterspreading between 1987-1991 is the equivalent of some £s 8,231,000.[8] This is about two-thirds of the development budget of the Ministry of Agriculture for the entire Eastern Region/State (£s 12,485,000). This same amount is over three times the development budget of the Kassala Department of Soil Conservation for the entire Eastern Region/State (£s 2,305,000) (table 7.2).[9] The 1990 mission, accordingly, concluded that MFEP/KADA was "[..] a very dominant factor in the development activities of the ministries [..]". This is even more so when we recall that its working area comprised only 2 of the total of 17 districts in the Eastern Region/State, and reached only 10 % of its population (*cf.* GOS/GON 1990,11). The WARK expenditures were separately financed by the National Council for Research with funds from the Ford Foundation (table 7.3). The total expenditures over 1989-1991 amounted to £s 411,000. This was the equivalent of $ 95,000 under the current official project exchange rate. The WARK expenditures are not included in tables 7.1 and 7.2.

The initial establishment costs and also the annual running costs of the pilot scheme were approximately £s 1,800 per ha (1988 £s) (table 7.3). Data collected for a typical 3 fd-landholding in the area of Hedadeib indicate that the net returns from crop production in the "with project" case were £s 3,000-5,000 per ha (1988 £s) depending on cropping intensities (table 7.4). If all intervention activities in Hedadeib were to be priced, cultivation under 1989-1991 research conditions, which included mechanized land preparation at subsidized rates would still in most cases be financially attractive when compared with a typical 3 fd-landholding in Um Safaree as "without project" case (table 7.4). Hafarat households were likely to incur additional marginal costs £s 189 (50 % cropping intensity) and £s 508 (75 % cropping intensity) to obtain a gross marginal benefit with the project of respectively £s 1,182 and £s 3,349.

---

8 This is the total of the MFEP/KADA Sudanese pounds budget, and of the Dutch guilder budget at the official exchange rate of 1 Dfl.= 3.75 £s. The bilaterally agreed rate for the CVF of MFEP/KADA was only 1 Dfl.= 1 £s. However, this CVF budget (*cf.* footnote 9) is revalued at the official rate, to arrive at figures that can be compared with the budgets of the regional government. Thus: £s 1,333,500 plus £s 861,428 multiplied by 3.75 equals £s 8,230,980.

9 Expenditures of the government of the Sudan contain two categories (i) the current budget. This includes "chapter 1 wages and salaries" and "chapter 2 other expenditures"; and (ii) the development budget for projects. Budgets (i) and (ii) are centrally approved by the Ministry of Finance (MFEP), but have been administered regionally between 1986-1991. New rules now apply in the new system of federal states of 1991, however. When international programmes contribute to government expenditures, a third budget called (iii) project budget is created. This contains a Counter Value Fund (CVF) and Foreign Exchange Component (FEC) in the case of programmes being financed by The Netherlands. The first is generated by commodity-aid, for which GOS pays in Sudanese pounds at low rates. The second is allocated to various bilateral GOS/GON projects in the country. Until 1991, it consisted of direct financial aid made available in Dutch guilders under yearly approval of the then bilateral programme between the countries.

Table 7.3 **WARK project expenditures in the Hedadeib pilot scheme over the two-year period 1989/90-1990/91, (in constant 1988 £s, rounded)**

Establishment costs (year 1)	50,000
Research, monitoring and evaluation costs	90,000
Total costs	140,000
Total costs per hectare	1,867
Running costs (year 2)	43,000
Research, monitoring and evaluation costs	90,000
Total costs	133,000
Total costs per hectare	1,773

Source: files MFEP/KADA and WARK. Note: costs per ha are based on a developed area in the pilot scheme of 75 ha, 1 £s 1988 equals $ 0.27.

No other standard techniques to evaluate the Hedadeib project costs and benefits can be applied. This is because still a substantial share of the MFEP/KADA overhead remains hidden at the level of the project.[10] The break-even point in the situation "with project" and a cropping intensity of 50 % is at a sorghum production of 1.8 sacks per fd (400 kg/ha) and a stalk production of 289 bundles per fd (2,064 kg/ha). In the situation of a 75 %-cropping intensity, these production levels of sorghum and stalks are 2.3 sacks (504 kg/ha) and 374 bundles (2,671 kg/ha) respectively.[11] According to outcomes of the L&E Survey for Hafarat (*cf.* appendix 6.1), these break-even levels have on average not been reached by land users in the Hedadeib pilot scheme in 1989.

Alternative crop production techniques for application in the Border Area include according to Mulder (1990) stone-reinforced contour embankments, stone bunds and improved *terus*. The cost-effectiveness of the pilot scheme, taking these alternative techniques into account, remains unknown. The 1990 evaluation mission considered the first alternative as financially not feasible, and the second was recommended for alternative development after 1990 (GOS/GON 1990,30). The mission did not comment on the potential of improved *teras* as an alternative for the ongoing government SWC interventions in the Border Area, however.

The earth dam

In the early 1980s, the overhead costs of the Department of Soil Conservation in Kassala formed some 70-80 % of the total budget (excluding the project budget, *cf.* footnote 9). This figure gradually decreased to a more favourable 10-20 % from 1986 onwards (Annual

---

[10] This applies to common discounting procedures, such as the internal rate of return.

[11] We assume that expenditures are equally divided over the production of grain and stalks. The calculations are furthermore based on a sorghum price of £s 200 per sack and a bundle price of £s 1.25 for stalks. The break-even sorghum production is calculated as: total operating expenditures equal gross grain production benefits; ½ (the share of grain production) x £s 1,114 = £s 200 x number of sacks of sorghum; where the number of sacks is 2.7 for 1.5 fd, and 1.8 for 1 fd cultivated. The same calculations can be made for stalks (Van Dijk 1991,53-56).

Reports SCLUWP, Kassala). A total staff of 41 is employed, of whom 10 in executive positions. The department controls a working area in the Kassala region of approximately 15,000 km². The team operating in the field in the framework of the Border Area Earth Dam programme usually consists of 2-3 technicians, 2-3 drivers and a supervisor. Although the balance between staff size and overhead costs must be considered as favourable, there is no balance with the set of tasks executed. This is because a significant underuse of manpower exists, as a result of transportation problems due to regular shortages of fuel and spare parts.

Table 7.4 **Crop budget of a typical 3 fd-landholding (in constant 1988 £s) with pilot scheme project and without pilot scheme project, situation 1989**

	WITHOUT PROJECT	WITH PROJECT		MARGINAL ANALYSIS	
Landholding size	3 fd	3 fd	3 fd		
Cropping intensity	50 %	50 %	75 %		
Cultivated area	1.5 fd	1.5 fd	2.25 fd		
	(1)	(2)	(3)	(2-1)	(3-1)
Production					
– grain	792	1,122	1,683		
– stalks	2,813	3,854	5,779		
Gross benefit	3,605	4,976	7,462	1,371	3,857
(Family) labour	450	497	743		
Seeds	4	4	6		
Tractor rent	330	90	90		
Implements	10	20	20		
Jute bags	330	330			
Taxation	79	112	168		
Transport	22	31	46		
Total operating expenditure	895	1,084	1,403	189	508
Net benefit	2,710	3,892	6,059	1,182	3,349
Net benefit per ha	2,151	3,089	4,809		

Source: Van Dijk (1991,56; adjusted). Note: the without project data are based on cultivation under wildflooding techniques in the area of Um Safaree.

The Border Area Earth Dam programme is not separately accounted for by the department. There is also no breakdown into its individual projects. A rough indication of costs can be obtained from similar interventions in the Red Sea province (WARK 1990c, files of SCLUWP office Port Sudan). The construction costs of the earth dams varied considerably in this area. This was between £s 50-600 (in constant 1988 £s) per hectare of cultivated land developed. The average costs (four documented projects) were £s 315 per ha (1988 £s). This is only 17 % of the same costs in the Hedadeib pilot scheme. However, the latter also include expenditures for research, monitoring and evaluation (*cf.* tables 7.3 and 7.5).

Table 7.5 **Costs of earth dam projects in the Red Sea province (in constant 1988 £s) implemented by the SCLUWP office Port Sudan**

Area	Year	Earth dam length (m)	Irrigated area (ha)	Total project costs (1988 £s)	Cost per ha (1988 £s)
Gowb	1987	1,000	295	13,149	45
Kilo 10	1987	215	30	16,236	541
Ashaf	1989	800	100	6,605	66
Hashatrebab	1990	400	21	12,759	608

Source: WARK (1990c), files SCLUWP office Port Sudan. Note: Gowb is located 14 km S.E. of Suakin, Kilo 10 at 75 km S.W. of Erkowit, Ashaf at 25 km S.W. of Sinkat, Hashatrebab at 8 km N.W. of Sinkat (cf. figure 3.6).

After construction, earth dams usually incur recurrent demands as to maintenance. Since damage is frequently in the form of full breaches, the level of the latter costs approaches the costs of constructing new dam sections. The cost-effectiveness of the earth dam, taking the alternative techniques mentioned by Mulder (1990) into account, remains unknown. However, the technical backstopping mission of MFEP/KADA made a strong plea to abandon this intervention programme, which viewpoint was shared by the joint Sudanese-Netherlands evaluation mission of 1990 (GOS/GON 1990,30).

7.1.2 Information from L&E Survey data

SIM (1990, 1991) distinguish several fields of direct and indirect impact of projects. We will discuss direct effects which enhance the economic position of the target group, and indirect effects of interaction with the physical and socio-economic environment. We also consider the project-sustainability and the potential of replication of government SWC interventions in the Border Area (after *Ibid.* 1990,7-10). Mainly L&E Survey data collected in Hafarat and Telkook are used to assess these effects not identified by the evaluation missions. The general concept of sustainability was discussed in section 1.1. Project-sustainability, more specifically, is defined here as the potential to deliver an appropriate level of benefits for an extended period of time after major financial, managerial and technical assistance from an external donor has been terminated (IOV 1993,132-134; IOV 1992,87).

• Impact: enhancing economic positions

The following questions are important when the direct effects of "enhancing economic positions" are used to evaluate government SWC intervention projects in the Border Area (SIM 1990,7; adapted):

1. *"Did the project result in higher production and in higher productivity, did it result in improved production methods ?"*

2. *"Did the project result in a change in income, income per unit labour, access to land resources, employment and organization of the target group ?"*

We first address the part of the second question regarding the subsidiary effects of government SWC interventions on income and employment locally. Several new activities in and around the pilot scheme have contributed to the household incomes in Hafarat from 1987 onwards. New employment opportunities were created in the form of surveillance duties for three households. One household is employed in the local operation of the meteorological station. Another household in the provision of domestic water to the Hedadeib camp. Incomes earned in these jobs ranged from £s 100-600 monthly for the years between 1987-1990. Labour was recruited in Hafarat for maintenance work on earth embankments in 1987 and 1988. This was at rates for casual labour of £s 5 per day. No local employment has been generated by the earth dam project in Telkook. However, one household is employed for surveillance duties in the guesthouse and equipment store that were later built in the village by the Department of Soil Conservation.

The impact of SWC interventions on production, productivity, land use and production incomes in Hafarat and Telkook is assessed by comparing "with project" and "without project" situations, and "before project" and "after project" situations. In the first, contributions of landholdings under pSWC are compared with those under all techniques other than pSWC. This group includes *teras*, brushwood panels, wildflooding and rainfed cultivation (these are referred to as nopSWC; *cf.* section 6.1, table 6.1). This comparison is made for the situation in the wet year 1988 and is repeated for the normal-to-dry year 1989. Only the contributions of local lands of the households are considered in order to make the cases of Telkook and Hafarat, where households differ in access to non-local lands, more comparable. We present the average household scores at the landholding level, unless indicated otherwise. In the second approach, the situations are compared for identical landholdings before and after implementation of the SWC project. The years should be comparable in agro-climatological respect. Given the data available in the L&E Survey, 1983 and 1989 were considered the best choice for this purpose. The situation in the deviant wet year 1988 is included in these presentations for comparison.

The selected characteristics for which in all cases the household average scores will be compared are presented in tables 7.6 and 7.7. These characteristics include total crop production, crop production per feddan, land use in terms of cropping intensities, the income from crop production per feddan, and the income from crop production per unit of labour time invested (for selected activities of maintenance, gap-filling, weeding and thinning over the growing season months June to and including October). Table 7.6 indicates that in Hafarat, generally higher scores can be found with project than without project in 1988, except for total production of sorghum (6.2 sacks against 7.6), of millet (4.3 sacks against 8.0), and for millet production per fd (1.6 sack against 1.7). The sorghum production per fd with project in this year equalled the level reached at other local landholdings. Also in Telkook, generally higher scores were obtained with project than without project in 1988, except for total production of sorghum (3.7 sacks against 4.3), of stalks (430 bundles against 590), and for production income per unit of labour time (£s 47 against £s 57). The overall level of production and productivity in Telkook was substantially lower than in Hafarat.

Table 7.6 **Selected crop production characteristics. Situation in 1988 without SWC project and with SWC project. Data are at the landholding level for nopSWC and pSWC landholdings respectively, all households, Hafarat and Telkook**

	HAFARAT				TELKOOK			
	1988 without project	N	1988 with project	N	1988 without project	N	1988 with project	N
TOTAL PRODUCTION								
sorghum (sacks)	7.6	52	6.2	20	4.3	9	3.7	11
millet (sacks)	8.0	49	4.3	15	1.0	1	--	
stalks (bundles)	762	79	801	22	590	5	430	10
PRODUCTION PER FD								
sorghum (sacks)	1.8	52	1.8	20	1.2	9	1.3	11
millet (sacks)	1.7	49	1.6	15	0.2	1	--	
stalks (bundles)	167	79	267	22	133	5	175	10
LAND USE								
cropping intensity (%)	77	98	86	23	77	11	89	17
PRODUCTION INCOME PER FD (constant 1983 £s)	520	90	754	23	209	11	307	13
PRODUCTION INCOME PER UNIT LABOUR TIME (1983 £s per man-hour selected activities [A])	111	64	190	19	57	4	47	6

Source: L&E Survey. Note: based on partly modelled data, [A] selected activities include maintenance, gap-filling, weeding and thinning for the growing season months June to and including October, -- means not cultivated.

Table 7.7 **Selected crop production characteristics. Situation in 1989 without SWC project and with SWC project. Data are at the landholding level for nopSWC and pSWC landholdings respectively, all households, Hafarat and Telkook**

	HAFARAT				TELKOOK			
	1989 without project	N	1989 with project	N	1989 without project	N	1989 with project	N
TOTAL PRODUCTION								
sorghum (sacks)	4.3	13	2.9	3	3.5	8	1.2	6
millet (sacks)	3.4	8	1.5	2	--	--		
stalks (bundles)	593	21	505	5	801	8	357	7
PRODUCTION PER FD								
sorghum (sacks)	1.0	13	0.8	3	2.3	5	0.3	6
millet (sacks)	0.8	8	0.4	2	--	--		
stalks (bundles)	139	21	128	5	533	3	713	7
LAND USE								
cropping intensity (%)	77	47	89	16	82	7	90	14
PRODUCTION INCOME PER FD (constant 1983 £s)	310	23	208	5	107	5	130	7
PRODUCTION INCOME PER UNIT LABOUR TIME (1983 £s per man-hour selected activities [A])	72	17	33	5	nd		57	7

Source: L&E Survey. Note: based on partly modelled data, [A] selected activities include maintenance, gap-filling, weeding and thinning for the growing season months June to and including October, -- means not cultivated, nd means no data available.

The question of the intervention impact on improved crop production methods can also be addressed with L&E Survey data for the situation in 1988. We consider the frequency of application of weeding and thinning as tentative indicators for this purpose. In Telkook, the average number of weeding rounds was, in 1988, higher at landholdings with project (1.92 times, N=13) than without project (1.65 times, N=17). The percentage of landholdings where thinning was practised was with 47 % (8 out of 17) also higher than without project (35 %, or 7 out of 20). A similar picture emerges in Hafarat. Here, the crops at 61 % of the landholdings with project (14 out of 23) were being thinned in 1988, against 36 % (36 out of 99) without project. However, the average number of weeding rounds was lower with project (1.67, N=24) than without project (1.82, N=90). This latter outcome is influenced by mechanized land preparation in the pilot scheme (*cf.* section 7.1.3).

The greater part of the more favourable performances in Hafarat and Telkook in 1988 with project are not again found for the situation in 1989 (table 7.7). In Hafarat, landholdings with project only favourably distinguish themselves from those without project by higher cropping intensities (89 % against 77 %). The situation was slightly better in this respect in Telkook. In 1989, the landholdings with project had a higher stalk production per fd (713 bundles against 533), higher cropping intensity (90 % against 82 %) and higher production income per fd (£s 130 against £s 107), than without project.

The same selected crop production characteristics can also be considered on identical landholdings in Hafarat and Telkook before the project (1983) and after the project (1989 and also 1988 for comparison) (tables 7.8 and 7.9).

Table 7.8 **Selected crop production characteristics. Situation in 1983 without SWC project and situations in 1988 and 1989 with SWC project. Data are at the landholding level for pSWC households and identical 1983-1989 pSWC landholdings only, Hafarat**

	HAFARAT					
	1983	N	1988	N	1989	N
TOTAL PRODUCTION						
sorghum (sacks)	6.8	11	6.2	20	2.9	3
millet (sacks)	3.2	6	4.3	15	1.5	2
stalks (bundles)	586	11	801	22	505	5
PRODUCTION PER FD						
sorghum (sacks)	1.5	11	1.8	20	0.8	3
millet (sacks)	0.5	6	1.6	15	0.4	2
stalks (bundles)	139	11	267	22	128	5
LAND USE						
cropping intensity (%)	99	12	86	23	89	16
PRODUCTION INCOME PER FD (constant 1983 £s)	281	13	754	23	208	5
PRODUCTION INCOME PER UNIT LABOUR TIME (1983 £s per man-hour selected activities [A])	37	8	190	19	33	5

Source: L&E Survey. Note: based on partly modelled data, [A] selected activities include maintenance, gap-filling, weeding and thinning for the growing season months June to and including October, -- means not cultivated.

In Hafarat, average scores on identical landholdings were consistently lower after the project in 1989, than before the project in 1983 (table 7.8). Generally higher scores after the project only occurred for the wet year 1988, except for total sorghum production (6.2 sacks against 6.8) and cropping intensity (86 % against 99 %). The same picture emerges in Telkook when the situations for 1983 and 1989 are compared (table 7.9). However, the number of valid cases is too small to draw meaningful conclusions here.

Government SWC interventions in Hafarat and Telkook did not influence access to cultivated lands. However, the government presence did arouse dispute over land use (*cf.* section 7.1.2 Impact: interaction with environment). In addition, participants in the Hedadeib pilot scheme gained preferential access to another resource namely mechanized land preparation. Ploughing by tractor was carried out by the Department of Soil Conservation either without charge, or at highly subsidized rates. This resulted in substantial financial benefits for the households in Hafarat (table 7.10). When the lands are prepared mechanically, the first manual weeding round can normally be skipped. The savings on this physically demanding, time-consuming and hence expensive activity generally were about 60 % of the total household land preparation costs in Hafarat over 1987-1991. The Department of Soil Conservation started to charge the households in Hafarat for this service in 1989. The fees were not collected in the very dry year 1991 on humanitarian grounds. In all years, anyhow, the costs of tractor hire remained lower than the first-round weeding costs. In a number of situations, this benefit alone is known to have been the main reason for households to participate in the pilot scheme.

Table 7.9 **Selected crop production characteristics. Situation in 1983 without SWC project and situations in 1988 and 1989 with SWC project. Data are at the landholding level for pSWC households and identical 1983-1989 pSWC landholdings only, Telkook**

	TELKOOK					
	1983	N	1988	N	1989	N
TOTAL PRODUCTION						
sorghum (sacks)	3.0	1	3.7	11	1.2	6
millet (sacks)	--		--		--	
stalks (bundles)	--		430	10	357	7
PRODUCTION PER FD						
sorghum (sacks)	3.0	1	1.3	11	0.3	6
millet (sacks)	--		--		--	
stalks (bundles)	--		175	10	713	7
LAND USE						
cropping intensity (%)	100	2	89	17	90	14
PRODUCTION INCOME PER FD (constant 1983 £s)	213	1	307	13	130	7
PRODUCTION INCOME PER UNIT LABOUR TIME (1983 £s per man-hour selected activities)	nd		47	6	57	7

Source: L&E Survey. Note: based on partly modelled data, [A] selected activities include maintenance, gap-filling, weeding and thinning for the growing season months June to and including October, -- means not cultivated.

The available data from the L&E Survey with respect to local organization are exclusively at the level of households.[12] The effect of SWC interventions executed at the landholding level can therefore not be assessed. Local organization was found to have increased in certain respects in the Hedadeib pilot scheme after 1989 (*cf.* section 7.1.1 Effectiveness *The pilot scheme*). Maintenance and repair was taken up after the use of brushwood panels had been propagated by the project. This work invariably was executed spontaneously on the own cultivated lands. However, work on structures of a communal character demanded more persuasiveness on the part of the village leaders for land users to engage in them (WARK 1990b,2). No communal activities have been reported on the cultivated lands under command of the earth dam in Telkook.

Table 7.10 **Total land preparation costs (in current £s) per feddan without SWC project and with SWC project, Hafarat area, 1987-1991**

	Tractor ploughing cost per fd		Weeding costs per fd			Total land preparation costs per fd [A]	
	Commercial rate (1)	SCLUWP rate (2)	1st round	2nd round (3)	3rd round (4)	Without project (1+3+4)	With project (2+3+4)
1987	50	0	25	20	20	90	40
1988	60	0	30	20	20	100	40
1989	110	30	50	25	25	160	80
1990	300	40	100	75	75	450	190
1991	450	0	200	150	150	750	300

Source: L&E Survey, Van Dam & Houtkamp (1992,67,72). Note: [A] calculated as costs of tractor ploughing plus costs of labour for a second and third weeding round.

• Impact: interaction with environment

The following questions are important when the indirect effects of "interaction with the physical and socio-economic environment" are used to evaluate government SWC intervention projects in the Border Area (SIM 1990,9; adapted):

1. *"Were there any external factors that influenced the economic position of the target group ?"*

2. *"Did the project have any negative consequences for the ecology, if so, was this recognized and were special measures introduced ?"*

3. *"Did the project have any negative effects on the position of disadvantaged groups who do not belong to the defined target group ?"*

---

[12] In Telkook, 17 out of 30 households (57 %) received support from voluntary workgroups in Telkook in 1983. This was 53 % (9 out of 17) in 1988. In Hafarat, the same figures were 75 % (6 out of 8 households) in 1983 and 69 % (9 out of 13 households) in 1988.

Three external and related developments have interacted over the years after the 1984-1985 drought with the impact of government SWC interventions in the Border Area. These include (i) the process of livelihood change triggered by this drought (*cf.* section 6.6.1); (ii) the occurrence of a new series of very dry years in 1990 and 1991 (*cf.* section 5.1); and (iii) the accomplishments of the movement of Ali Betai (*cf.* section 4.3.3). With respect to the first development, the drought resulted in substantial losses of livestock in the Border Area. This forced households to venture into other subsistence means. In Hafarat and Telkook, these mainly included, besides crop production, labour migration and off-farm employment. With respect to the second development, the very dry years 1990 and 1991 were unfavourable for local cultivation. The combined effect of these two developments can be expected to have been disadvantageous for SWC interventions in the Border Area. These projects implicitly assume the presence of land users for maintenance and repair and are dependent on local (regional) rainfall. With respect to the third development, the movement of Ali Betai received considerable government and international support (*cf.* section 4.3.3). This supposedly positively influenced the economic position of households in Telkook. The movement also secured tenancy titles for these households in the GDAC irrigation scheme (but these were not considered in the L&E Survey).

The pilot scheme

The second evaluation mission of 1990 concluded that no harmful environmental effects had resulted from the MFEP/KADA programme in general (GOS/GON 1990,10). However, a clear negative effect caused by the interventions in Hedadeib is gully erosion. At the margins of the pilot scheme, spillwater freely drains. Erosion has been common here since the scheme was established in 1987. Different measures were taken to control this process locally, including the application of sand bags and brushwood. Still, the lengths of gullies continued to increase annually at rates of some 50-100 % (Van Dijk 1990). An important negative intervention effect on other groups is the upsurge of local land dispute. The Sabdera, being mainly downstream land users in Hedadeib, receive less floodwater as a result of the embankments. The water which reaches their lands also arrives in a highly modified flow. This caused the Sabdera to reconsider their participation in the pilot scheme already in the first year of implementation.

The earth dam

Increased erosion as a result of the SWC project was also found on the cultivated lands of Ungwateit in Telkook. No measures were taken by the Department of Soil Conservation, or by local land users, to reduce these negative effects. The department merely assumed the role of project-executor. The households in Telkook have never been involved in the project and leave all responsibilities to the department. No negative effects of the SWC intervention have been assessed in Telkook on non-participating households.

7.1.3 Sustainability and potential of replication

The following questions are important when project-sustainability and the potential of replication of government SWC interventions in the Border Area are considered (SIM 1990,10; adapted):

1. *"Do the activities initiated by the project continue to be carried out ?"*

2. *"Are there any signs of project activities and results being sustainable in financial terms; is there an adequate level of participation that warrants the continuation of the activities after completion of the project ?"*

3. *"Can the project be replicated; if so, is this process taking place spontaneously ?"*

4. *"Was the intervention innovative and have elements been replicated elsewhere?"*

5. *"Is the counterpart in a position to continue the operations after the donor and government funding decreases or terminates ?"*

The backgrounds and objectives of the pilot scheme and earth dam interventions differ substantially. This has consequences on how the above questions regarding sustainability and replication must be appreciated. The pilot scheme was a combined Sudanese-Dutch initiative. The Sudanese government acted through the regional Ministry of Agriculture and its Department of Soil Conservation formed the counterpart in MFEP/KADA. In the programme's dual objective of development *cum* institution-building, the department itself was considered as the first and higher-level target group. This department, in turn, was assumed to address the second and lower-level target group of households in the Border Area. The continued use of the pilot scheme, project-sustainability and potential of replication must therefore be evaluated from this same dual perspective. However, the earth dam project was an entirely Sudanese activity. The objectives which ruled this intervention were based on considerations of development and economic growth. The target group consisted of households in the Border Area, but this group was not defined in detail beforehand. The project-sustainability of this intervention in Telkook can be meaningfully considered, but its potential of replication can not. This is because the dam was never meant to be replicated by the households in the first place.

The pilot scheme

In the course of the first half of 1991, the activities in the Border Area of MFEP/KADA and WARK were both stopped.[13] However, already in the late 1990s the Department of

[13] The WARK project was completed in May 1991. The MFEP/KADA programme was terminated in July 1991.

Soil Conservation had decided to reallocate its resources to two new pilot schemes in Kawateib and Shellalob. This followed the recommendations made by the second evaluation mission and technical backstopping mission (El Faki 1991). These recommendations had been made because Hedadeib had by then clearly demonstrated to be an unfavourable location for SWC interventions (*cf.* section 4.2.3). Termination of government interventions in the pilot scheme was therefore not a direct consequence of the withdrawal of international support. Still, the activities here are unlikely to have been project-sustainable in the way defined in section 7.1.2. This can be illustrated by comparing the department's annual development budgets (table 7.2) and the pilot scheme project costs in "adjusted" form (table 7.3). In 1989-1990, the department spent respectively £s 380,000 and £s 945,000 (current £s) for the entire eastern Sudan. The project costs in these same years in Hedadeib were respectively £s 140,000 and £s 133,000 (1988 £s). When these budgets of the Department of Soil Conservation are expressed in the same 1988 £s (£s 362,000 for the budget of 1989, and £s 583,000 for 1990), the single Hedadeib pilot scheme would account respectively for some 40 % and 25 % of the regional budgets. This is unlikely to have been manageable for the department.

The project-replication potential for households as second target group, must equally be considered to have been low. The annual net crop production returns per hectare of £s 3,000-5,000 (table 7.4) favourably compare with the annual establishment costs and running costs of the scheme of about £s 1,800 per hectare (table 7.3). The latter even include a substantial share of research, monitoring and evaluation costs. Still, these household returns can be expected to have been too insecure to risk the engagement. In 1989, for example, the break-even levels required were on average not attained in Hafarat (*cf.* section 7.1.1 Efficiency). At present, land users continue to cultivate in Hedadeib without external support. They rely on indigenous techniques to repair the breached embankments. However, insufficient floodwater was received between 1991 and the date of our last report from the area of August 1993, to bring substantial areas under cultivation.

Finally, replicability of certain components of the MFEP/KADA programme can also be evaluated from the perspective of the Department of Soil Conservation. These concern training and institution-building. The potential of replication of these activities was high as shown by the department's autonomous initiatives in this field in other pilot schemes in the Border Area (*cf.* section 7.1.1 Effectiveness *The pilot scheme*). A range of SWC projects is currently being implemented in eastern Sudan. The Department of Soil Conservation and other departments participate in these, and the experiences gained at Hedadeib will directly contribute here.[14]

---

[14] Current SWC intervention programmes with involvement of the Kassala Department of Range and Pasture Management include: (i) "Development of pasture and irrigated fodder crops (E. Region) Sudan (UNO/SUD/009/EOA, TCP/SUD/4525). This FAO programme is implemented in the Gash Dai area. It includes a floodwater harvesting component to improve fodder production in a 150,000 fd area. The Kassala Department of Soil Conservation participates in: (ii) Area Development Scheme (ADS) El Butana, Eastern Region (SUD/90/006a/01/99). This UNDP programme includes, i.a., the introduction of improved SWC techniques in the Butana Rural Council. Finally, the two departments are also involved in projects in (iii) the Red Sea province where different types of water management systems

Several activities propagated in the pilot scheme have also spontaneously been adopted by the households of Hafarat for application in other farming zones. These include (i) measures for improved erosion-control using wattled brushwood panels; (ii) new crops such as okra, rosella, watermelon and the forage legume *Dolichos lablab* (coll. *lubia*); and (iii) improved crop husbandry practices, such as more frequent thinning and gap-filling, and the use of seed-dressings to prevent pests.

The earth dam
The lands in Ungwateit are still being cultivated by the households of Telkook, but the breached earth dam contributes little to regulate the supply of floodwater. The last repair works were made by the Department of Soil Conservation in 1985/86. The project has since then been considered as completed. The department's development budget is insufficient to take up the required additional maintenance and repair works. There is also no international support to restore the dam. The intervention, therefore, was not project-sustainable. New crops of okra and rosella were introduced by the department locally as part of the Earth dam programme interventions. These have been adopted and are also being cultivated by the households of Telkook in other farming zones.

7.1.4 Particulars of impact of SWC intervention

The participants in the Hedadeib pilot scheme were, according to initial findings by Van Dam & Houtkamp (1992), not representative for the population of Hafarat as a whole. The bias would be towards households in the higher-income groups (*Ibid.* 1992,62-63). This is confirmed by the L&E Survey data where for this purpose not only household incomes, but also livestock wealth and land entitlements were compared. In Hafarat, these are all higher for pilot scheme participants (pSWC households) than for non-participants (nopSWC households). In Telkook, only the household income of earth dam participants is higher than for non-participants (table 7.11). The relationships between participation in SWC projects and these characteristics were examined by means of the Chi-square measure. However, none proved statistically significant.[15]

When the scores on these same variables are ranked and compared for all households in the four research villages, by all sub-groups discerned in the L&E Survey, the following pattern emerges (table 7.12). The target group of government SWC interventions (pSWC households) has the highest scores in terms of household income, but lower scores for entitlement to cultivated land and livestock wealth. The mean rank of 2.7 also places them

---

for crop production are being developed. These projects are supported by the Norwegian Red Cross, OXFAM, UNICEF, WHO, EuroAction ACCORD and EEC (*cf.* OXFAM 1990,14-15).

[15] Household sub-group is a dichotomous variable (yes/no member). The three variables representing household wealth were recoded to make them dichotomous. The median was used to divide high and low scores. For all relationships holds that P>0.05, which means that they are not significant at the 95 % confidence level.

at an intermediate position on the list. When we finally consider the outcomes presented in the two tables together, we conclude that the composition of target groups of government SWC intervention only tended to favour groups of higher household wealth. This was more so in Hafarat than in Telkook. This equally was more pronounced and more consistently the case in terms of household income than in other respects. Household wealth in terms of land and livestock ownership is only in Hafarat higher for the target group than among other households.

Table 7.11 **Average household wealth in terms of total land entitlement (in fd), livestock ownership (in TLU) and total household income (in 1988 £s) for selected sub-groups, Hafarat and Telkook, situation 1988**

HOUSEHOLD Sub-group	LAND (fd)	N	LIVESTOCK (TLU)	N	INCOME (1988 £s)	N
Hafarat pSWC	15.3	26	7.10	26	8,437	25
Hafarat nopSWC	12.6	34	6.70	31	7,933	34
Telkook pSWC	5.6	22	5.81	21	7,635	19
Telkook nopSWC	7.4	14	6.17	21	6,108	32

Source: L&E Survey. Note: based on partially modelled data. pSWC means project SWC households, nopSWC means no project SWC households, sub-group definitions are given in section 6.1.

Table 7.12 **Average household wealth in terms of total land entitlement (in fd), livestock ownership (in TLU) and total household income (in 1988 £s) for selected sub-groups and their rank number (r) over the sub-groups, Ilat Ayot, Telkook, Um Safaree, Hafarat, situation in 1988**

HOUSEHOLD Sub-group	LAND (fd)	r	LIVESTOCK (TLU)	r	INCOME (1988 £s)	r	MEAN rank r
iSWC	14.3	2	6.27	4	5,101	4	3.3
pSWC	10.5	4	6.46	3	8,036	1	2.7
nopSWC	12.8	3	12.91	2	6,589	2	2.3
noSWC	14.4	1	17.12	1	6,224	3	1.7

Source: L&E Survey. Note: based on partially modelled data. iSWC means indigenous SWC households (hh), pSWC means project SWC hh, nopSWC means no project SWC hh, no SWC means no SWC hh. All sub-groups listed are mutually exclusive, sub-group definitions are given in section 6.1.

Selected indicators

The impact of government SWC interventions on the economic positions of the households in Hafarat and Telkook can also be assessed more quantitatively. We consider the scores for the with project and without project situations, and before project and after project situations of crop production income, allocation of labour time, and cropping intensity. These data are available at the level of households and landholdings. The same reference years 1983, 1988 and 1989 as in section 7.1.2 "Impact: enhancing economic positions" are used. The score distributions for the with and without situations are tested in the Mann

Whitney U-test. The same score distributions of the before and after situations are tested in the Wilcoxon Matched-pairs Signed-ranks test (*cf.* section 2.3.3 for their applications). A positive (+) sign in tables 7.13 and 7.14 indicates an association of occurrence of government SWC intervention and, for the Mann Whitney U-test, higher mean rank scores on concerned variables for the group with intervention than for the group without intervention. With respect to the Wilcoxon Matched-pairs Signed-ranks test, this positive sign means an association of occurrence of government SWC intervention and higher mean rank scores on the concerned variables for identical landholdings and households after intervention (1988, 1989), than before intervention (1983). Since the Wilcoxon test, unlike the test of Mann-Whitney, considers only scores over matching research units, the first test is usually based on a smaller number of valid cases. We again consider only landholdings on the local lands in Hafarat and Telkook (*cf.* section 7.1.2 Impact: enhancing economic positions).

Table 7.13 presents the outcomes of the comparisons, firstly, for the with and without situations in 1988 and 1989. Most signs in this table are positive for the wet year 1988. Generally higher mean rank scores are found in Hafarat and Telkook with the project than without the project for this year, except for crop production income at the landholding level. For the normal-to-dry year 1989, all signs are negative except for the variable cropping intensity. For this year, generally lower mean rank scores are found in Hafarat and Telkook with the project than without the project.

Table 7.13 **Impact of government SWC intervention comparing the with project and without project situations. Direction of relationship (sign) and observed significance levels in Mann Whitney U-test (P) for selected variables. At the landholding level, nopSWC landholdings are compared with pSWC landholdings. At the household level, nopSWC households are compared with pSWC households. "+" indicates associations of high mean rank scores and intervention occurrence. "-" indicates associations of low mean rank scores and intervention occurrence, Telkook and Hafarat, 1988 and 1989**

	LANDHOLDING LEVEL A		HOUSEHOLD LEVEL	
	Sign	P	Sign	P
Situation in 1988				
Crop production income	–		+	
Labour-time allocation	+		+	
Cropping intensity	+	.0430	+ B	
Situation in 1989				
Crop production income	–		–	
Labour-time allocation	–	.0089	–	
Cropping intensity	+		+ B	

Source: L&E Survey partly based on modelled data. Note: only P≤0.05 is shown, [A] local lands only, [B] mean cropping intensity of local lands. Appendix 7.1 lists the mean rank scores behind the signs.

Table 7.14 presents the outcomes of the comparison of situations before the project (1983) and after the project (1988 and 1989), respecting identical landholdings and households. The table shows exclusively positive signs when 1988 is compared with 1983. This indicates higher mean ranks for all variables after intervention. However, this cannot be entirely attributed to the intervention impact, because the agro-climatological conditions were also considerably more favourable in 1988 than in 1983. When the situation in 1983 more appropriately is compared with 1989, this relationship changes for the majority of variables. At the landholding level, exclusively negative signs appear. These indicate lower mean ranks after the project in 1989, than before the project in 1983. This means that the interventions made little impact in this respect on the lands in Hafarat and Telkook. At the household level, two out of three signs are positive. This indicates that higher mean ranks are found after the project for the situation in 1989, than before the project in 1983, with the exception of cropping intensity. This means that the households were doing better in terms of crop production income, and generally also allocated more labour time to this sector after the project than before. However, since the signs for these variables differ between the two levels of landholding and household, the interventions may not have made a substantial contribution to these better performances in 1989. The mean ranks of the cropping intensities at both the landholding level and household level are lower for the situation after the project than before. At the household level, this is the only statistically significant relationship found (based on the mean cropping intensity of all local landholdings of the households).

Table 7.14 **Impact of government SWC intervention comparing the before project and after project situations. Direction of relationship (sign) and significance levels in Wilcoxon Matched-pairs Signed-ranks test (P) for selected variables. Identical pSWC-landholdings and pSWC-households are compared only. "+" indicates a mean signed-rank in 1988, respectively 1989, higher than in 1983, "-" indicates a mean signed-rank in 1988, respectively 1989, lower than in 1983, Hafarat and Telkook, 1983, 1988 and 1989**

	LANDHOLDING LEVEL [A]			HOUSEHOLD LEVEL		
	Sign	P	N	Sign	P	N
Situation in 1983 and 1988						
Crop production income	+	0.0029	12	+	.0001	20
Labour-time allocation	+		14	+	.0121	24
Cropping intensity	+		20	+ [B]		26
Situation in 1983 and 1989						
Crop production income	–		4	+		8
Labour-time allocation	–		11	+		16
Cropping intensity	–		14	– [B]	.0277	19

Source: L&E Survey partly based on modelled data. Note: only P≤0.05 is shown, [A] local lands only, [B] mean cropping intensity of local lands. Appendix 7.1 lists the mean signed-rank scores behind the signs.

Developments over time

Figures 7.1 to 7.4 illustrate for Hafarat and Telkook the developments over the years in selected crop production characteristics, by cultivation technique. We consider for this purpose cropping intensity, allocation of labour time per fd (selected activities of maintenance, gap-filling, weeding, thinning for the growing season months June to and including October), crop production income per fd, and crop production income per man-hour of allocated labour time (returns from selected activities of maintenance, gap-filling, weeding and thinning for the growing season months June to and including October). The figures are based on average scores at the landholding level for all households in Hafarat and Telkook. Indigenous SWC includes *teras* and brushwoods panels, project SWC includes the earth dam in Telkook and the pilot scheme in Hafarat. There were no interventions in 1983 which means also no pSWC landholdings existed in this year. NoSWC includes wildflooding and rainfed cultivation. Finally, the group Gash region includes irrigation in the GDAC scheme and cultivation under flood-recession on the Gash margin. However, the number of landholdings in the Gash region is small: 10 in 1983, 13 in 1988, 13 in 1989 and 22 in 1990. The high incomes from crop production in 1990 (*cf.* figures 7.3 and 7.4) resulted from the abnormal price developments in this very dry year. The average grain price per sack in 1990 was £s 990 for sorghum and £s 1,300 for millet. The average stalk price reached in this year £s 4.5 per bundle. Van Dam & Houtkamp (1992,77-78) present details on the development of the sorghum prices in Hafarat. They report successive rises, starting from £s 280 per sack in December 1989, and finally reaching £s 1,600 per sack in December 1990.

Figure 7.1 **Average cropping intensity. All households in Hafarat and Telkook, landholding level, 1983, 1988-1990. Note: iSWC means indigenous SWC, pSWC means project SWC, noSWC means no SWC techniques applied. Source: L&E Survey**

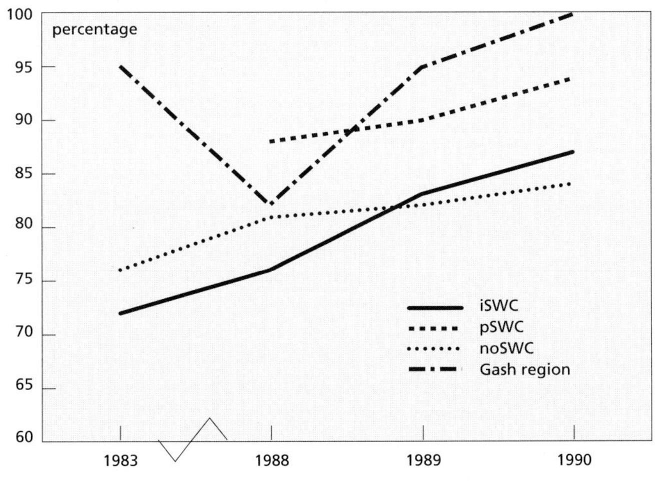

The cropping intensities in the intervention villages of Telkook and Hafarat (figure 7.1) tended to continuously increase over the period 1983-1990, except on landholdings in the Gash region (N over individual years 1983 and 1988-1990 for the Gash region is respectively 3, 8, 5, 11). Cropping intensities in the Gash region were usually the highest of all techniques, but there was a clear dip in the wet year 1988. This can be explained from the fact that when conditions for local cultivation are favourable, there is less need for households to fully engage in other more distant regions, such as the Gash. Besides the Gash, also landholdings under pSWC showed consistently higher cropping intensities than those under other cultivation techniques. A possible explanation is that payments of mechanized land preparation (Hedadeib pilot scheme) and water fees (GDAC irrigation scheme) compelled a more intensive use of land in order to make the investments profitable. iSWC landholdings had among the lowest cropping intensities listed, except in the normal-to-dry year 1989, and very dry year 1990.

Figure 7.2 **Average labour-time allocation per feddan (selected activities). All households in Hafarat and Telkook, landholding level, based on partly modelled data, 1983, 1988-1990. Note: iSWC means indigenous SWC, pSWC means project SWC, noSWC means no SWC techniques applied. Source: L&E Survey**

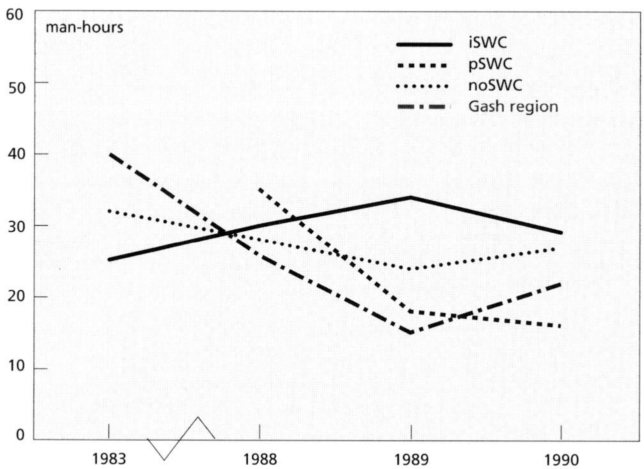

The allocations of labour-time per feddan in Telkook and Hafarat (figure 7.2) are variable over the years. However, these fluctuations seem to have occurred within a certain bandwidth of approximately 15-40 man-hours. This can be understood as follows. Most households simultaneously engage in different techniques as was described in the village presentations (*cf.* sections 6.3 and 6.5). However, this mutual substitution of labour between techniques is constrained by the total labour-time budget of the household, which is itself a function of household size and other factors. Figure 7.2, accordingly, suggests that this budget was on a per feddan basis in the mentioned range of 15-40 man-hours.

Landholdings under pSWC received, on average, the maximum allocation of labour time in the wet year 1988, but received among the lowest in the other dryer years. The landholdings under iSWC to some extent show the reverse picture. The allocations of labour time here were higher in the dryer years 1989 and 1990, but were lower in the wet year 1988. The pattern in the Gash region (N over 1983 and 1988-1990 is respectively 3, 8, 5, 11) is variable. The highest allocations of labour time were made in 1983, the lowest in 1989. Intermediate levels were recorded in the Gash region for the other years.

Figure 7.3 **Average crop production income per feddan. All households in Hafarat and Telkook, landholding level, based on partly modelled data, 1983, 1988-1990. Note: iSWC means indigenous SWC, pSWC means project SWC, noSWC means no SWC techniques applied. Source: L&E Survey**

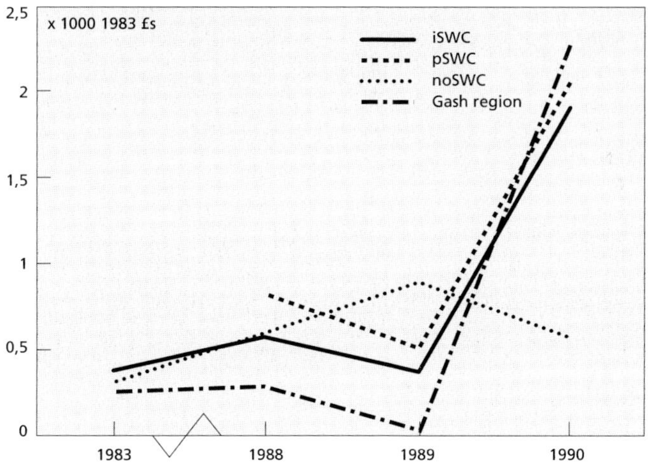

Figure 7.3 displays the developments over the years in average crop production incomes per feddan cultivated and harvested. These incomes steeply increased in 1990 as a result of the mentioned anomalous price developments. The pattern also shows higher incomes in the wet year 1988, and lower incomes in the normal-to-dry years 1983 and 1989. This is according to expectation. Only landholdings which were not cultivated under SWC techniques (noSWC) show the highest average crop production income in another year, namely 1989. The most important technique in this category is wildflooding. Khors, in the Border Area frequently with large and non-local catchments, apparently provide a comparative advantage in such situation, but not in a very dry year like 1990. Landholdings under pSWC provided high returns, in addition to the anomalous 1990, also in the wet year 1988. However, their contributions to the crop production income were more modest in the dryer 1989. Landholdings under iSWC take an intermediate position. They provided the highest returns of all techniques in 1983, but the lowest in 1990. Landholdings in the Gash region (N over 1983 and 1988-1990 is respectively 1, 5, 5, 5)

showed the lowest scores of all, except for the anomalous 1990. The number of valid cases is too small to draw any meaningful conclusions from them, however.

Figure 7.4 **Average crop production income per manhour (selected activities). All households in Hafarat and Telkook, landholding level, based on partly modelled data, 1983, 1988-1990. Note: iSWC means indigenous SWC, pSWC means project SWC, noSWC means no SWC techniques applied. Source: L&E Survey**

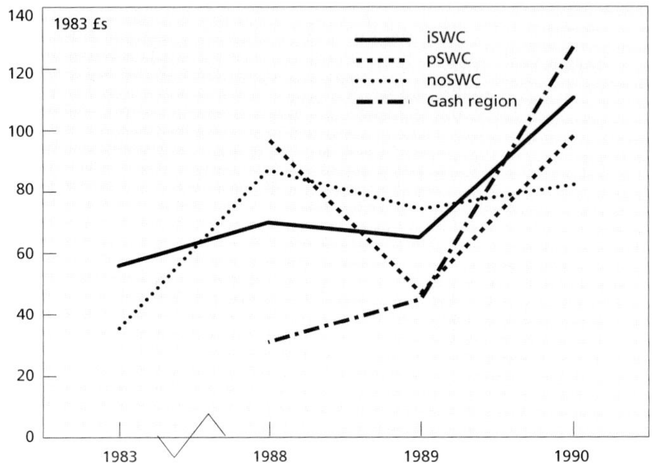

Figure 7.4 shows the crop production incomes per unit of household labour time allocated in Telkook and Hafarat. This is the most variable pattern of all. The 1990 anomalous price developments are again shown. A second peak occurred in the wet year 1988, but lower scores prevailed in the dryer years 1983 and 1989. Landholdings under pSWC provided the highest crop production incomes per man-hour in the wet year 1988. However, they provided lower returns in the dryer years conforming with the general pattern in figure 7.4. Landholdings under iSWC, on the contrary, provided among the lowest returns per man-hour in the wet year 1988, but the highest in the normal-to-dry year 1983 (over two techniques only). They also provided the one but highest average returns in the normal-to-dry year 1989 and in the very dry year 1990. The noSWC techniques showed the best performances in 1989, probably for the reason given above. Finally, landholdings in the Gash region provided the lowest returns per man-hour allocated in all years, except in the very dry 1990. The number of valid cases over 1988-1990 (no data available for 1983) is, with respectively 4, 5 and 3 households, small. A comparison of these hourly returns from crop production with returns in other livelihood categories in the Border Area will be made in section 8.2.

Figure 7.5 **Average cropping intensity. Sub-group iSWC and pSWC households in Hafarat and Telkook, landholding level, 1983, 1988-1990. Note: iSWC means indigenous SWC, pSWC means project SWC. Source: L&E Survey**

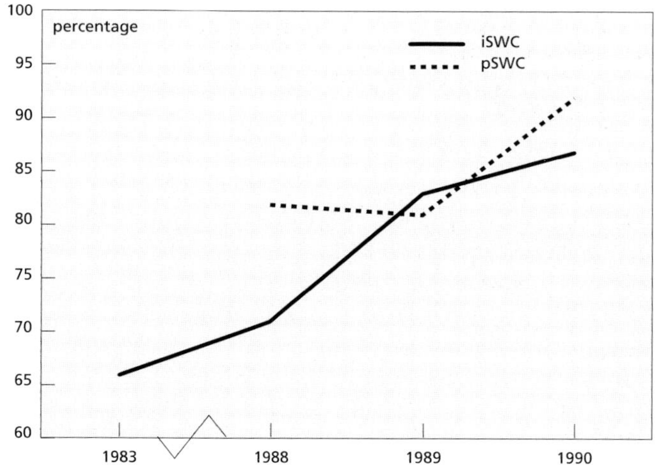

Figure 7.6 **Average labour-time allocation per feddan (selected activities). Sub-group iSWC and pSWC households in Hafarat and Telkook. Data are at the landholding level and partly modelled, 1983, 1988-1990. Note: iSWC means indigenous SWC, pSWC means project SWC. Source: L&E Survey**

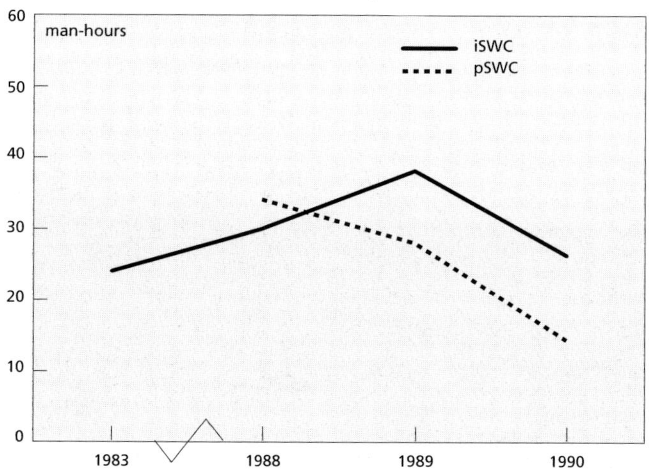

Two exercises
Two additional exercises are of interest to this study. In the first, we address the possible effects of government SWC interventions on the performances on landholdings under iSWC. In the second, we address the effect of mechanized land preparation in Hafarat on the allocation of labour time to other cultivation techniques. Figures 7.5 to 7.8 show the same crop production characteristics over the years as have been discussed above. The difference is that this time, we consider only Hafarat and Telkook households who were using both project and indigenous SWC techniques (sub-group "iSWC and pSWC households", Telkook N=5, Hafarat N=16, pSWC over 1988-1990 only).

The first figure 7.5 shows that average cropping intensities varied over the years. These resulted in mirror images in the years 1988, 1989, and 1990 after the date of implementation of government SWC interventions, which was in 1985 and 1987 in Telkook and Hafarat respectively. This may indicate that operations in either group of techniques are associated, for example by relations of imposing ceilings to the amount of land that can be cultivated by the household under each, in any one year. The allocations of labour time per fd (figure 7.6) show more comparable patterns for iSWC and pSWC techniques in Hafarat and Telkook. Landholdings under iSWC received lower allocations of labour time in the wet year 1988 and higher allocations in the dryer years 1989 and 1990, than those under pSWC.

Figure 7.7 **Average crop production income per feddan. Sub-group iSWC and pSWC households in Hafarat and Telkook. Data are at the landholding level and partly modelled, 1983, 1988-1990. Note: iSWC means indigenous SWC, pSWC means project SWC. Source: L&E Survey**

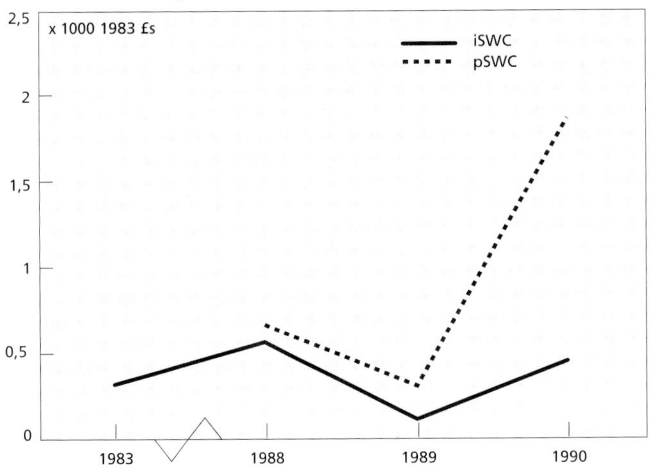

The financial returns per feddan from crop production of households in Telkook and Hafarat, were in all years higher for landholdings under pSWC than under iSWC (figure 7.7).

Figure 7.8 **Average crop production income per manhour (selected activities). Sub-group iSWC and pSWC households in Hafarat and Telkook. Data are at the landholding level and partly modelled, 1983, 1988-1990. Note: iSWC means indigenous SWC, pSWC means project SWC. Source: L&E Survey**

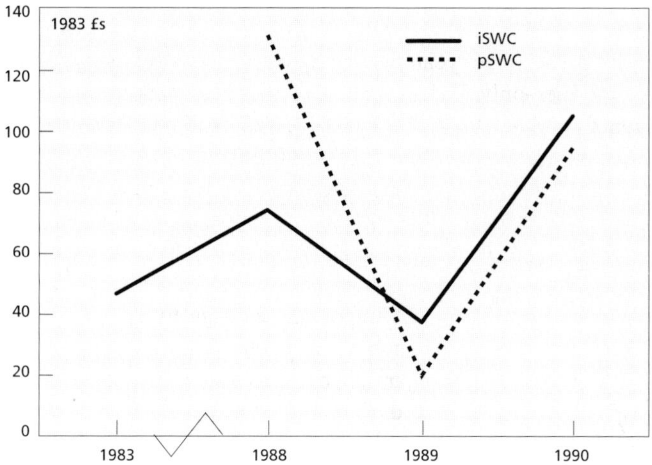

Finally, the crop production returns per man-hour of labour time allocated (selected activities of maintenance, weeding, thinning and gap-filling and for the growing season months June to and including October) indicate that landholdings under pSWC produced higher returns in the wet year 1988 (figure 7.8). However, in the dryer years 1989 and 1990, landholdings under iSWC performed better in this respect.

The findings illustrated in figures 7.5-7.8 allow the following two conclusions to be drawn. Firstly, there are no significant effects of household participation in pSWC on the application of iSWC techniques locally. The only exception to this may be the straightforward limitation to the amount of land that can be cultivated per year (*cf.* figure 7.5). Secondly, these findings suggest that iSWC and pSWC techniques may have their own comparative advantages for different types of years. This advantage would lie with pSWC techniques in the wetter years (although the indications presented here are based on one wet year 1988 only). This advantage, alternatively, would lie with iSWC techniques in the dryer years. Obviously, the limited number of years studied in detail in the Border Area cannot provide any decisive proof of this proposition.

For the purpose of the second exercise, only the sub-group of pSWC households in Hafarat (N=26) is selected[16] (no mechanized land preparation was carried out in Telkook). The data of the Gash region have been excluded because these concern in this sub-group in Hafarat one household only. Figure 7.9 shows that the household allocation of

[16] This sub-group includes households who also apply other techniques (iSWC, noSWC) without however having been selected on the basis of this particular characteristic (*cf.* section 6.1, table 6.1).

labour time (selected activities of maintenance, weeding, thinning and gap-filling for the growing season months June to and including October) for 1988-1990 was the lowest on landholdings under pSWC. This allocation, in addition, also continued to decline over these same years. Landholdings under noSWC and iSWC techniques of the same households in Hafarat consistently received higher allocations of labour time. There is also a slight tendency for these amounts to increase after 1989. It is plausible to assume that households who saved labour time as a result of mechanized land preparation in the pilot scheme introduced in 1987, have allocated this to their landholdings under other cultivation techniques. This extra labour time may also have been allocated to other livelihood activities, but the available data cannot be used to consider this alternative. Mechanized land preparation, therefore, positively contributed to improve the economic positions of households in Hafarat. However, the mere availability of government tractor services for ploughing was also a main reason among these households to give a positive evaluation of the SWC intervention. In the words of one respondent interviewed in Hafarat by Van Dam & Houtkamp (1992,109) "[..] the (floodwater harvesting) system is good, because the government takes care of the people [..]". Participation in the SWC project, therefore, was largely based on enjoyed fringe benefits, possibly including benefits of labour time, instead of recognized usefulness of the introduced SWC technique.

Figure 7.9 **Average labour-time allocation per feddan (selected activities). Sub-group pSWC households in Hafarat. Data are at the landholding level and partly modelled, 1983, 1988-1990. Note: iSWC means indigenous SWC, pSWC means project SWC, noSWC means no SWC techniques applied. Source: L&E Survey**

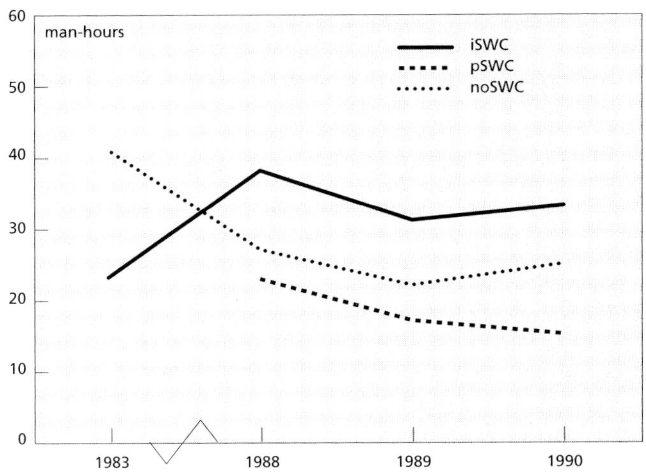

7.1.5 Concise presentation of findings

The government SWC intervention in Telkook is less well documented than the intervention in Hafarat. Still, the evidence available indicates in main lines comparable outcomes for the two projects. In terms of accordance with policy, the earth dam and pilot scheme objectives tallied with the respective Sudanese (GOS) and Netherlands government (GON) policy formulations. However, these policies themselves were mutually opposing. The executive government institutions of GOS and GON assumed in the Border Area a role of project-implementor and project-facilitator, respectively. Either approach was not participatory in the sense defined in section 1.3. The factual outcome was a Transfer-of-Technology situation. This was, under the GOS objectives and approaches, an accepted principle. However, under the GON objectives and approaches, this was an undesired consequence. The target groups in Telkook and Hafarat have factually been determined by the patterns of household entitlement to cultivated lands that came to be developed. (Other ways of target group delineation, however, would have been difficult to carry out because the introduced techniques are of a communal character). The effectiveness of government SWC interventions, in terms of the succesful development of operational floodwater-harvesting systems, was low. No elaborate assumptions lied at the base of the development of earth dams. Also no significant changes were pursued to improve their performance after implementation. The underlying assumptions of the pilot scheme technique proved wrong in different respects. However, flexibility was maintained to experiment with new approaches, and to adjust to newly emerging conditions. The research and training objectives of the pilot scheme project have been achieved.

It is difficult to draw conclusions on the SWC intervention outcomes for the Border Area in terms of efficiency. There is a lack of data on this subject. In general, it holds that the intervention objectives have been reached with the resources earmarked. However, the cost-effectiveness of these projects cannot be assessed. The costs to develop one hectare in the pilot scheme were at least six times higher than the average costs for land development in earth dam projects. However, the latter type of intervention in Telkook was in the end also less effective because of high breach damage. Alternative allocations of resources for SWC development in the Border Area are available in the field of indigenous SWC support and improvement, particularly of the *teras* technique.

The impact of the earth dam on the economic positions of households in Telkook was low. A positive contribution of the pilot scheme in Hafarat, in this respect, could only be demonstrated for the wet year 1988. Several improved cultivation methods have been taken up by land users at both locations. The returns to labour increased in the case of the pilot scheme. Expectedly, this must to a large extent be attributed to the beneficial role of mechanized land preparation, not to the system of floodwater harvesting alone. Household access to cultivated lands did not change as a result of the earth dam and pilot scheme projects. Under both interventions, only lands were developed which were already in use under traditional tribal rules. There was no local organization of households cultivating the lands commanded by the earth dam in Telkook. In Hafarat, efforts were made to increase the involvement of the local households during the later development phases of the pilot scheme. This eventually resulted in a modest revival of elementary forms of local organization.

The impact of government SWC interventions in terms of interaction with the environment included negative effects on the physical environment in the form of increased erosion. Negative effects on the local community took the form of disputes over land and, in Hafarat, of deprivation of downstream land users. There were also distinct external factors which influenced the developments in the Border Area in the 1980s and early 1990s. These included livelihood change, drought years and religious organization.

Both the pilot scheme and the earth dam interventions were not project-sustainable. In Telkook, the effective use ceased after government involvement stopped. In Hafarat, the use of the pilot scheme in its original design also ceased, but the withdrawal of international support was no relevant factor in this. Still, the high costs make it unlikely that the Department of Soil Conservation would be able to sustain the intervention without it. In the earth dam project in Telkook, no responsibilities had been delegated to the participants. The intervention was a public service, whose standard could not be improved due to insufficient government means and resources. In the pilot scheme project in Hafarat, on the other hand, participants were supposed to take up the activities themselves. However, this replication objective was not attained.

A comparison of selected crop production characteristics in Hafarat and Telkook with and without SWC intervention (tables 7.6 and 7.7) shows that, commonly, better performances were obtained with the SWC project for the wet year (1988). This, however, was not the case for the normal-to-dry year (1989). A similar comparison of situations before and after SWC intervention (tables 7.8 and 7.9) shows that, commonly, the 1989 performances after the project were also less favourable than before (1983). (However, the number of valid cases for the 1983 "before project" situation in Telkook is small). Largely identical outcomes arise when other variables of crop production income, allocation of labour time and cropping intensity are used for the purpose of this comparison (tables 7.13 and 7.14). Their score patterns were considered at the two levels landholdings and households. The comparison of with and without SWC intervention situations (table 7.13) again shows that the with project performances were generally more favourable than the without project performances in 1988, but not in 1989. (However, the 1988 crop production income at the landholding level, and 1989 cropping intensities at the landholding and household levels show a deviant pattern in this respect). The comparison of the before and after SWC intervention situations (table 7.14), shows that the performances were invariably more favourable after the project in 1988. However, they were less favourable after the project in 1989. (The 1989 crop production income, and 1989 allocation of labour time at the household level show a deviant pattern in this respect).

The developments in selected characteristics of crop production over the years (figures 7.1-7.8) suggest that possibly comparative advantages exist for individual cultivation techniques in agro-climatologically different types of years. Government SWC projects show better overall performances in the wet year 1988. However, indigenous SWC shows better overall performances in the dryer years.

Mechanized land preparation in the pilot scheme saved household labour time. This is likely to have been allocated to other cultivation techniques (figure 7.9). Probably, this was also allocated to other livelihood activities. Ploughing, therefore, positively contributed to improve the economic positions of households in Hafarat. However, it had negative effects on participation. This participation is likely to have become largely based on enjoyed financial fringe benefits, instead of on the recognized usefulness of the introduced technique.

7.2 The importance of indigenous SWC and crop production

The central premise in our research framework was discussed in section 2.1. We expect that a positive relationship exists between distance to the regional urban centre Kassala, and the importance of crop production and indigenous SWC in household livelihoods. This relationship is primarily based on economic distance. More in particular, the effects of labour-opportunity costs play a role in this (*cf.* section 2.1.2). This first group of theories postulates that iSWC importance is likely to be higher at a greater distance to the town of Kassala, and lower at a smaller distance to this town. A second group of theories postulates that urban-rural relations are also controlled by characteristics of the urban centre, or town, itself (*cf.* section 2.1.2). These characteristics generate relations where, on balance, the net transfer of wealth is either towards the urban centre (parasitic relations), or towards the rural areas (generative relations). Finally, there is an overlap, so-to-speak, in the reach of both groups of theories for the situation near the urban centres. In this particular situation, the first group would postulate a generally lower importance of crop production and iSWC. This is also the case for the second group of theories, provided that urban-rural relations are of a parasitic type. In the case of prevailing generative relations, also a higher crop production and iSWC importance would be expected in this second group of theories. This means that, according to our research framework, we can assess and compare scores of the dependent phenomena (L&E research variables indicating crop production and iSWC importance) for situations at a greater and smaller distance to urban centres in the Border Area. However, findings near the urban centre in the end cannot be explained from either group of theories. Suffice to say that identification of the prevailing pattern in the Border Area, as we attempt here, may already contribute to improve policy-making in the field of local SWC interventions.

Until sofar, we implicitly assumed that the Border Area could be regarded as a uniform landscape in environmental and socio-economic terms. This is not true. Two factors were discussed in section 6.6.1 to possibly interfere with the outcomes in the relationships discussed above. These are the local gradient in aridity and the tribal-signature of Hadendowa and Beni Amer groups of the Border Area. It was discussed in this same section that what we subsequently will call the "distance-to-Kassala" pattern, in fact measures the effect of economic distance, plus the aridity-gradient, plus tribal-signature. Recent socio-economic research in the Gash delta (HVA 1993) showed that largely similar employment opportunities as found in the Kassala town also exist in

Aroma. This major town in the Gash delta had a population of 10,000 in 1990. It lodges the management of the GDAC irrigation scheme. Besides the distinct urban pull-factor in the form of Aroma town and its service apparatus, the delta also generates a considerable rural pull in the form of labour opportunities in the irrigation scheme. Part of its seasonal labour force is drawn from the neighbouring villages in the Border Area (*Ibid.* 1993, *cf.* chapter 6). Unlike access to GDAC tenancy titles, access to urban and rural jobs in the Gash region is not controlled by tribal membership.

Figure 7.10 **Research villages in the Border Area and distance to Kassala and Aroma (Gash)**

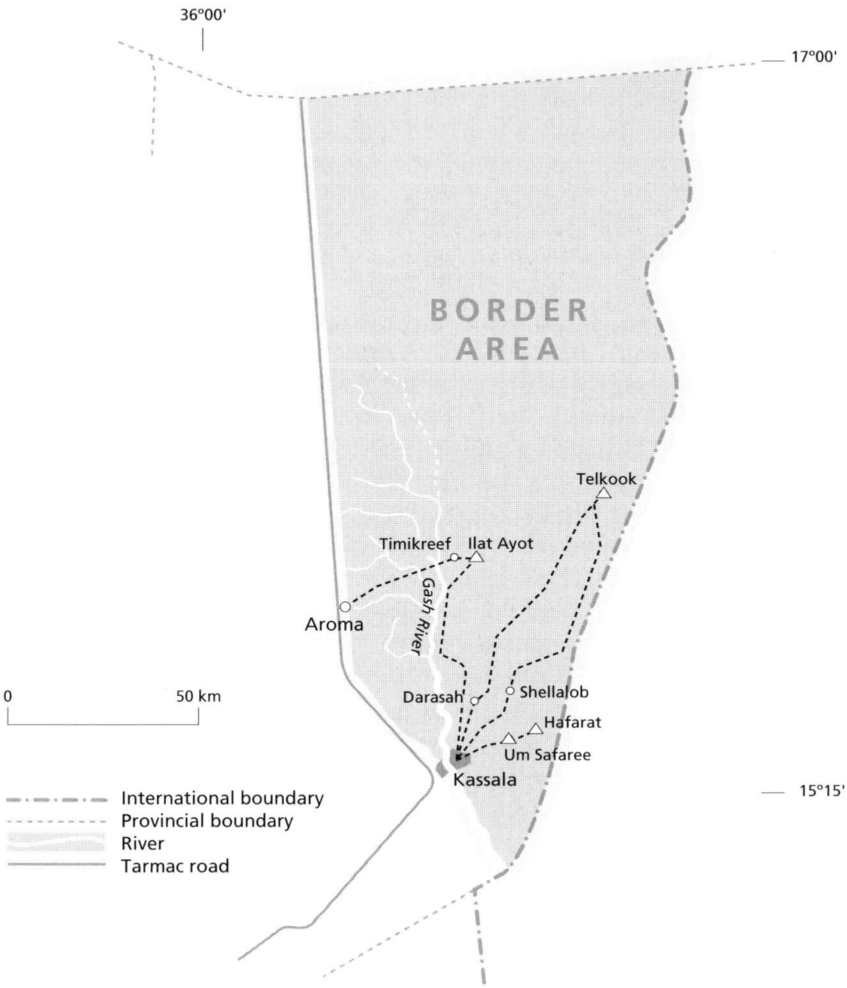

Table 7.15 **Travel effort to cover distances between the research villages in the Border Area and the towns of Kassala and Aroma (Gash delta region), in geographical distance (km) and travelling time (minutes), various routes**

	ILAT AYOT		TELKOOK		UM SAFAREE		HAFARAT	
	km	min	km	min	km	min	km	min
KASSALA town								
– shortest route	70	n-s	88	330	15	45	20	60
– longest route	—	—	115	360	—	—	—	—
Seasonally weighed		n-s		333	45		60	
AROMA town								
– shortest route	40	240	153	480	80	195	85	210
– longest route	135	330	180	510	—	—	—	—
Seasonally weighed	280		485		195		210	

Source: fieldwork and 1:100,000 topographical maps. Note: Telkook to Kassala shortest route is via Darasah, longest route is via Shellalob police station. Ilat Ayot to Aroma shortest route crosses the dry Gash riverbed, longest route is via Kassala town. Telkook to Aroma shortest route is via Darasah and Kassala, longest route is via Shellalob police station and Kassala. 60 minutes travelling time are added for a change of bus in Kassala when required. Assumed average velocity of travel by camel and donkey (route 1 Ilat Ayot to Aroma) is a conservative 10 km/hr. Travel from all villages faces impediments after khor discharges in the rainy season. The longer Telkook route by consequence must be used for 1-7 days immediately following localized discharges. The longer Ilat Ayot route must be used during consecutive periods of up to 1 month after Gash discharges. Seasonal weighing for Ilat Ayot is 7 months shortest route and 5 months longest route; ibid. for Telkook is 11 months shortest route and 1 month longest route. n-s means night-stop, — means no other route available.

An analysis of this "distance-to-Gash" pattern, accordingly, would conveniently measure the influence of economic distance and its effects of labour-opportunity costs without the intervening effects of aridity gradient and tribal-signature. For the purpose of comparing relations in this second pattern, we consider for practical reasons the distance between the research villages in the Border Area, and the town of Aroma which represents the Gash region.[17] At first sight, the best combination would be to compare the scores of the research variables of the L&E Survey in Ilat Ayot and Um Safaree (small distance to Aroma and Gash) with those in Telkook and Hafarat (great distance to Aroma and Gash). However, travelling effort in the Border Area is mainly determined by time instead of distance. According to table 7.15 and figure 7.10, a comparison of scores in Telkook (great distance to Aroma and Gash), with those in Ilat Ayot, Um Safaree and Hafarat combined (small distance to Aroma and Gash) would be the best choice on grounds of travelling time. This is also the combination we will use in the next section. Hafarat, however, is left out of the group of three to obtain more comparable numbers of households, while tribal diversity over groups at greater and smaller distance is at the same time maintained.[18]

---

[17] Commonly, all travel to rural employment opportunities in the Gash delta passes through Aroma town.

[18] The alternative combinations of villages have also been checked. Ilat Ayot, Um Safaree and Hafarat together as a group at a small distance to Aroma (Gash) produce a pattern of signs exactly similar to that presented in table 7.16 (although no significant relationship can be found for % indigenous SWC in crop production income). Ilat Ayot and Hafarat together only produce one different sign (% crop production in household income for 1983 is negative instead of positive), and one additional significant

The importance of crop production and indigenous SWC is assessed in terms of household income, allocation of labour time, allocation of cultivated land, and land user perception (*cf.* appendix 2.1). Table 7.16 presents the scores together in the distance-to-Kassala and distance-to-Gash patterns. These are based on the mean ranks of households in the different groups of villages compared. The households who participate in government SWC projects have been excluded from these calculations to avoid possible disturbing effects (the studied sub-group is therefore nopSWC households). A positive (+) sign in table 7.16 indicates a positive relationship between distance and the variable, a negative sign (-) indicates a negative relationship. All relationships are tested at a 95 %-significance level (two-tailed, $\alpha=0.05$) in the Mann-Whitney U-test. The table shows besides the sign, also the observed probability for relationships with $P\leq0.05$. However, this value of P cannot be used to evaluate the outcomes of specific variables in terms of the strength of relationships they represent.[19] The mean rank scores behind these signs and the number of valid cases are separately listed in appendix 7.2.

7.2.1 The patterns of distance-to-Kassala and distance-to-Gash

The distance-to-Kassala pattern represents effects of economic distance and also includes the effects of an aridity-gradient and tribal-signature. The scores in table 7.16 show that the contributions of crop production income to total household livelihood income tend to be higher at a greater distance to Kassala town (positive signs for CROPLIV8, CROPLIV3). However, these generally contain a smaller iSWC component (negative signs for SWCCROP8, SWCCROP3, SWCLIVE8). An exception is the situation in 1983 for the iSWC contribution to the total household livelihood income (positive sign SWCLIVE3). It is shown in the same way that allocations of labour time to iSWC (SWCTIME8, SWCTIME3) and iSWC land shares in the total household entitlements to cultivated land of local and non-local lands combined (SWCTOTA8, SWCTOTA3), all are smaller at a greater distance to Kassala. However, the latter shares in the household entitlements to local lands are greater (SWCBORD8, SWCBORD3). The scores of Potential crop production time *C*, which indicate the potential household labour availability for crop production (POTTIME8, POTTIME3), are also higher at a greater distance to Kassala. The priority rankings of perceived iSWC importance (SWCRANK8, SWCRANK3) and crop production importance (CROPRAN8, CROPRAN3), are all smaller at a greater distance to Kassala town. Statistically significant relationships are listed separately in the table with their value of P. These are for the distance-to-Kassala pattern in all but one case (CROPLIV3) of a negative sign. This outcome of prevailing negative significant relationships is precisely opposite to what we were expecting on the basis of distance and labour-opportunity cost relations alone (*cf.* section 2.1.2). Given these outcomes, we reject H_0 in favour of the alternative hypothesis H_1 which states that scores have different

relationship (% SWC-time S in total labour time in 1988; P=.0404).

[19] P also depends on the sample size, which is not constant for the different variables and research years used.

averages for households in villages located at either smaller or greater distance to Kassala, when P≤0.05. This is the case for the following relationships:[20]

The dimension Income

% iSWC in total household crop production income for 1988 and 1983; where the importance of iSWC income contributions to household crop production incomes is smaller among households in villages located at a greater distance to the town of Kassala.

% Crop production in total household livelihood income for 1983; where the importance of crop production income contributions to household livelihood incomes is greater among households in villages located at a greater distance to the town of Kassala.

% iSWC income in total household livelihood income for 1988; where the importance of iSWC income contributions to household livelihood incomes is smaller among the households in villages located at a greater distance to the town of Kassala.

The dimension Land

% iSWC land in total household land entitlement for 1983; where the importance of iSWC land in local and non-local cultivated lands of the households combined is smaller in villages located at a greater distance to the town of Kassala.

The dimension Perception

Rank iSWC priority in crop production for 1988 and 1983; where the perceived importance by land users of iSWC in the crop production entreprise is smaller among households in villages located at a greater distance to the town of Kassala. However, only four valid observations in Hafarat contributed to the analysis (*cf.* section 6.5.5).

Rank crop production priority in household livelihood for 1988 and 1983; where the perceived importance by land users of crop production in household livelihood is smaller among households in villages located at a greater distance to the town of Kassala.

---

[20] No rejection of H_0 for other relationships does not mean that the null-hypothesis, which states that equal averages can be found in villages at either small or great distance to Kassala, is true. It only means that these can not be rejected as being quite unlikely.

Table 7.16 **Direction of relationship (sign) and observed significance levels (P) in Mann-Whitney U test (only P≤.05 is shown) for selected variables by factors distance to Kassala town and distance to Aroma town (Gash). "+" indicates associations of high mean rank scores and great distances to Kassala and Aroma (Gash). "-" indicates associations of low mean rank scores and great distances to Kassala and Aroma (Gash). Data are at the household level for nopSWC households, 1983 and 1988**

	DISTANCE-TO-KASSALA		DISTANCE-TO-GASH	
INCOME	sign	P	sign	P
% iSWC in total household crop production income				
– 1988 SWCCROP8	–	.0002	+	.0465
– 1983 SWCCROP3	–	.0009	+	
% Crop production in total household livelihood income				
– 1988 CROPLIV8	+		–	
– 1983 CROPLIV3	+	.0198	+	
% iSWC in total household livelihood income				
– 1988 SWCLIVE8	–	.0229	–	
– 1983 SWCLIVE3	+		+	
TIME				
% iSWC-time S in total household labour time				
– 1988 SWCTIME8	–		+	
– 1983 SWCTIME3	–		+	
Score C for the potential household labour time available [A]				
– 1988 POTTIME8	+		+	
– 1983 POTTIME3	+		–	
LAND				
% iSWC land in total household local land entitlement				
– 1988 SWCBORD8	+		+	.0104
– 1983 SWCBORD3	+		+	
% iSWC land in total household land entitlement				
– 1988 SWCTOTA8	–		+	.0026
– 1983 SWCTOTA3	–	.0436	+	
PERCEPTION [B]				
Rank iSWC priority in crop production				
– 1988 SWCRANK8	–	.0087 [C]	+	.0253
– 1983 SWCRANK3	–	.0258 [C]	+	.0189
Rank crop production priority in household livelihood				
– 1988 CROPRAN8	–	.0039	–	
– 1983 CROPRAN3	–	.0005	–	

Source: L&E Survey. Note: distance-to-Kassala compares scores in Telkook and Ilat Ayot (great distance to Kassala) with scores in Um Safaree and Hafarat (small distance to Kassala). Distance-to-Gash compares scores in Telkook (great distance to Gash) with scores in Ilat Ayot and Um Safaree (small distance to Kassala). [A] Over the number of economically active members in the household. [B] High mean rank scores for variables in the dimension Perception indicate low importance, contrary to other variables listed. Their sign in this table is chosen such, that associations of high perception importance (low mean ranks) and great distance to Kassala and Aroma (Gash) are represented by positive signs. Appendix 7.2 lists the mean rank scores behind the signs. [C] Only 4 households in Hafarat.

The second pattern of distance-to-Gash relationships is not importantly disturbed by the intervening effects of the aridity-gradient and tribal-signature. These outcomes indicate that the importance of indigenous SWC and crop production is only in a few cases smaller at a greater distance to the town of Aroma (Gash) (negative signs for CROPLIV8, SWCLIVE8, POTTIME3, CROPRAN8, CROPRAN3). However, the greater part of these variables show a higher score at a greater distance to the town of Aroma (Gash) (positive signs for SWCCROP8, SWCCROP3, CROPLIV3, SWCLIVE3, SWCTIME8, SWCTIME3, POTTIME8, SWCBORD8, SWCBORD3, SWCTOTA8, SWCTOTA3, SWCRANK8, SWCRANK3). All significant relationships in this distance-to-Gash pattern are also of a positive sign. This outcome, in turn, is in agreement with the relationship between indigenous SWC importance and distance we were expecting (*cf.* section 2.1.2).

The main differences with the first distance-to-Kassala pattern include: (i) a more frequent change of sign of variables when scores for the situations in 1988 and 1983 are compared (3 changes in sign against 1). This is likely to be due to seasonality between years. The effect is therefore more pronounced in the Gash region economy with an important rural component, than in the Kassala urban economy of the first pattern; (ii) The number of significant relationships is smaller. These also occur more often for the situation in 1988 than for 1983 (5 relationships for SWCCROP8, SWCBORD8, SWCTOTA8, SWCRANK8 and SWCRANK3, against 9 in the distance-to-Kassala pattern). Given these outcomes, we reject H_0 in favour of the alternative hypothesis H_1 which states that scores have different averages for households in villages located at either smaller or greater distance to Aroma (Gash), when $P \leq 0.05$. This is the case for the following relationships:

The dimension Income

% iSWC in total household crop production income for 1988; where the importance of iSWC income contributions to household crop production incomes is greater among households in villages located at a greater distance to the town of Aroma (Gash).

The dimension Land

% iSWC land in the total household local land entitlement for 1988; where the importance of iSWC land shares in local lands of households is greater in villages located at a greater distance to the town of Aroma (Gash).

% iSWC land in the total household land entitlement for 1988; where the importance of iSWC land shares in the total of local and non-local lands combined of households is greater in villages located at a greater distance to the town of Aroma (Gash).

The dimension Perception

Rank iSWC priority in crop production for 1983 and 1988; where the perceived importance by land users of iSWC in the crop production entreprise is greater among households in villages located at a greater distance to the town of Aroma (Gash).

7.2.2 Effects of aridity and tribal characteristics

We finally consider table 7.16 again to see if we can identify any of the assumed effects of the aridity-gradient and tribal-signature. We discuss for this purpose the significant relations in the affected distance-to-Kassala pattern only. The pattern shows that the contributions of iSWC to the household crop production income are smaller at a greater distance to Kassala, which is at higher and slightly dryer latitudes (negative signs for SWCCROP8, SWCCROP3). Aridity is most likely to have influenced this outcome. The production levels under iSWC are normally highly responsive to differences in the annual supply of rainfall as exist between 1983 and 1988. The failure of these differences to appear is likely to indicate that these areas at a great distance to Kassala are located at the environmental margins with respect to water. The high contribution of crop production incomes in 1983 at the same greater distance to Kassala (positive sign for CROPLIVE3) is partly based on production in the GDAC irrigation scheme, and is therefore not in contradiction with this. Finally, another indication of the role of aridity is that at a greater distance to Kassala, the mentioned lower iSWC income shares (negative signs for SWCCROP8, SWCCROP3) have also been obtained precisely with higher allocations of iSWC land (positive signs for SWCBORD8, SWCBORD3) (table 7.16).

The following observations are in support of the existence of a tribal-signature effect. The use of iSWC generally requires the building and maintenance of structures to control run-off. These activities are, in eastern Sudan, associated with longer sedentary habitats and lifestyles (*cf.* section 4.3.4). The four research villages in the Border Area differ in this respect along lines of respectively Hadendowa and Beni Amer settlements (*cf.* section 6.6.1). The smaller iSWC at a greater distance to Kassala in the Hadendowa villages of Ilat Ayot and Telkook, may therefore also be explained from a different phasing in sedentarization histories.[21] This smaller iSWC importance, furthermore, may also relate to the Hadendowa preferential access to tenancies in the GDAC irrigation scheme and non-local lands on the Gash margin. Cultivation in the Gash region is later than in the Border Area (*cf.* section 5.2.1 *Domains and techniques*), but still some competition is likely to occur. The latter lands in the Border Area are economically speaking usually less competitive than the first (*cf.* appendix 6.1), resulting in smaller iSWC importance.

---

[21] We discussed in section 6.6.1 that these differences only exist with respect to village settlement dates, not household settlement duration in these villages.

Finally, all perception variables in the distance-to-Kassala pattern show significant relationships of a negative sign (SWCRANK8, SWCRANK3, CROPRAN8, CROPRAN3). This indicates that consistently lower priority is given to iSWC and crop production at a greater distance to the town. We argue that it is more plausible to interpret these subjective outcomes as largely being influenced by factors which are directly perceived and continuously experienced by land users, such as aridity and tribal characteristics, rather than by more abstract effects of distance.

7.2.3 Concise presentation of findings

Two different sets of relationships between distance and indigenous SWC and crop production importance were compared for the Border Area in table 7.16. The greater number of significant relationships in the first pattern is of a negative sign (8 negative against 1 positive). This means that the importance of indigenous SWC and crop production is smaller at a greater distance to the town of Kassala. We interpret this to be the result of economic distance and its associated effects of labour-opportunity costs. However, the intervening effects of aridity and tribal characteristics also exercise their influence in this first pattern. All significant relationships in the second pattern have a positive sign. The effects of distance and labour-opportunity costs are less distorted. However, the best possible combination of villages used still has shortcomings due to different sample sizes and incomplete control over effects of tribal-signature. (The optimal combination would have been to compare the group Ilat Ayot and Um Safaree with the group Telkook and Hafarat. However, this would not have been in accordance with considerations of travelling distance in terms of time).

We finally conclude that (i) the importance of indigenous SWC and crop production in household livelihoods can be assessed by using different types of operational variables; (ii) that for the case of the Border Area, we cannot determine the precise effect exercised by the factor distance to the regional urban centre. This is due to an excessive large variation in other environmental and socio-economic variables; (iii) that it is more likely for three reasons that in the Border Area the positive relationship between distance, on the one hand, and importance of indigenous SWC and crop production, on the other, prevails. The first reason is that this is the outcome of the distance-to-Gash pattern which is the least influenced by known intervening factors. The second reason is that a positive relationship can also be better understood from the second group of theories of urban-rural relationships. In the Sudan, towns with parasitic traits prevail (*cf.* section 2.1.2).[22] This results in a low iSWC importance near the urban centre (*viz.* a positive relationship between iSWC importance and the factor distance). The third reason is that in the only other regional reference known to us which addresses this matter, this distance relationship

[22] There are no studies made on this for Kassala town. The relatively high sectoral average incomes in Um Safaree close to Kassala (*cf.* figure 6.15) only point to a high level of urban-rural interaction. However, it provides no evidence of the type of relationship because the level of benefits for Kassala town remains unknown.

is of a negative sign. This particular study, which discusses the relationship between distance to urban centres and crop production occurrence in general, was made in the desert climate area of the Red Sea province (ERGO 1990, appendix 4, tables 4.56a, 4.56b). It gives us reason to believe that it is in fact aridity which is an important factor behind the same negative signs in the distance-to-Kassala pattern. This, in turn, supports our notion that positive relationships better represent the situation in the greater part of the Border Area, which has only a semi-arid climate.

8

Soil and water conservation: conclusions and policy considerations

8.1 The Border Area in perspective

The Border Area was depicted as a dryland region in eastern Sudan which is marginal in environmental and economic terms. The traditional mainstay of the Beja population was nomadic and transhumance livestock husbandry, until their transition to more settled lifestyles. Most communities have been cultivating some subsistence crops since at least the late 19th century. The first accounts of crop production in the Border Area date from 1902. The early techniques used were mainly of the wildflooding type. The first reference to government SWC interventions dates from 1941. This concerns a dam in the area of Um Safaree and Hafarat. Finally, indigenous SWC, such as the use of *teras* and brushwood panels, is likely to have been adopted by the Beja with their later transition to more settled lifestyles. This process mainly took place between the 1920s-1950s, while the first evidence on the application of *teras* in the Border Area dates from 1953.

The most plausible history of *teras* adoption is that its rudimentary forms were developed in the Nile region. Run-off use in *hafirs* was already known in the Nubia of the last millenium B.C. (Mokhtar 1990). These techniques later spread over the entire central Sudan. They seem to have mainly been used by riverine dwellers with a tradition in crop production, such as the communities at the time of the *Funj* Kingdom (1504-1820). Since the earliest times, the nomadic Beja frequented these regions, and must have been acquainted with SWC. Apparently, they saw little need in its use then, because their livelihoods were still predominantly livestock-oriented. The factual introduction of *teras* from the Nile region to the Kassala region outside the Border Area could well have been made by West-African pilgrims crossing the Sudan. These pilgrims settled in the immediate surroundings of Kassala town from 1900 onwards. Unlike the Beja, their livelihoods were to a greater extent oriented to crop production. Occasionally, they also used *teras* for this purpose.

In the course of the following decades, the social, economic and political positions of the Beja and West-Africans would drastically change. The second group gradually became richer, while the Beja at the same time impoverished (Kuhlman 1994,182). Before the 1950s, the Beja still would hire West-Africans to work in crop production for them. They also sub-letted their tenancy titles to them in the GDAC irrigation scheme. However, the tide was turned by a series of drought years in the late 1940s. The Beja suffered great losses of livestock, and many were forced to sedentarize. However, also several other factors of which the influences were felt both before and after the 1950s contributed to their sedentarization in the Border Area. For the situation before the 1950s, these include infrastructural development and market integration, the development of irrigation schemes, pacification and the rise of religious movements. After the 1950s, also government sedentarization policies and barriers to the free movement of livestock in the region contributed to this. Finally, also the demonstration effect of early SWC applications in the region by West-Africans, the same demonstration effect of a few early government SWC interventions in the Border Area, and the favourable wet period of the 1950s are likely to have induced the adoption of SWC techniques by the Beja locally (Van Dijk & Reij 1995). The earliest actual evidence on this is based on 1953 aerial photographs. These show lands cultivated under *teras* in the area of Um Safaree. In the meantime, most government interventions in water resources development had been made in the field of irrigation in the inland deltas of eastern Sudan. Large-scale irrigation schemes were established in the Gash and around Tokar in the early decades of the 20th century. The main crop was cotton for export. This policy focus on high potential areas continued to prevail in the independent Sudan of the 1960s and 1970s. From this period onwards, the rainfall conditions in central Sudan changed unfavourably. This resulted in generally shorter and less predictable developments of the rainy season, and mainly hit rainfed production in the low-potential areas. These developments eventually resulted in a change of government policy favouring these same low-potential areas from the early 1980s onwards. This change of policy was furthermore facilitated by the new system of regional administration which had brought Beja politicians with an electorate in these low-potential areas to high ranks in the Ministry of Agriculture. The ministry approved a programme of the Department of Soil Conservation to develop floodwater-harvesting techniques in the Border Area in 1982. This Earth Dam programme reached the height of its performance in the years following the great 1984-1985 drought. The same drought subsequently attracted international attention. It resulted, in addition to relief aid, in projects to improve crop production by means of introduced SWC techniques. The MFEP/KADA programme was the main facilitator of these interventions in the Border Area from 1987 onwards. Its starting point was to down-scale the earth dam techniques already in use. This resulted in the development of a system based on low earth embankments built on the terrain contour. Viewed against this background from the then current paradigm of Integrated Rural Development, progress has been made in changing the heavy top-down approach of the Department of Soil Conservation. However, looking back from the recent perspective of Participatory Approaches, also these interventions must now be said to have been mainly governed by a Transfer-of-Technology approach.

8.2 The effects of SWC interventions and importance of indigenous SWC

Government SWC interventions
When the livelihoods in the Border Area of the 1980s and early 1990s are examined at the aggregate village level, the outcomes of the L&E Survey indicate that the decline of livestock importance was compensated by an increasing dependence on non-livestock incomes. In Ilat Ayot and Um Safaree without government SWC intervention, these were mainly earned in off-farm employment and labour migration. In Telkook and Hafarat with government SWC intervention, these depended to a greater extent on crop production. When at this same village level the total household incomes are compared, the increase between 1983 and 1988 was the smallest in the non-intervention village Ilat Ayot (from £s 3,200 in 1983 to £s 3,780 in 1988 in constant 1983 prices, or a 18 %-increase). It takes an intermediate position in the non-intervention village Um Safaree (from £s 8,980 to £s 14,100, or a 57 % increase). The greatest increase was found in the village with the most elaborate, and internationally supported, intervention in SWC which is Hafarat (from £s 2,900 to £s 5,090, or a 76 % increase). The intervention in Telkook takes an intermediate position in this respect (from £s 2,700 to £s 4,170, or a 54 % increase). These outcomes may at first sight be interpreted to indicate a positive impact of government SWC interventions. However, this proves not to be the case when the data are examined at the level of households and landholdings (see below). Also in terms of the common criteria for project evaluation, the intervention achievements in the Border Area were shown to have been small.

In terms of *accordance with policy,* the objectives of the earth dam project in Telkook and the pilot scheme project in Hafarat tallied with policy formulations of the Sudanese and Netherlands governments. In the field, however, two different approaches were being followed. The Department of Soil Conservation was a project-implementor in Telkook, but was supposed to be project-facilitator in Hafarat. The factual outcome, in both cases, was a Transfer-of-Technology situation. This resulted in projects which could not be sustained after government and international support had ended.

The *effectiveness* of these interventions was low. No floodwater-harvesting systems have been developed which fulfill the objectives stated in the respective projects. The poor performances of the earth dam in Telkook were mainly caused by the application of an inappropriate technique. The gradual breakdown of the Sudanese "Water Organization" since the 1970s is debet to this. The Department of Soil Conservation, by consequence, came to rely on a technique from which certain vital components had to be left out for lack of experience with its implementation. This made the dams vulnerable to breach. Poor performances in the pilot scheme were caused by two main factors. Firstly, its design was inappropriate. The harshness of the environment and the complexity of local livelihoods were underestimated. Secondly, the implementation in Hedadeib was imperfect due to conflict over land use and the relocation of the project into an unfavourable zone. Both in Telkook and Hafarat, therefore, a technique was developed which proved a pale shadow of the original system. The latter might have worked, technically speaking, if it had been properly executed. This also illustrates that the two techniques were too sophisticated for implementation by the Department of Soil Conservation in the Border Area.

The *efficiency* of the interventions can not adequately be assessed due to a lack of data. Comparable earth dam projects in the Red Sea province indicate that their establishment costs are low. We know from the Border Area that the recurrent maintenance and repair demands are high. The pilot scheme showed slightly better performances in this latter respect. Maintenance and repair can be reduced by partly relying on indigenous techniques. However, the pilot scheme's establishment costs (including research, monitoring and evaluation) were also six times higher than of the earth dams projects. The financial returns to pilot scheme participants were attractive partly because of subsidization. However, the production level required to break even when all inputs were priced was usually not attained by them.

The *impact on household economic positions* was generally low in Telkook. A high positive impact could only be demonstrated in Hafarat for the wet year 1988, but not for any of the dryer years. Several improved cultivation methods were taken up by land users in Telkook and Hafarat as a result of the intervention. The interventions had no influence on the access of households to cultivated lands.

The *impact on the physical and socio-economic environment* included negative effects of increased erosion in Telkook and Hafarat. It also included deprivation of downstream land users, which was only reported in Hafarat. Equally, only in Hafarat were counter measures taken in the form of erosion control. The negative effect on downstream land users was not addressed here, but was evaded by changing the location of the project. Generally speaking, three interrelated external factors have also exercised their influence on the SWC intervention outcomes in the Border Area of the 1980s and 1990s. These include a change in household livelihoods as a result of the 1984-1985 drought, the occurrence of a second series of very dry years in 1990 and 1991, and local organization based on religious movements. The most influential was the movement of Ali Betai. This group succeeded in securing new tenancy titles for Hadendowa households in the GDAC irrigation scheme.

The *project-sustainability* of interventions in Telkook and Hafarat was low. The earth dam and pilot scheme no longer function after the termination of respectively government and international support for construction and maintenance. However, in the case of Hafarat this was not a result of the withdrawal of international support, but of a policy choice of the Department of Soil Conservation. On grounds of costs, however, the project-sustainability of the pilot scheme in Hafarat must be considered as low.

The *potential of replication* by the households of Hafarat of the pilot scheme technique is expected to be low. The same potential of replication by the Department of Soil Conservation of the project's institution-building and training component proved favourable.

Project particulars
(i) The outcomes of the L&E Survey indicate that the participants in the pilot scheme, and to a lesser extent in the earth dam project, are characterized by a greater household wealth than non-participants. The differences are not significant, however; (ii) the developments in selected crop production characteristics over the years suggest that project SWC, indigenous SWC and non-SWC techniques may be ruled by comparative advantages for agro-climatologically different types of years; (iii) the simultaneous involvement of households in project SWC and indigenous SWC in Telkook and Hafarat showed a few small mutual effects. Mechanized land preparation in the pilot scheme freed household labour. This could be allocated to local crop production under other techniques, including indigenous SWC, and to other livelihood activities (no such land preparation was carried out in the earth dam project in Telkook). This is a negative effect from the narrow viewpoint of SWC intervention, for the benefits did not accrue to the project. Tractor-ploughing, in addition, acted as an improper incentive to raise popular participation. This working method furthermore strengthened the conception among the households in Hafarat that interventions were made for them, and not with them.

The importance of indigenous SWC
A first approximation of the distribution of indigenous SWC in the Border Area was made by examining satellite imageries in a GIS. A total area of about 1,600 ha was identified as likely to be under *teras*. This is 0.04-1.05 % of the total area covered by the imageries for the northern and southern part of the Border Area respectively. The outcomes of the L&E Survey for the 1980s and early 1990s indicate that, generally speaking, the indigenous techniques of *teras* and brushwood panels were important in local crop production. However, this crop production sector itself was only of modest importance to the household livelihoods in the Border Area. These livelihoods can be typified, according to definitions based on these same sectoral compositions of the household economies, as either agro-pastoral, agricultural, or mixed. For the purpose of a more general reference, we speak of multi-resources economies. When the survey outcomes of all four research villages in the Border Area together are considered for 1983 and 1988, indigenous SWC contributed an average 74 %, respectively 77 %, to the total household crop production incomes in these years (based on the sub-group iSWC households). Crop production income contributed an average 56 %, respectively 51 %, to the grand total household incomes in these years (sub-group all households). Finally, were the indigenous SWC incomes to be expressed as shares of the latter grand total household incomes directly, these figures would be an average 31 % for 1983, and 25 % for 1988 (sub-group iSWC households). The share of labour time allocated to indigenous SWC (for selected land preparation, cultivation activities and months of the growing season) was an average 87 %,

respectively 89 %, of total labour time allocated to these same activities for the household crop production sector as a whole. (The value of another measure used Potential crop production time C has no meaning in the absolute sense discussed here). The share of cultivated lands allocated to indigenous SWC by the households in the four research villages, was an average 78 % of the total household entitlement to the local lands. This was 73 % when the same figure is expressed as a share in the total household entitlement to the local and non-local lands combined. Finally, the perceived importance of indigenous SWC in crop production for the situations in 1983 and 1988 was given an average rank of 1.32, respectively 1.27, on a list of five cultivation techniques. The importance of crop production in the household livelihood was similarly given an average rank of 1.62 and 1.41 over five livelihood categories in 1983 and 1988 respectively.

When the pattern of these scores over the individual research villages is considered, the survey outcomes indicate that the importance of indigenous SWC and crop production is usually lower at a greater distance to the town of Kassala. However, the expected effects of economic distance and labour-opportunity costs in this could not unequivocally be assessed. Also a gradient in the level of aridity, and effects of certain tribal characteristics influence these outcomes. They, moreover, do so in precisely the same direction as the distance factor does. When, alternatively, the effect of distance to the town of Aroma in the Gash region is examined to control these intervening effects to the best possible extent, precisely the opposite pattern of relationships arises. The importance of indigenous SWC and crop production, accordingly, is usually higher at a greater distance to the urban area of Aroma. It was argued that this latter positive relationship between, on the one hand, importance of indigenous SWC and crop production, and distance to the regional urban centre on the other, better represents the effects of economic distance and labour-opportunity costs for the case of the Border Area.

The crop production income return to labour per man-hour invested, as was illustrated in figures 7.4 and 7.8, shows that government SWC projects are attractive in this respect in the wet year, while indigenous SWC is more attractive in the dryer years. These figures were based on labour allocations to a range of selected land preparation and cultivation activities. Unfortunately, we do not have at our disposal the data to compare these outcomes with true returns to crop production labour, and engagements in other livelihood activities of households in the Border Area. However, a rough estimate can still be made by using a list of average figures presented in Bokkers & Dabloub (1986), De Leeuw (1987) and Ausenda (1987). Table 8.1 is based on these average figures and L&E Survey data. The data, firstly, show the effect of inflation and deteriorating incomes in constant prices. They, secondly, illustrate that indigenous SWC is a relatively unprofitable option when judged by its returns to labour alone. This is not only the case when indigenous SWC returns are compared with returns from other cultivation techniques, but partially also when compared with returns from other livelihood activities. Among these, labour migration and local off-farm employment are the most rewarding.

Table 8.1 **Household returns to labour on a daily and monthly basis (in constant 1983 £s) for the main livelihood engagements in the Border Area (Ilat Ayot, Telkook, Um Safaree, Hafarat), 1983 and 1988, (indicative figures are between brackets)**

	INCOME DAILY BASIS		INCOME MONTHLY BASIS	
	1983	1988	1983	1988
CROP PRODUCTION [A]				
– Indigenous SWC	1.2	2.4		
– Project SWC	na	3.7-5.7		
– Av. local techniques	3.9	4.8		
– Av. non-local techniques	7.2	6.1		
LIVESTOCK (indicative) [B]				
– Husbandry own herd	(42)	(23)		
LIVESTOCK-RELATED [C]				
– Selling produce	0.5-1	3.8-13		
– Animal rent	6.3-7.5			
LABOUR MIGRATION [C]				
– Kassala private sector	3-25	5-19		
– Kassala public sector		3.1		
– Gash delta region			30-250	25-125
– New Halfa irrigation scheme			20-150	95-190
– Port Sudan urban employment				190
– Contract-herding eastern Sudan region			150-200	65-220
LOCAL OFF-FARM EMPLOYMENT [C]				
– Water collection		4.4-6.3		
– Wood-cutting	8-48	0.8-18.8		
– Charcoal making			350	155
– Shop	4-6.5	5-9.4		
NETWORKING (indicative) [D]				
– Gifts	(2.2)	(1.3)		

Source: L&E Survey, Bokkers & Dabloub (1986), De Leeuw (1987), Ausenda (1987). Note: na means not applicable. [A] Based on household average scores on L&E Survey variables. The returns are calculated by multiplying the average grain and stalk production with the current prices of produce (cf. table 3.4) and dividing the outcome by the cultivated area multiplied by the average labour demands per area (cf. table 5.5 and Ausenda 1987,140 for the Gash delta). Labour demands of project SWC are based on wildflooding techniques using 1 and 2 weeding rounds in Hafarat and Telkook respectively, taking account of fringe benefits from mechanized land preparation in the first village. [B] Based on household average scores on L&E Survey variables of livestock wealth. The returns are calculated in its simplest form by multiplying the household livestock wealth with the current livestock prices (cf. table 3.4) which outcome is divided by the number of days in the year, 365. This is an indicative figure only. [C] Ranges are taken from the L&E Survey data. Labour migration incomes from the Gash, New Halfa and Port Sudan, from contract-herding, and from charcoal-making only exist on a monthly basis. [D] Based on household average scores on L&E Survey variables. Available data of annual totals are expressed as daily income, but normally receipts are weekly or bi-monthly. This is an indicative figure only.

The indicative returns from livestock husbandry, in the way defined here, are also relatively high.[1] The figures in table 8.1 agree in broad lines with those collected for the Kassala area by Kuhlman. He presents daily labour returns (in current 1987 prices) of £s 1-4 for traditional cultivation, £s 5-6 for wood-cutting in rural areas, £s 6-7 for urban employment, mechanized farming and horticulture, and daily returns of £s 15-25 for artisans who possess their own tools and equipment (*Ibid.* 1994,255-257).

A challenging question raised by Kuhlman for the Kassala region, and also by Little for dryland Kenya, is why households under marginal conditions continue to pursue crop production. The answer of Kuhlman is that they have no alternative (*Ibid.* 1990,91). Little refers to the unreliability of grain markets, the favourable interactions between crop production and livestock sectors, and the common ambiguity surrounding pastoral rights to land in general (*Ibid.* 1992,98-104). We believe that the latter arguments also apply to the situation in the Border Area of the 1980s and early 1990s. The unpredictable development of grain prices, for example, was illustrated for the situation in 1990 in the figures of section 7.1.4. The possibility of temporarily relying on even a small stock of subsistence grain clearly is a bonus in such events. The interactions between crop production and livestock in Little's study refer to the building up of grain funds to purchase or barter additional livestock, the opportunities of reducing the need to sell or barter own animals for household subsistence needs, and the opportunities of avoiding sales of livestock when markets are devaluated as a result of high supply. The latter is a typical situation during droughts. No data have been collected in the L&E Survey on these particular matters for the Border Area. However, such motives have at various occasions been raised by the respondents during our interviews. Finally, encroachment on pastoral lands is common in the Border Area region (*cf.* section 4.3.3). Seasonal land preparation, for example, in the form of raising *teras* bunds also serves in the area the particular purpose of confirming entitlement to (cultivated) land (Van Dijk & Ahmed 1993).

We believe that for the case of the Border Area the following additional motives to proceed with cultivation under marginal conditions are equally important. Crop production is usually considered to add an extra category to the range of activities simultaneously pursued to sustain the livelihood. The wider and more diverse this range, the better are the opportunities to be successful in any one of them. The livelihood risk, in other words, is being spread. Individual activities, like crop production, as part of such behaviour can be tolerated even if they provide low returns. This is because the sum of the contributions of all activities of individual household members together is what counts. This sum is usually sufficient to sustain the livelihood in normal years, and to survive in dry years. It was also discussed that indigenous SWC in the Border Area is likely to have comparative advantages over other techniques in these dry years. The role of these indigenous

---

[1] *Cf.* note in table 8.1. Other calculations of income returns from livestock husbandry can be made by specifying all sales of livestock and produce (*cf.* Abdullahi 1990,98-105). Another possibility is to consider the average herd increase from growth and reproduction, and multiply the outcome by livestock prices (*cf.* Little 1992,95). Either approach remains indicative, however.

techniques in household survival may therefore be of particular importance. This collection of motives is sometimes referred to as the strategy of "livelihood diversification" (Reitsma *et al.* 1992). This same strategy is also pursued at other levels of scale. Rainfall is highly variable over time and place in the Border Area. Cultivation is therefore more likely to provide a return when the range of locations of individual landholdings is as wide and diverse as possible. This is achieved by dispersing them geographically. Finally, the same diversification strategy is also reflected in indigenous procedures of land allocation in the Border Area. Land users have access to different farming zones. Within these farming zones, they cultivate different landholdings, at different positions in the direction of the slope and across the width of the khor bed (*cf.* section 3.2 *Beja land and water resources access*).

Diversification and risk-spreading behaviour in land use were studied from the data of the L&E Survey by comparing the locational characteristics of landholdings with patterns of allocation of labour time. Risk-spreading behaviour was defined for this purpose as the occurrence of maximum allocations of household labour time to landholdings which are sub-optimal. The latter were defined according to subjective characteristics based on standards freely selected by the land users themselves, and objective ones based on levels of grain production at the individual landholdings. The findings indicate, firstly, that a difference exists between the research villages in the number of landholdings available per household, and hence in risk-spreading opportunities. This seems to be associated with the age of the village. The greatest number of landholdings per household (2.8) and the highest occurrence of risk-spreading behaviour (36-29 % according to subjective and objective standards respectively), was found in the oldest village Hafarat. Alternatively, the smallest number of landholdings (1.3) and the lowest occurrence of risk-spreading behaviour (8-11 % respectively), was found in the more recent Telkook. There are obvious limitations to the strategy of livelihood diversification. Its successful application entails the continuous management of a collection of different landholdings, the seasonal preparation of part of these landholdings for cultivation, and finally the harvesting of only those which in the end proved to have received sufficient rainfall and run-off. The chief limiting factor is availability of labour. The significance of this strategy is probably to be appreciated in the light of the willingness of households to incur costs in terms of labour opportunities and income foregone, in order to buy a greater overall household subsistence security. The factors in this decision-making process are not exclusively economic in character in the way we have mainly described them in this study. These may equally cover any of the non-financial social, religious and political costs.

This diversification behaviour, together with the adoption of selective mixes of subsistence and market orientations, has been identified as one of the outstanding operational principles of livelihood strategies in dryland regions (Dietz *et al.* 1992). In terms of the typology of livelihood strategies proposed in this study, the application of indigenous SWC use in the Border Area can now be labelled as a short-term preservation or "recovery strategy". Its general goal is "[..] the solving of seasonal stress, adaption to sudden changes and survival of crises [..]". Finally, the strategy also aims at self-rescue.

The application of indigenous SWC can also be labelled as a long-term strategy of "structural improvement". The general objective of this strategy is "[..] the amelioration of manoeuvring networks, and the improvement and accumulation of activities and resources [..]" (*Ibid.* 1992,41-42). The small contributions of individual categories of livelihood activities, including indigenous SWC, as were found in this study for the Border Area should therefore not be regarded as unimportant to subsistence. On the contrary, these should be appreciated on their merits of being part of a greater strategy of livelihood diversification.

The dynamics of physical aspects in dryland environments have now been relatively well-documented (*cf.* Hall *et al.* 1979). However, similar dynamics of socio-economic aspects in these areas, which are important to understand how people survive in such harsh lands, have been much less documented. These dynamics, furthermore, have important research implications. Firstly, data collected at a high level of aggregation, such as the village level as was shown above, are likely to provide misleading information. These hide the caleidoscope of activities which are part and parcel of the diversification strategies. Even when the household is chosen as lowest level for study, these dynamics can only be appreciated when due consideration is given to its individual members and their activities at different locations. Several research techniques of "Rapid Rural Appraisal" frequently applied under Participatory Approaches are inappropriate when they fail to register this diversity.[2]

8.3 The role of technical and organizational constraints

The Border Area was found to show generally better performances than other dryland regions in SSA with respect to a set of common constraints to crop production and government SWC intervention.

Water availability

The most important areas of SWC application are located in the region of 250-500 mm annual rainfall. Water conservation is usually not essential in areas receiving above 500 mm, while the supply of water is generally insufficient for crop production in areas below 250 mm. *Teras* SWC is still practised under 150 mm rainfall conditions in the Border Area, however. The following physical conditions facilitate this by contributing to high levels of run-off despite low levels of annual rainfall. (i) The extreme northern part of the Border Area is already under influence of the Red Sea Desert climate regime. This generates small amounts of winter rainfall precisely at a time when the losses through evaporation are low. Water use is therefore highly efficient; (ii) the Border Area is flanked by foothills which in Eritrea reach heights of 1,500 m above local ground level. These mountain ranges reinforce the typical mechanisms of a dryland climate of localized, high-

[2] For example, methods of direct observation, group walks, group discussions, a study of landscape-transects and interviewing key-informants.

intensity and short-duration rainfall; (iii) water supply, particularly to landholdings under *teras*, is greatly increased by certain soil-physical properties. Micro-lamination, surface-crusting and surface-sealing by micaceous minerals contribute in undisturbed soils to high volumes of surface run-off and low rates of infiltration. These are favourable conditions for catchment areas of the *teras*; (iv) these foothills, in addition to the ground level collection areas, also provide supplementary natural rock catchments. In the smaller ones, up to 60 % of the captured rainfall may eventually become available on the cultivated lands as floodwater; (v) this can be tapped on the alluvial flats of the braided drainage system which is typical for the area. The predominant light-clay to loamy soil textures facilitate a rapid infiltration, especially when the surface crust is broken by land preparation. The resulting levels of soil moisture are favourable to sustain crop production under local conditions. *Teras* SWC can, for all these reasons, be used in the Border Area under lower total rainfall than is normally the case in SSA.

Soil nutrient status

The soil nutrient balance is a key concept for the study of nutrient flows in land use systems. Nutrient depletion is the situation where, on balance, more nutrients leave the system than return into its reserves. This depletion is an important causal factor in dryland degradation. Certain types of SWC not only harvest rainwater and floodwater, but also nutrients stored in animal droppings, sediment and debris which are all captured with this water. The extent to which this gain and other imports can compensate the loss of nutrients by removal of plant material (harvesting) and soil erosion remains subject of debate. Typical subsistence land uses in SSA proved to be vigorous soil miners. This process is usually not accounted for by land users. If this were to be priced as a positive income from negative "environmental costs", as Van der Pol (1992) did in terms of costs of purchasing equivalent amounts of chemical fertilizer to compensate this loss, about 40 % of the farmer income would be derived from "soil mining". Smaling (1993) concluded that under current systems of land use in SSA, virtually all soils are being exhausted in this way. A case study made in Ilat Ayot, however, indicated that the soil nutrient status is more favourable with *teras*, than without *teras*. This is therefore a remarkable finding (*cf.* Niemeijer 1993). The merits of the *teras* are based on a favourable combination of the technique itself and the local environment. In central Sudan of the flat clayplains, where the environment is less dynamic in terms of continued sediment deposition than the Border Area, soil exhaustion also takes place when *teras* is applied (El-Dishouni 1989).

Labour availability

The labour demands of SWC in a world-wide inventory range from an annual 10 to 1,800 man-days per hectare. The average is some 300 man-days (*cf.* section 1.3). Only about 6-16 man-days are required per year, per hectare, for *teras* construction in the Border Area. Annually, between 3-18 man-days per hectare would be required for average seasonal maintenance and repair work. These ranges are largely set by the working period and local soil characteristics. The highest demands apply to building in the dry season, and on heavier clay soils. The lowest, alternatively, when building is during the wet season on the lighter soils. However, the labour demands are occasionally higher in the northern part of the Border Area. Brushwood cores are used here in *teras* bunds to trap windblown sand. A maximum of 60 man-days should in this case be added to the labour demands for construction per hectare. The bulk of this labour results from travel to areas where wood-cutting is allowed under tribal agreement. These figures, in any case, are low when compared with those in the inventory. Indigenous SWC, including *teras*, has low returns to labour in absolute terms and in relative terms when this is compared with other techniques (*cf.* table 8.1). However, it was found to produce in the Border Area among the highest returns per man-hour invested in dry years.

Support institutions for SWC interventions
The institutional constraints to SWC intervention under Participatory Approaches pertain to the higher level of relations between the donor, and the government institutions as target group. For the Sudanese situation, a top-down approach in public administration is still the rule. The constraints in this field are also difficult to remove (*cf.* El Moula 1985,26-43). However, at the lower level of relations between the government and the target group of users of the SWC techniques, such constraints are less pronounced. Despite the decline in local traditional binding-mechanisms like voluntary workgroups, religious movements gain in importance in the rural Border Area. These provide the obvious counterpart for local interventions in SWC in the future. The movement of Ali Betai is the best organized (*cf.* section 4.3.3). It exercises a direct influence in the process of SWC adoption through its call on the Beja to sedentarize and engage in crop production. It also exercises an indirect influence through the socio-economic and political support it provides to the households in the Border Area. The movement secures the livelihoods in adverse years when crop production and other livelihood activities fall short and functions as a safety net for destitutes. Similar religious organizations involved in the propagation of SWC techniques were also found among other Beja groups in the Red Sea province (Tilley & McEwan 1988).

There are still other potentially constraining factors to crop production and SWC intervention in the Border Area which have not been studied. These include *inter alia* human health, the occurrence of pests and diseases and the standard of land preparation and cultivation activities.

8.4 The contribution of indigenous SWC to the development of drylands

The SWC techniques introduced by the Department of Soil Conservation and MFEP/KADA in the Border Area in the 1980s are not the appropriate vehicles to support crop production locally. The most tangible evidence for this is that the projects cannot be sustained when major external financial, managerial and technical assistance is stopped. Recent reports of SWC projects in Africa (*cf.* IFAD 1992) clearly illustrate that the experiences in the Border Area are not an isolated incident. Introduced SWC techniques are generally difficult to manage locally. The relations presented in figure 8.1 may help to understand why the government interventions have failed for the particular case of the Border Area. They also illustrate why indigenous techniques, and we particularly refer in this last section to the *teras*, are likely to be more successful in this respect.

Figure 8.1 **SWC techniques, water volumes, organization types, and rates of success**

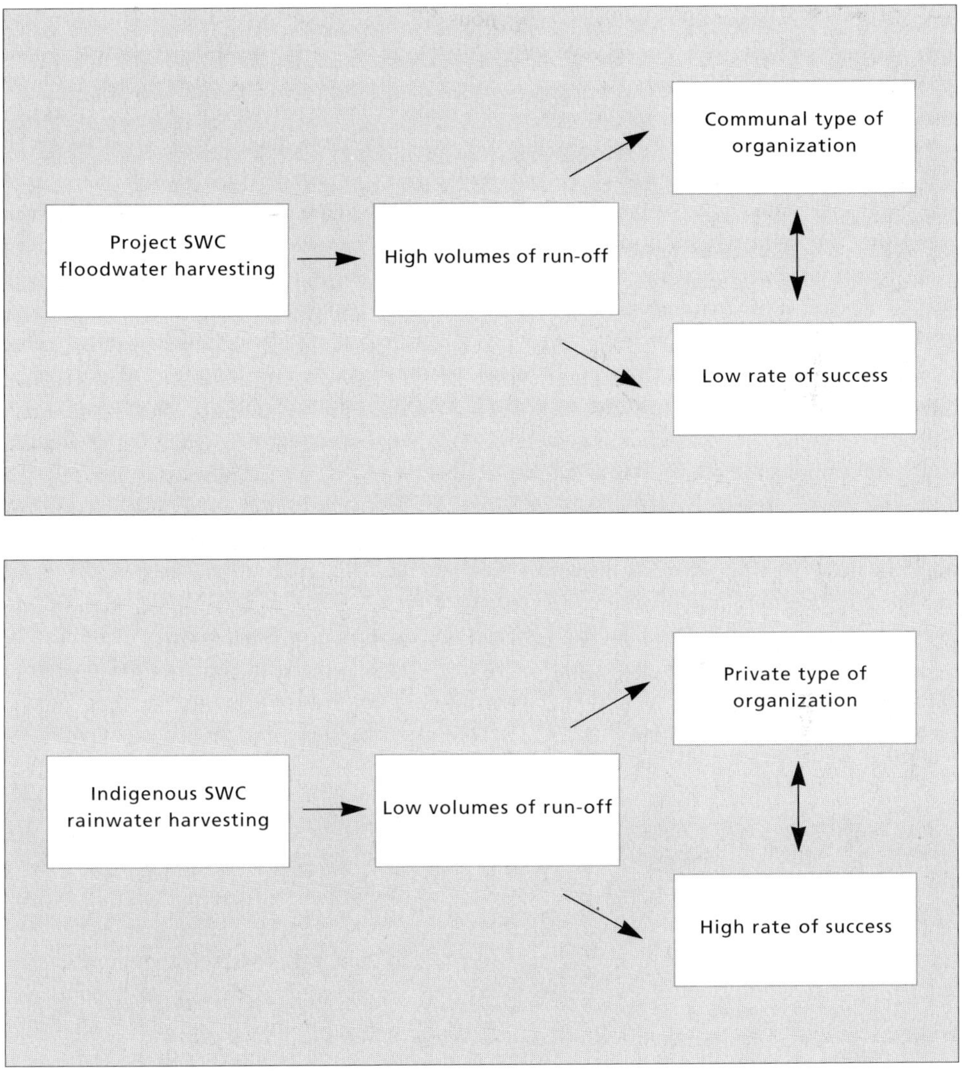

The selection of a type of SWC technique simultaneously determines a number of other elements of the system. Several of the presented relationships in figure 8.1 are obvious and are usually also accounted for in the context of conventional irrigation (*cf.* Barrow 1993). Others are less obvious, and are usually not considered, particularly not in SWC projects. In the Border Area, all introduced SWC techniques are of the floodwater-harvesting type. This implies that land users have to manage relatively large volumes of run-off. This normally demands a communal type of organization, because of the bulkiness of the resource. The large volumes of run-off also explain the high rate of failure of the projects due to erosion and breaching. Finally, the SWC projects have, to varying degrees, assumed some elementary form of communal organization to exist at the level of the target group of users of the technique. This would be required for the organization of maintenance and repair works at dams and embankments. This assumption proved false. In reality, even precisely the opposite situation of dispute over land use was encountered in the field. This further undermined the local intervention achievements. The indigenous *teras* is basically a rainwater-harvesting technique, although commonly also additional floodwater is collected. It implies that, generally, land users have only to manage smaller volumes of run-off. This allows a private type of organization, because the resource can be handled and controlled by individual users. Smaller volumes of run-off, in turn, are less likely to cause a failure in the system. Finally, private organization also reduces the chance of conflict and failure, because the responsibilities of land users are defined.

Three more arguments also plead in favour of propagating the application of private instead of communal SWC techniques in the Border Area. Firstly, Beja livelihoods are characterized by a high level of absenteeism from the village area for parts of the year. Communal obligations are therefore difficult to adhere to. Secondly, regularly changing water courses require continuous adjustments in the execution of SWC works. This usually can be better done on landholdings held in private, than in communal types of organization. Thirdly, private techniques allow a better delineation of target groups. This is of relevance when improvements in SWC techniques are supported in the framework of rural development projects.

Developments on the short term

The *teras* technique, and its land use, can still be improved on the following four points. (i) When the base bunds are better positioned on the (near) terrain contour, the efficiency of run-off collection can be increased. This, in turn, will also reduce the hazard of overtopping water and the amount of damage caused to the earth bunds. Although this may seem a relatively sophisticated engineering objective, simple tools such as a spirit level and hose level have already succesfully been applied in SSA for this purpose (*cf.* Chleq & Dupriez 1988); (ii) maintenance and erosion protection of *teras* bunds can be improved. Compaction of newly raised earth works is hardly practised in the Border Area. Already a considerable improvement can be achieved in this respect, when earth works are tamped after construction by simply walking them. Several cheap biological measures for the purpose of protection, which were tested in the pilot scheme, proved ineffective under local conditions. However, simple mechanical means of protection such as brushwood, sand bags and stones are favourable alternatives; (iii) excessive rates of sedimentation in the *teras* must be avoided. Although sediment harvesting improves the nutrient status of local soils, too high rates of sedimentation will distort the hydraulics of irrigation inside the bunded area. The measures required to tackle this problem lie in the field of common property resources management. These include the improvement of the upstream vegetative cover, and underline the importance of the various reafforestation initiatives already underway in the area (*cf.* chapter 6); (iv) the standard of land preparation and cultivation in the *teras* can still be improved. This could entail the promotion of more frequent weeding, thinning and pest control. It also can be sought in the field of introducing suitable crop varieties and new crops.

The experiences in the Border Area also provide a number of lessons. Programmes which aim to support and improve the application of *teras* and indigenous SWC should consider the following points.

1. *Despite evident manifestations in rural areas of landholdings under indigenous SWC, the amount of labour allocated to these landholdings, and the income contributions which such individual landholdings make to the household livelihood, may still be modest in terms of time and money.*

Indigenous SWC is usually the product of labour investments which have been made over a series of years. This, however, does not imply that labour will also be allocated by the households for this particular purpose every single year. On the contrary, since livelihood diversification strategies are common in drylands, household labour is usually short in supply. It therefore will not be allocated to a "project" which aims to improve the indigenous technique, or introduce a new technique for that matter, unless its returns can reasonably be expected to outweigh those from competing allocations. These frequently exist in local crop production, but also in local and non-local non-farm activities.

2. *The relative importance of indigenous SWC in household livelihoods is, for the situation in the Border Area, likely to increase with distance to urban centres.*

This greater importance of indigenous SWC was found for the Border Area over the east-west distance to the town of Aroma in the Gash delta. However, it was not found over the north-south distance to the town of Kassala due to intervening factors. The conclusion presented under 2 must therefore remain tentative. The lesser indigenous SWC importance near urban centres implies a greater contribution of other activities to the household livelihood locally. This, in turn, results in a greater competition of labour-allocation opportunities and in higher labour-opportunity costs (*cf.* point 1). The policy implication is that particularly in areas close to urban centres, a high priority should be given to control the conditions which influence these crucial returns to labour.

3. *The adoption of introduced SWC techniques in itself is no proof of the appropriateness of the techniques proposed.*

Innovations are sometimes evaluated without considering the real motives behind their adoption. The case of the Hedadeib pilot scheme illustrates that these motives (mechanized and subsidized land preparation, the presence of government and international organizations which "look after the people") may be contrary to the notions which are held by the institutions involved in the intervention. The policy implication is that the appropriateness of techniques can only be assessed by continuous and detailed monitoring and evaluation. It implies, more in particular, that the notion of adoption by "participation" should be reconsidered.[3]

4. *Pressure of population on scarce natural resources is not in all situations a general principle which governs technological change and the adoption of SWC techniques.*

The population density in the Border Area and the carrying capacity of its natural resources vary independently. This means that, for a given location, the degree of pressure of population on scarce resources can not meaningfully be calculated. However, it is unlikely that critical densities have played a role in the adoption of SWC techniques in the Border Area. Firstly, against expectation, we find in its more populous southern part the highest average entitlement to cultivated land per household recorded in the survey. Secondly, against expectation, we find in its less populous northern part several of the more elaborate examples of indigenous SWC application and experimentation. Also the greatest expansion of indigenous SWC in relative terms was recorded in the 1980s for this northern part, namely in Telkook. This is not in accordance with observations for example in Machakos, Kenya, by Tiffin *et al.* (1994,271) who state that low levels of population density are "[..] beyond doubt inimical to technology change [..]" (*cf.* section 4.3.4, this book).

5. *Adoption of SWC measures is also found in settings which are not primarily governed by a continuous exposure to commercial market influences.*

This point also is given lengthy consideration by Tiffin *et al.* (1994). They underline the importance of commercial zones for their beneficial effects on access to markets, multiplication of technical knowledge, and adoption of new techniques. However, the findings in the Border Area show that similar beneficial effects can also be achieved under non-commercial infrastructure and institutions which are based on religion.

Developments in the long term

The indigenous *teras* is a promising technique for crop production under dryland conditions. The physical environment and socio-economic setting under which this *teras* is

---

[3] *Cf.* Oakley (1987) for an extensive review of people's participation in soil and water conservation.

presently applied in the Border Area are also particularly favourable. However, the outlook in the long term is none too bright. The system of land use receives little or no input of fertilizers other than nutrients captured from animal droppings, sediment and debris. A declining level of soil fertility is therefore likely to result, and will ultimately constrain sustainable crop production in the manner defined by Van Reuler & Prins (1993) (*cf.* section 1.1). At this point in time, two paths are open to the households in the Border Area. The first ideally leads to a specialization in land use. This "specialization option" is sometimes referred to in terms of a paradigm of economic geography called "regional comparative advantages". It implies that regions should specialize in those functions in which they are, economically speaking, the best. Its advocators point out that pursuing comparative advantages is the only viable way to attain the targets set in rural development programmes in Africa (*cf.* Balacs 1985). Other commentators add that this is the "realistic option" and frequently also what households in these areas themselves do (Reitsma & Dietz 1992,180-181). When we consider the climate since the mid 1960s (*cf.* section 1.1.1), the comparative advantages of the Border Area are more likely to lie in livestock husbandry than in crop production. However, specialization in livestock husbandry requires that one important precondition to sustain a sufficient number of people is met. This precondition is that the "calorific terms of trade" for livestock against grain (Dietz 1993) is favourable. Research is currently being carried out into this subject in Kenya and Burkina Faso, as part of the Livelihood and Environment programme (Zaal 1994). The second path ideally leads to a greater combination of, on the one hand, different land use functions and, on the other hand, various local and non-local, non-farm activities. Under this "combination option", resources available from the latter can be allocated to increase the level of sustainability of crop production in the long term. Typically, this would mean that organic and inorganic fertilizers are purchased with incomes which have been earned in non-farm sectors. Another possibility would be that rangelands are specifically managed for the purpose of generating nutrients for local crop production. This entails a higher intensity of relations in the crop-livestock systems (Stroosnijder 1994).

Reshaping rural development programmes for the drylands within the general framework of either the specialization or combination option requires certain favourable prior conditions. We mention two. Firstly, a minimum level of market integration of the village economies is required for livestock-grain exchange, and the supply of necessary inputs to sustain crop production in the long term. Secondly, there must be sufficient potential for the scaling-up of the specialization or combination options in order to reach meaningfull levels of impact. In the end, the decision to favour either one of these options is a political one.

Still, irrespective of the prior conditions and the policy option chosen, indigenous SWC has an important contribution to make in the drylands now. One far from modest achievement would be in the words of Harrison (1990), who was quoted in the first section of this book, "[..] to gain time [..] in order to break through in the battle for land and food [..]". This time should be used to make the necessary preparations along lines of either specialization or combination to facilitate the transformation of rural livelihoods away from too great a dependency on the drylands' whims.

Appendices

Appendix 2.1 Names of major research variables

The dimension Income

SWCCROP3
: Percentage iSWC income in total household crop production income, 1983

SWCCROP8
: Percentage iSWC income in total household crop production income, 1988

CROPLIV3
: Percentage crop production income in total household livelihood income, 1983

CROPLIV8
: Percentage crop production income in total household livelihood income, 1988

SWCLIVE3
: Percentage iSWC income in total household livelihood income, 1983

SWCLIVE8
: Percentage iSWC income in total household livelihood income, 1988

The dimension Time

SWCTIME3
: Percentage labour time allocated to iSWC techniques in total household labour time (selected activities of maintenance, gap-filling, weeding and thinning), 1983

SWCTIME8
: Percentage labour time allocated to iSWC techniques in total household labour time (selected activities of maintenance, gap-filling, weeding and thinning), 1988

POTTIME3
: Indicator score C for the potential household labour time available for local crop production (over the number of economically actives in the household), 1983

POTTIME8
: Indicator score C for the potential household labour time available for local crop production (over the number of economically actives in the household), 1988

The dimension Land

SWCBORD3
Percentage land shares under iSWC in total household entitlement to local land, 1983
SWCBORD8
Percentage land shares under iSWC in total household entitlement to local land, 1988
SWCTOTA3
Percentage land shares under iSWC in total household entitlements to local and non-local lands combined, 1983

SWCTOTA8
Percentage land shares under iSWC in total household entitlements to local and non-local lands combined, 1988

The dimension Perception

SWCRANK3
Rank number given to the perceived priority of iSWC in crop production, 1983
SWCRANK8
Rank number given to the perceived priority of iSWC in crop production, 1988
CROPRAN3
Rank number given to the perceived priority of crop production in household livelihood, 1983
CROPRAN8
Rank number given to the perceived priority of crop production in household livelihood, 1988

Appendix 2.2 **Remote-sensing materials**

Coverage of the Border Area by satellite imageries (full-scenes) and black and white aerial photographs

VILLAGE	DATE	LANDSAT TM track-frame	SPOT XS path-row	AERIAL PHOTOGRAPHS	PHOTO SCALE
Telkook	05/1964			B85 108-111	40,000
Telkook	12/1966			E21 147-156	40,000
Telkook	02/1979			V17 159-169	75,000
Telkook	05/1979			V32 001-007	80,000
Telkook	10/1987	171-049			
Telkook	08/1988		130-318		
Ilat Ayot	?? /1963			B32 021-027	40,000
Ilat Ayot	12/1966			E17 074-079	40,000
Ilat Ayot	12/1966			E20 074-077	40,000
Ilat Ayot	12/1966			E22 058-062	40,000
Ilat Ayot	02/1979			V17 113-121	75,000
Ilat Ayot	02/1979			V17 164-165	75,000
Ilat Ayot	05/1979			V31 089-091	80,000
Ilat Ayot	05/1979			V32 178-180	80,000
Ilat Ayot	12/1986			SD86 04 001-005	20,000
Ilat Ayot	12/1986			SD86 04 071-075	20,000
Ilat Ayot	10/1987	171-049			
Ilat Ayot	08/1988		130-318		
US & HA	12/1953			series 138	15,000
US & HA	?? /1963			B32 008-013	40,000
US & HA	12/1966			E17 087-090	40,000
US & HA	12/1966			E22 069-075	40,000
US & HA	02/1979			V17 109-113	75,000
US & HA	05/1979			V31 095-096	80,000
US & HA	05/1979			V32 173-175	80,000
US & HA	12/1986			SD86 05 076-078	20,000
US & HA	12/1986			SD86 05 139-155	20,000
US & HA	12/1986			SD86 05 211-214	20,000
US & HA	10/1987	171-049			
US & HA	08/1988		130-319		

Note: US & HA means Um Safaree and Hafarat village areas.

Appendix 3.1 **Selected climatic parameters for the Border Area based on observations at the Kassala meteorological station, 1941-1970**

(cf. next page)

(Legend to appendix 3.1)

A: extreme maximum temperature (°C)
B: mean daily maximum temperature (°C)
C: mean temperature (°C)
D: mean daily minimum temperature (°C)
E: extreme minimum temperature (°C)
F: sunshine duration (h)
G: sunshine duration (%)
I: relative humidity 06.00 hrs (%)
J: relative humidity 12.00 hrs (%)
K: relative humidity 18.00 hrs (%)
L: rainfall (mm)
M: rainfall no. of days 0.1-1 mm
N: rainfall no. of days 1-10 mm
O: rainfall no. of days over 10 mm
P: maximum daily rainfall (mm)
Q: windspeed (miles/h)
R: wind direction
S: Class-A pan evaporation (mm)

Source: TNO (1982).

	JAN	FEB	MAR	APR	MAY	JUN	JUL	AUG	SEP	OCT	NOV	DEC	YEAR
A	41.5	34.1	44.8	45.2	46.5	45.7	42.3	40.3	42.3	43.0	41.3	40.0	46.5
B	34.3	35.4	38.6	40.7	41.7	39.9	35.6	33.7	36.3	38.9	37.5	35.0	37.3
C	25.1	25.9	28.9	31.4	33.3	32.5	29.5	28.1	29.7	31.1	29.1	26.0	29.2
D	16.0	16.4	19.2	22.1	24.8	25.2	23.4	22.6	23.0	23.2	20.6	17.0	21.1
E	5.4	6.2	9.4	11.1	15.0	18.4	18.7	18.7	18.0	12.5	10.5	7.4	5.4
F	10	11	10	11	11	10	8	8	10	10	10	10	10
G	90	90	85	86	82	77	65	66	80	88	88	88	82
H	61	54	45	35	35	47	65	72	62	49	48	57	53
I	27	23	18	17	18	25	41	50	39	25	23	26	28
J	41	35	28	27	29	38	57	67	59	43	38	42	42
K	0	0	1	4	12	27	97	110	51	9	3	0	314
L	--	--	0.2	0.9	1.6	3.2	9.4	10.7	5.8	1.5	0.4	--	34
M	--	--	0.2	0.8	1.3	3.0	8.4	9.6	5.1	1.3	0.4	--	30
N	--	--	--	0.1	0.5	0.9	3.1	3.6	1.5	0.4	0.1	--	10
O	1	8	6	20	39	56	78	96	91	25	15	--	96
P	3	3	3	3	3	5	6	5	5	2	2	3	3.6
Q	NNE	NNE	NNE	NNE	S	S	S	S	S	S	S	NNE	NNE
R	190	230	280	320	320	300	220	180	210	250	220	90	2900

Appendix 6.1

Average household production of sorghum (in sacks per fd), 1983 and 1988-1990

	Ilat Ayot				Telkook			
	1983	1988	1989	1990	1983	1988	1989	1990
Local lands								
– Teras	0.81	1.88	1.70	2.45	2.82	1.26	2.33	1.33
– Brushwood	2.11	2.44	8.00	6.00	na	1.67	0.20	0.20
– Wildflooding	1.34	2.18	1.68	3.10	0.90 [A]	1.62 [A]	0.31 [A]	1.48 [A]
– Pilot scheme	na	na	na	na	na	na	na	na
Non-local lands								
– GDAC scheme	4.59	3.80	5.70	7.17	4.23	4.93	2.08	4.00
– Gash margin	5.04	4.00	3.33	3.33	na	na	na	na

	Um Safaree				Hafarat			
	1983	1988	1989	1990	1983	1988	1989	1990
Local lands								
– Teras	1.50	2.92	0.98	1.27	1.55	2.36	0.91	0.92
– Brushwood	1.16	1.98	1.14	na	1.41	1.32	na	1.00
– Wildflooding	1.93	4.00	2.44	1.35	2.37	3.36	1.17	1.15
– Pilot scheme	na	na	na	na	na	2.11	0.78	0.25
Non-local lands								
– GDAC scheme	na	1.00	1.00	2.00	1.25	na	2.20	6.00
– Gash margin	na	na	na	na	na	na	na	na

Source: L&E survey. Note: [A] 1988 production on wildflooding land in Telkook includes project SWC land (earth dam), na means not applicable.

Average household production of millet (in sacks per fd), 1983 and 1988-1990

	Ilat Ayot				Telkook			
	1983	1988	1989	1990	1983	1988	1989	1990
Local lands								
– Teras	0.57	0.80	na	na	na	na	na	na
– Brushwood	na	na	na	na	na	na	na	na
– Wildflooding	na	0.18	na	na	0.10 [A]	0.17 [A]	na	na
– Pilot scheme	na	na	na	na	na	na	na	na
Non-local lands								
– GDAC scheme	na	na	na	na	na	na	na	na
– Gash margin	na	na	na	na	na	na	na	na

	Um Safaree				Hafarat			
	1983	1988	1989	1990	1983	1988	1989	1990
Local lands								
– Teras	1.33	1.33	0.31	0.50	1.68	2.53	1.50	1.13
– Brushwood	na	1.13	na	0.40	0.68	1.82	0.40	0.37
– Wildflooding	1.37	0.88	na	na	1.11	1.87	0.77	0.59
– Pilot scheme	na	na	na	na	na	1.47	0.40	0.25
Non-local lands								
– GDAC scheme	na	na	na	na	na	na	na	na
– Gash margin	na	na	na	na	na	na	na	na

Source: L&E survey. Note: [A] 1988 production on wildflooding land in Telkook includes project SWC land (earth dam), na means not applicable.

Average household production of stalks (in bundles per fd), 1983 and 1988-1990

	Ilat Ayot				Telkook			
	1983	1988	1989	1990	1983	1988	1989	1990
Local lands								
– Teras	61	98	200	58	na	156	533	667
– Brushwood	269	272	400	55	na	93	115	40
– Wildflooding	164	237	91	368	120 [A]	221 [A]	64 [A]	262 [A]
– Pilot scheme	na	na	na	na	na	na	na	na
Non-local lands								
– GDAC scheme	291	304	263	350	583	425	438	54
– Gash margin	220	272	150	133	na	na	na	na

	Um Safaree				Hafarat			
	1983	1988	1989	1990	1983	1988	1989	1990
Local lands								
– Teras	222	259	116	89	160	241	204	98
– Brushwood	106	306	168	123	179	155	na	103
– Wildflooding	255	279	153	266	168	222	95	185
– Pilot scheme	na	na	na	na	na	275	148	133
Non-local lands								
– GDAC scheme	123	100	102	200	na	na	400	500
– Gash margin	na	na	na	na	na	na	na	na

Source: L&E survey. Note: [A] 1988 production on wildflooding land in Telkook includes project SWC land (earth dam), na means not applicable.

Average household cropping intensity (in % of cultivated land), 1983 and 1988-1990

	Ilat Ayot				Telkook			
	1983	1988	1989	1990	1983	1988	1989	1990
Local lands								
– Teras	75	72	100	70	83	75	77	83
– Brushwood	51	67	20	25	na	100	100	100
– Wildflooding	79	77	66	65	87 [A]	87 [A]	99 [A]	73 [A]
– Pilot scheme	na	na	na	na	na	na	na	na
Non-local lands								
– GDAC scheme	88	86	35	38	77	89	77	84
– Gash margin	100	84	71	66	na	na	na	na

	Um Safaree				Hafarat			
	1983	1988	1989	1990	1983	1988	1989	1990
Local lands								
– Teras	58	63	75	73	80	75	89	84
– Brushwood	62	68	88	91	100	84	86	89
– Wildflooding	61	55	79	70	77	76	84	79
– Pilot scheme	na	na	na	na	na	91	90	88
Non-local lands								
– GDAC scheme	20	83	40	20	40	60	40	40
– Gash margin	na	na	na	na	na	na	na	na

Source: L&E survey. Note: [A] 1988 production on wildflooding land in Telkook includes project SWC land (earth dam), na means not applicable.

Appendix 6.2

The importance of SWC in crop production. Average household scores of SWC-time S as % of total household labour-time allocations to crop production, 1989 and 1990 (iSWC households)

	Ilat Ayot		Telkook		Um Safaree		Hafarat	
	1989	1990	1989	1990	1989	1990	1989	1990
Teras	71	64	100	100	61	49	53	50
Brushwood	26	28	100	100	55	54	22	21
Teras and brushwood	0	0	0	0	66	57	85	74
Total iSWC								
– (teras/brushwood)	64	58	100	100	66	56	55	53
Project pSWC	na	na	100	97	na	na	42	31

Source: modelled data based on L&E Survey. Note: labour time spent on maintenance, gap-filling, weeding and thinning for growing season months June to and including October, na means not applicable.

The importance of SWC in crop production. Average household acreage in % of total land entitlement to local lands, 1989 and 1990 (iSWC households)

	Ilat Ayot		Telkook		Um Safaree		Hafarat	
	1989	1990	1989	1990	1989	1990	1989	1990
Teras	67	61	100	100	68	68	47	51
Brushwood	100	100	100	100	57	55	48	47
Teras and brushwood	0	0	0	0	85	88	82	83
Total iSWC								
– (teras/brushwood)	72	75	100	100	76	71	56	61
Project pSWC	na	na	100	100	na	na	32	30

Source: L&E Survey.

The importance of SWC in crop production. Average household acreage in % of total land entitlement to local and non-local lands combined, 1989 and 1990 (iSWC households)

	Ilat Ayot		Telkook		Um Safaree		Hafarat	
	1989	1990	1989	1990	1989	1990	1989	1990
Teras	58	52	100	100	68	68	47	52
Brushwood	81	92	100	100	57	55	45	46
Teras and brushwood	0	0	0	0	85	88	82	83
Total iSWC								
– (teras/brushwood)	61	69	100	100	76	71	55	62
Project pSWC	na	na	89	91	na	na	32	30

Source: L&E Survey.

Appendix 7.1

Impact of government SWC intervention comparing the with project and without project situations. Mean rank scores of selected variables in the Mann-Whitney U-test. At the landholding level, nopSWC landholdings are compared with pSWC landholdings. At the household level, nopSWC households are compared with pSWC households, Hafarat and Telkook, 1988 and 1989

	LANDHOLDING LEVEL [A]			HOUSEHOLD LEVEL		
Situation in 1988	nopSWC	pSWC	N	nopSWC	pSWC	N
Crop production income	75.74	66.67	146	41.17	45.06	85
Labour-time allocation	73.99	78.55	149	41.17	45.06	87
Cropping intensity	76.62	91.77	160	42.63 [B]	45.46 [B]	87
Situation in 1989	nopSWC	pSWC	N	nopSWC	pSWC	N
Crop production income	25.44	21.67	48	20.00	17.00	36
Labour-time allocation	38.22	26.02	64	27.50	19.87	43
Cropping intensity	47.84	58.48	101	27.67 [B]	34.46 [B]	62

Source: L&E Survey partly based on modelled data. Note: [A] local lands only, [B] mean cropping intensity of local lands. pSWC means project SWC, nopSWC means no project SWC, cf. table 7.13.

Impact of government SWC intervention comparing the before project and after project situations. Mean signed-rank scores of selected variables in the Wilcoxon Matched-pairs Signed-ranks test. Identical pSWC landholdings and pSWC households are compared only, Hafarat and Telkook, 1983 and 1989

	LANDHOLDING LEVEL [A]			HOUSEHOLD LEVEL		
Situation in 1983 and 1988	–	+	N	–	+	N
Crop production income	1.00	7.00	12	0.00	10.50	20
Labour-time allocation	4.50	5.14	14	7.93	13.78	24
Cropping intensity	2.00	5.00	20	10.42 [B]	11.23 [B]	26
Situation in 1983 and 1989	–	+	N	–	+	N
Crop production income	2.67	2.00	4	3.67	7.00	8
Labour-time allocation	8.00	4.86	11	7.50	9.50	16
Cropping intensity	3.33	2.50	14	10.00 [B]	8.79 [B]	19

Source: L&E Survey partly based on modelled data. Note: [A] local lands only, [B] mean cropping intensity of local lands. "+" ranks are defined as scores for 1988, respectively 1989, higher than 1983, "-" ranks are defined as scores for 1988, respectively 1989, lower than 1983, pSWC means project SWC, cf. table 7.14.

Appendix 7.2

Mean rank scores of selected variables in the Mann-Whitney U-test by factors distance to Kassala town and distance to Aroma town (Gash) (great versus small distance). Data are at the household level for nopSWC households, 1983 and 1988

	DISTANCE-TO-KASSALA			DISTANCE-TO-GASH		
INCOME	great	small	N	great	small	N
% iSWC in total hh crop production income						
– 1988 SWCCROP8	31.81	49.08	84	50.00	35.62	73
– 1983 SWCCROP3	19.25	33.40	58	37.50	27.64	55
% Crop production in total hh livelihood income						
– 1988 CROPLIV8	73.31	71.80	144	55.58	56.45	123
– 1983 CROPLIV3	63.06	48.89	109	53.33	49.78	99
% iSWC in total hh livelihood income						
– 1988 SWCLIVE8	34.78	47.25	84	31.57	37.58	73
– 1983 SWCLIVE3	30.50	29.12	58	36.00	27.70	55
TIME						
% iSWC-time S in total hh labour time						
– 1988 SWCTIME8	42.10	46.08	88	50.00	38.46	78
– 1983 SWCTIME3	35.14	40.86	77	45.00	34.85	69
Score C for the pot. hh labour time available [A]						
– 1988 POTTIME8	91.79	88.11	179	80.93	77.18	155
– 1983 POTTIME3	82.61	78.39	160	67.33	70.64	139
LAND						
% iSWC land in total hh local land entitlement						
– 1988 SWCBORD8	54.39	49.64	102	64.50	43.39	90
– 1983 SWCBORD3	46.71	46.40	92	58.50	39.52	79
% iSWC land in total hh land entitlement						
– 1988 SWCTOTA8	47.53	54.06	102	69.00	42.89	90
– 1983 SWCTOTA3	38.57	50.15	92	62.00	39.43	79
PERCEPTION [B]						
Rank iSWC priority in crop production						
– 1988 SWCRANK8	53.18	40.62	91 [C]	27.50	45.67	87
– 1983 SWCRANK3	49.41	38.51	85 [C]	23.50	43.70	83
Rank crop production priority in hh livelihood						
– 1988 CROPRAN8	79.12	63.43	142	73.58	66.56	134
– 1983 CROPRAN3	70.76	51.40	121	73.50	57.64	117

Source: L&E Survey. Note: [A] over the number of economically active members in the household. [B] High mean rank scores for variables in the dimension Perception indicate low importance contrary to scores of the other variables listed. [C] Only 4 households in Hafarat, cf. table 7.16. Variable names are described in full in appendix 2.1.

Archival sources

National Records Office (NRO) in Khartoum, Sudan.

NRO ALLEN (1929), *Letter from the manager of the Gash Board to the Governor of Kassala*, 10 October 1929.

NRO ARBER 57/A.9, *Handing Over Notes of the Kassala District Commissioner*, 1946.

NRO BLACKLEY KD/20.H.10, *Letter from the DC Kassala to the Governor of Kassala*, 1 November 1936.

NRO BLACKLEY KD/8.B.2, *Letter from the DC Kassala to the Governor of Kassala*, 21 March 1937.

NRO CARLISLE BD/19.B.4, *Letter from the DC Beja to the Governor of Kassala*, 18 June 1949.

NRO CUMMING 57.A.9, *Handing Over Notes of Kassala*, not dated (late 1930s, JvD).

NRO DINGWALL BD/2.F, *Letter from the DC Beja to the Governor of Kassala*, 12 January 1947.

NRO DINGWALL BD/2.F.1, *Letter from the DC Beja to the Governor of Kassala*, 22 November 1947.

NRO DINGWALL BD/2.F.6, *Letter from the DC Southern District to the Governor of Kassala*, 14 September 1947.

NRO DINGWALL BD.19.B.2, *Letter from the DC Beja to the Governor of Kassala*, 29 March 1949.

NRO LEA GRDC/22.C.1, *Letter from the DC Gedaref to the Governor of Kassala*, 18 September 1945.

NRO OWEN BD/66.K, *Report of Hadendowa Annual Meeting 1935*.

NRO OWEN DCH/66.K1, *Report on tour of Maman-Odi area*, March 1933.

NRO SANDISON KD/2.N.1, *Letter from the DC Kassala to the Governor of Kassala*, 1 September 1941.

NRO SUDAN GOVERNMENT KD/57.A.1, *Kassala Province Annual Report*, various years 1928-1952.

NRO SUDAN GOVERNMENT KP/19.B.2, *Letter from the Kassala Governor's Office to the Sudan Famine Relief Committee Port Sudan*, 5 June 1949.

NRO WEEKLY REPORT No.6 (1949), *Weekly Report on Famine Conditions, Governor's Office Kassala*, 15 June 1949.

NRO WEEKLY REPORT No.8 (1949), *Weekly Report on Famine Conditions, Governor's Office Kassala*, 1 June 1949.

Durham Sudan Archive (DSA) of the University of Durham, England

DSA BARTER 448/1, *Hadendowa District Annual Report 1930 and Tamarein Meeting March 1931*.

DSA BARTER 646/2, *Trek notebook 1930*.

DSA CARLISLE 725/3, *Handing Over Notes ADC Hadendowa District*, 1 January 1951.

DSA GASH BOARD 499/1, *Letter from Divisional Engineer Projects to the Director of Irrigation*. Wad Medani, date unknown, 1944.

DSA MACGREGOR 499/1, *Letter from the Irrigation Advisor Sudan Government to the Financial Secretary Sudan Government*. London: 8 September 1925.

DSA OWEN 414/5/15-20, *Letters home*, 25 February 1933.

DSA OWEN 414/5/21 *Sudan days*, mimeo, not dated, (written in 1960-1961, JvD).

DSA SAVILE 427/3, *Diary of tribute collecting tour in N. Kassala*, April 1903.

DSA SAVILE 427/5, *Diary 1902*.

DSA SAVILE 427/6, *Diary 1903* (part 2).

DSA SMITH 498/6, *Jebel Haura Water Storage Project. Preliminary note following reconnaisance*, 10 September 1937.

DSA THOMSON 402/6, *Diary 1916-1917*.

DSA TOTTENHAM 185/1, *Irrigation from the Gash river, memorandum from the Inspector-General of Sudan Irrigation Service*, 1913.

DSA WINGATE 112/3, *Memorandum on irrigation development in Gezira*, 8 February 1913.

References

ABDULLAHI, A.M. (1990) *Pastoral production systems in Africa. A study of nomadic household economy and livestock marketing in central Somalia.* Farming Systems and Resource Economics in the Tropics Vol.8. Kiel: Wissenschaftsverlag Vauk (Ph.D. thesis).

ABEL ATI, H.A. (1991) *The developmental impact of NGOs' activities in the Red Sea Province, Sudan.* Red Sea Area Programme Technical Papers, No. 3. Khartoum: RESAP/University of Khartoum, University of Bergen.

ABU SIN, M.A.A. (1970) *The regional geography of the Butana north of the railway.* Khartoum: University of Khartoum (M.A. thesis).

ABU SIN, M.A.A. (1991) "Urban process and resource conservation and management in Sudan". In: *Second International Sudan Studies Conference Papers Sudan: Environment and People held at University of Durham 8-11 April 1991*, Vol.2, pp.8-27. Durham: University of Durham.

ADAMS, W.M. (1990) "How beautiful is small? Scale, control and succes in Kenyan irrigation". *World Development*, Vol.18, No.10, pp.1309-1323.

ADAMS, W.M. (1992) *Wasting the rain. Rivers, people and planning in Africa.* London: Earthscan Publications Ltd.

AD HOC CHINESE EXPERTS GROUP (1989) *Investigation report on feasibility of utilization of seasonal river water resources in Eastern Region of the Sudan.* Beijing: Ministry of Foreign Economic Relations and Trade.

AHMED, A.G.M. (1976) *Some aspects of pastoral nomadism in the Sudan.* Khartoum: The Sudan National Population Committee and the Economic and Social Research Council.

AHMED, T. H. (1966) "Economic aspects of production in the Gash scheme, Eastern Sudan". In: D.J. Shaw, ed., *Agricultural development in the Sudan. Proceedings of the 13th annual conference, 3-6 December Khartoum*, Vol.2, pp.285-314. Khartoum: Philosophical Society of The Sudan in conjunction with The Sudan Agricultural Society.

AHMED, M. H. (1987) *Earth embankments in the Border Rural Council and their effect on soil conservation and regeneration of the natural resources. Paper prepared for the seminar on earth embankments, November 1987.* Kassala: Hans Siedel Foundation (in Arabic).

ALMKVIST, H. (1884) "Die Bischari Sprache Tū Bedāwie in Nordost Afrika". *Nova Acta Regiæ Societatis Scientiarum Upsaliensis Seriei Tertiæ*, Vol.XIII.

AL-NAQAR, U. (1972) *The pilgrimage tradition in West Africa, an historical study with special reference to the nineteenth century.* Khartoum: Khartoum University Press.

ANGLADETTE, A. & L. DESCHAMPS (1974) *Problèmes et perspectives de l'agriculture dans les pays tropicaux.* Paris: Maisonneuve et Larose.

ARMBRUSTER, C.H. (1965) *Dongolese-Nubian Lexicon.* Cambridge: Cambridge University Press.

ARNTZEN, J.W., G.M. MOOLHUYZEN & A.P.R. VISSER (1993) *Duurzame ontwikkeling als leidraad voor ontwikkelingsplanning*, Reprint Series Development Cooperation No.11. 's Gravenhage: DGIS.

ATTIYA, A.B.O. & F.G. NIVOSE (1966) "Rainwater conservation to supply rural areas of the Sudan". In: M.S. NUR, ed., *Engineering and development in the Sudan. 14th Annual Conference December 6-9 1966, Khartoum*, Vol.2 Written contributions, pp.7-15. Khartoum: Philosophical Society of The Sudan in conjunction with the Sudan Engineering Society.

AUSENDA, G. (1987). *Leisurely nomads: the Hadendowa (Beja) of the Gash Delta and their transition to sedentary village life.* Ann Arbor: UMI Microfilm, (Ph.D. thesis).

BAASHER, M.M. (1992) *Gash Die grazing rehabilitation and irrigated forage production proposed project.* Khartoum.

BAKKER, E.J. (1992) *Rainfall and risk in India's agriculture. An ex-ante evaluation of rainfall insurance.* Groningen Thesis in Economics, Management and Organization. Groningen: Wolters-Noordhof (Ph.D. thesis).

BALACS, P. (1985) *Rural development in Africa: a synthesis of project experience.* Report EV 400. London: ODA.

BARBOUR, K.M. (1961) *The Republic of the Sudan. A Regional Geography.* London: University of London Press.

BARROW, C. [1987](1993) *Water resources and agricultural development in the tropics.* Harlow: Longman Development Studies.

BASCOM, J.B. (1990a) "Border pastoralism in Eastern Sudan". In: *The Geographical Review,* Vol.80, No.4, pp.416-431.

BASCOM, J.B. (1990b) "Food, wages and profits: mechanized schemes and the Sudanese State". In: *Economic Geography,* Vol 2, pp.109-123.

BEAUVILAIN, A. (1992) *Nord-Cameroun, crises et peuplement.* Tome 1 et 2. Rouen: L'Université de Rouen (Ph.D. thesis).

BELL, H. (1956) *Irrigation by gravity from the river Atbara part 1. A preliminary report on a dam at Kashm El Girba and the area it can command.* Khartoum: Ministry of Irrigation and Hydro-Electric Power.

BELSHAW, D.G.R., P.M. BLAIKIE, & M.A. STOCKING (1991) "Identifying key land degradation issues and applied research priorities". In: J.T. WINPENNY, ed., *Development research: the environmental challenge,* pp.66-91. London: ODI.

BENNET, H.H. (1939) *Soil conservation.* Berkeley: McGraw-Hill.

BLAIKIE, P. [1985](1989) *The political economy of soil erosion in developing countries.* Harlow: Longman Development Studies.

BLAIKIE, P. & H. BROOKFIELD (1987) *Land degradation and society.* London: Routledge.

BLOKHUIS, W.A. (1993) *Vertisols in the Central Clay Plain of the Sudan.* Wageningen: Grafisch Service Centrum Landbouwuniversiteit (Ph.D. thesis).

BLOKLAND, A. & F. VAN DER POL (1992) "Conclusions and policy recommendations". In: A. BLOKLAND & F. VAN DER STAAIJ, ed., *Sustainable development in semi-arid Sub-Saharan Africa,* pp.73-81. The Hague: DGIS DVL/OS.

BOKKERS, J. & SAYYED DABLOUB (1986) *Subsistence agriculture in the Kassala Rural District.* Kassala: KADA/DHV Consulting Engineers/Ministry of Agriculture.

BONSU, M. (1981) "Assessment of erosion under different cultural practices on a savanna soil in the northern region of Ghana". In: R.P.C. Morgan, ed., *Soil conservation: problems and prospects,* pp.24-253. Accra: Soil Research Institute.

BORN, M. (1965) *ZentralKordofan. Bauern und Nomaden in Savannengebieten des Sudan. Marburger Geographische Schriften, Heft 25.* Marburg: Geographisch Institutes der Universität Marburg.

BORN, M., D.R. LEE & J.R. RANDELL (1971) "Ländliche Siedlungen im nordöstlichen Sudan". In: *Arbeiten aus dem Geographischen Institut, Band 14,* pp.78-112. Saarbrücken: Universität des Saarlandes.

BOSERUP, E. (1965) *The conditions of agricultural growth: the economics of agrarian change under population pressure.* London: Allen & Unwin.

BOSERUP, E. (1981) *Population and technology.* Oxford: Blackwell.

BREMAN, H. & N. DE RIDDER (1991) *Manuel sur les pâturages des pays sahéliens.* Wageningen: ACCT-CTA-KARTHALA, CABO-DLO.

BROOKFIELD, H. (1991) "Environmental sustainability with development: what prospects for a research agenda ?" Papers from the 6th EADI General Conference, Oslo 27-30 June 1990. In: *The European Journal of Development Research,* Vol.3, No.1, pp.42-66.

BRUINS, J.H., M. EVENARI, & U. NESSLER (1986) "Rainwater-harvesting agriculture for food production in arid zones: the challenge of the African famine". In: *Applied Geography,* Vol.6 No.1, pp.13-32.

BRYCESON, D.F. (1993) *De-agrarianization and rural employment generation in Sub-Saharan Africa: process and prospects.* Working Paper, Vol.19. Leiden: Afrika-Studiecentrum.

BURCKHARDT, J.L. [1819] (1987) *Travels in Nubia.* London: Darf Publishers Ltd.

BURRA, R. and K. VAN DEN HEUVEL (1987) *Traditional irrigation in Tanzania. Vol.1: historical analysis of irrigation tradition and government intervention.* Dar es Salaam.

CARLSTEIN, T. (1982) *Time resources, society and ecology. On the capacity for human interaction in space and time. Volume 1: preindustrial societies.* London: Allen & Unwin.

CARTER, D.C. & S. MILLER (1991) "Three years experience with an on-farm macro-catchment water harvesting system in Botswana". In: *Agricultural Water Management* 19, pp.191-203.

CASLEY, D.J. & K. KUMAR (1988). *The collection, analysis, and use of monitoring and evaluation data.* World Bank Publication. Baltimore: The John Hopkins University Press.

CHAMBERS, R. [1983] (1991) *Rural development: putting the last first.* London: Longman.

CHAMBERS, R. (1990) *Microenvironments unobserved.* IIED Gatekeeper Series 22. London: International Institute for Environment and Development.

CHAMBERS, R., A. PACEY & L. ANN THRUPP, eds., (1991) *Farmer first. Farmer innovation and agricultural research.* London: Intermediate Technology Publications.

CHISHOLM, M. (1973) *Rural settlement and land use. An essay in location.* London: Hutchinson University Library.

CHLEQ, J.L. & H. DUPRIEZ (1988) *Vanishing land and water. Soil and water conservation in dry lands.* London: MacMillan Publishers with Terres et vie.

CLIFFE, L. (1989) "The impact of war and the response to it in different agrarian systems in Eritrea". In: *Development and Change*, Vol.20, pp.373-400.

COLE, R. (1989) *Measuring drought and drought impacts in Red Sea Province.* Port Sudan: OXFAM.

COLVIN, R.C. (1939) *Agricultural survey of Nuba mountains.* Khartoum: Government of Sudan.

COSIJN, G.P.A. & J.A. VAN DIJK (1989) *Waterspreading in the Hedadeib Pilot Scheme, Kassala, Sudan.* Kassala: KADA/DHV Consulting Engineers BV.

COX, K.R. (1972) *Man, location and behavior.* New York: John Whiley & Sons Inc.

CRAIG, G.M., ed., (1991) *The agriculture of The Sudan.* Oxford: Oxford University Press.

CRAMER ROBERTS, F.W. (1923) *Report on Tokar with proposals for the more efficient utilization of the khor Baraka and its flood waters.* Cairo: Ministry of Public Works, Government Press.

CRAWFORD, O.G.S. [1951](1978) *The Fung Kingdom of Sennar with a geographical account of the Middle Nile Region.* New York: AMS Press Inc.

CRITCHLEY, W. (1990) "How Africa's pastoralists catch the rain. Rainwater harvesting is nothing new to the people of the sub-Saharan drylands". In: *CERES* 125 pp.41-45.

CRITCHLEY, W. (1991b) *Looking after our land. New approaches to soil and water conservation in dryland Africa.* Oxford: RITA/ALIN-OXFAM/IIED.

CRITCHLEY, W. & K. SIEGERT (1991) *Water harvesting. A manual for the design and construction of water harvesting schemes for plant production.* Rome: FAO AGL/MISC/17/91.

CRITCHLEY, W., C. REIJ, & A. SEZNEC (1992) *Water harvesting for plant production. Part II: Case studies and conclusions for Sub-Saharan Africa.* Washington: World Bank Technical Paper 157.

CTFT (1979) *Conservation des sols au Sud du Sahara.* Paris: Ministère de la Cooperation.

CULLIS, A. & A. PACEY (1992). *A development dialogue. Rainwater harvesting in Turkana.* London: Intermediate Technology Publications.

CUMMING, D.C. (1937) "The history of Kassala and the Province of Taka, Part I". In: *Sudan Notes and Records*, Vol.20, p.1-54.

DAHL, G. & A. HJORT (1976) *Having herds. Pastoral growth and household economy.* Stockholm Studies in Social Anthropology 2. Stockholm: Department of Social Anthropology, University of Stockholm.

DAHL, G. (1991) "The Beja of Sudan and the Famine of 1984-1986". In: *Ambio*, Vol.20, No.5, pp.189-191.

DANA TOMLIN, C. (1990) *Geographic Information Systems and cartographic modeling.* New Jersey: Prentice Hall.

DAVIS, J.C. (1986) *Statistics and data analysis in geology.* New York: John Wiley & Sons.

DAVIE, W.A. (1924) *The cultivated crops of the Sudan including cotton.* Khartoum: Sudan Government, Department of Agriculture and Forests.

DAVIES, H.R.J. (1991) "Irrigation development programmes". In: G.M. CRAIG, ed., *The agriculture of The Sudan*, pp.339-364. Oxford: Oxford University Press.

DE BRUIJNE, G.A. (1989) "WOTRO and the geography of developing countries". In: *Tropical research in development. WOTRO 1964-1989.* The Hague: Netherlands Foundation for the Advancement of Tropical Research NWO, pp.48-52.

DE GRAAFF, J. (1993) *Soil conservation and sustainable land use. An economic approach.* Amsterdam: Royal Tropical Institute.

DE HAAN, L.J. (1988) *Overheid en regionale integratie van de savanne in Togo 1885-1985. Infrastructuur, marktintegratie, arbeidsmigratie en de lokale bestaanswijze.* Amsterdam: Netherlands Geographical Studies 79 (Ph.D. thesis).

DE LEEUW, M. (1987) *Activities, structures and bottlenecks within the traditional rainfed agriculture, a case study of El Marya.* Kassala: KADA/DHV Consulting Engineers BV.

DECARP (Desert Encroachment Control and Rehabilitation Programme) (1976) *Sudan's Desert Encroachment Control and Rehabilitation Programme.* Khartoum: Ministry of Agriculture, Agricultural Research Council, National Council for Research, UNEP, UNDP and FAO.

DE VADJA, A. (1966) "The use of water in Sudan's agriculture". In: D.J. SHAW. ed., *Agricultural development in The Sudan. Proceedings of the 13th Annual Conference, December 3-6, 1965, Khartoum.* Vol.2, pp.73-80. Khartoum: Philosophical Society of The Sudan in conjunction with The Sudan Agricultural Society.

DGIS (1991) *Slotdocument Kassala Area Development Activities.* 's Gravenhage: Ministerie van Buitenlandse Zaken, DGIS DAF/NF.

DGIS (1992) *Duurzaam landgebruik. Sector- en themabeleidsdocumenten van Ontwikkelingsdocumenten 2.* 's Gravenhage: Ministerie van Buitenlandse Zaken, DGIS.

DHV (1991) *KADA completion report. Report to DGIS/DAF/NF.* Amersfoort: DHV Consulting Engineers BV.

DHV & IES (1989) *Environmental Profile Kassala Province, Eastern Region, Sudan. Draft Main Report.* Amersfoort: DHV Consultants Repro.

DIEMER, G. (1990) *Irrigatie in Afrika. Boeren en ingenieurs, techniek en kultuur.* Amsterdam: Thesis Publishers (Ph.D. thesis).

DIETZ, A.J. (1987) *Pastoralists in dire straits. Survival strategies and external interventions in a semi-arid region at the Kenya/Uganda border: Western Pokot, 1900-1986.* Amsterdam: Netherlands Geographical Studies 49 (Ph.D. thesis).

DIETZ, A.J. (1993) "The state, the market, and the decline of pastoralism: challenging some myths, with evidence from western Pokot in Kenya/Uganda". In: J. MARKAKIS, ed., *Conflict and the decline of Pastoralism in the Horn of Africa*, pp.83-99. London: The Macmillan Press Ltd and ISS The Hague.

DIETZ, A.J., P. DRUIJVEN & D.W.J. FOEKEN (1992) "Coping mechanisms and livelihood strategies; a summary of concepts". In: H. REITSMA, A.J. DIETZ & L. DE HAAN, eds., *Coping with semiaridity. How the rural poor survive in dry-season environments*, pp.37-44. Amsterdam: Netherlands Geographical Studies 146.

DOORENBOS, J. & W.O. PRUITT (1977) *Guidelines for predicting crop water requirements.* Rome: FAO Irrigation and Drainage Paper 24.

DUMONT, H. & P. STEVENS (1987) *Soedan. Een wetenschappelijk en archeologisch avontuur.* Den Haag: Uitgevery Simon.

EICKELMAN, D.F. (1981) *The Middle East: an anthropological approach.* New Jersey: Prentice Hall.

EISA, M.H. (1978) *Smuggling in the eastern Sudanese boundaries.* Khartoum: University of Khartoum/IAAS (M.A. thesis).

EISSA, E.S. (1990) *Analysis of revenue and expenditure performance at local governments. Case study of Eldwim Shabasha Area Council.* Khartoum: University of Khartoum/DSRC (Diploma thesis).

EL ARIFI, S.A. (1971) *Regional economic development in the Sudan; a geographical analysis of the emergent regional pattern.* Ann Arbor: UMI Microfilm, (Ph.D. thesis).

El BERIER, O.M. (1982) *National public policy and natural hazards in developing countries: analysis of absorbtive capacity in Kassala city. Sudan.* New York: Saracuse University (M.Sc. thesis).

EL-DISHOUNI, S. (1989) *Tradition and modernization in Sudanese irrigated agriculture: lessons from experience*, DSRC Monograph Series No.36. Khartoum: Univeristy of Khartoum/DSRC.

EL FAKI, A. (1991) *Report on experiments with stone bunds in Shellalob 1991/1992 season*, Kassala: Department of Soil Conservation, Land Use and Water Programming (in Arabic).
EL HASSAN, A.M. & W.A. EL SEED (1989) *Crop assessment in Hedadeib area*. Kassala: Ministry of Agriculture and Natural Resources, Department of Soil Conservation, Land Use and Water Programming.
EL HASSAN, H.M. (1987) *Hedadeib water spreading. Innovation versus traditional practices*. Kassala: Department of Soil Conservation, Land Use and Water Programming.
EL HASSAN, H.M. (1990) *Rural land use systems in Red Sea Province*. Kassala: Department of Soil Conservation, Land Use and Water Programming.
EL HASSAN, I.S. (1991) *Religious institutions at work: the role of khalwa during the drought in the Red Sea area*. RESAP Technical Papers. Khartoum: Khartoum University Press.
EL HASSAN, H.M. & J.A. VAN DIJK (1988) *Research waterspreading pilot scheme*. Project proposal submitted to the National Council for Research and The Ford Foundation. Kassala: Department of Soil Conservation, Land Use and Water Programming.
EL JAHLI, T. (1973) *Inspection of El Fasher Water Spreading Schemes 11-14 March 1973*. Khartoum: Planning Section Soil Conservation Administration.
EL KARRAR, E. (1976) *The economics of agricultural production under flood conditions in Eastern Sudan: a case study of the Gash and Tokar deltas*. Khartoum: University of Khartoum (M.Sc. thesis).
EL MAHDI, S.M.A. (1979) *Introduction to the Land Law of The Sudan*. Khartoum: Khartoum University Press.
El MOSBAH, A.H. (1987) "Planning and management of streamwater in the Border Area through water spreading". *Paper for the Regional Symposium on Remote Sensing Applications in Hydrology and Natural Resources, Royal Jordanian Geographic Center, 22-26 October 1987, Amman.*
EL MOSBAH, A.H., O. EL MUBARAK, & A.M. EL HASSAN (1988) *Crop assessment in Hedadeib area*. Kassala: Ministry of Agriculture and Natural Resources, Department of Soil Conservation, Land Use and Water Programming.
EL MOULA, M.E.A. (1985) *On the problem of resource management in the Sudan*. Environmental Monograph Series No.4. Khartoum: University of Khartoum/IES.
EL SAMMANI, M.O. (1988) *Baseline survey for Darfur Region*. Khartoum: Government of Darfur for MFEP/UNDP and IES.
EL TOM, M.A. (1975) *The rains of The Sudan, mechanism and distribution*. Khartoum: Khartoum University Press.
ERGO (1990) *Integrated livestock surveys of Red Sea Province, Sudan*. Oxford: Environmental Research Group Oxford for OXFAM.
EUROCONSULT & NEWTECH (1987) *Socio-economic study of the Gash delta, Sudan. Final report.* Arnhem: Euroconsult/Newtech.
EVANS, H.E. & P. NGAU (1991) "Rural-urban relations, household income diversification and agricultural productivity". In: *Development and Change*, Vol.22, pp.519-545.
EVENARI, M. (1975) "Ancient desert agriculture and civilizations: do they point the way to the future ?" In: Y. MUNDLAK & S.F. SINGER, eds., *Arid zone development: potentialities and problems*, pp.83-97. Massachusetts: Ballinger Publishing Company.
FADL, M.S. (1988) *Hameskoreib the oasis of the Koran in Eastern Sudan*. Kassala: The Printing Press (in Arabic).
FALLOUX, F. & A. MUKENDI, eds., (1988) *Desertification control and renewable resource management in the Sahelian and Sudanian Zones of West Africa*. World Bank Technical Paper No.70. Washington: World Bank.
FAO (1973) *Water laws in Muslim countries*. Irrigation and Drainage Paper No.20/1. Rome: FAO.
FAO (1981) *Arid zone hydrology for agricultural development*. Irrigation and Drainage Paper No.37. Rome: FAO.
FAO (1985) *Rehabilitation of agriculture in Africa. Project proposals. Annex 6: Chad*. Rome: FAO.
FAO (1986a) *African agriculture: the next 25 years. Annex II, The land resource base*. Rome: FAO ARC/86/3.
FAO (1986b) *African agriculture: the next 25 years. Annex IV, Irrigation and water control*. Rome: FAO ARC/86/3.

FAO (1986c) *Report to the Government of The Sudan of the FAO/WFP multidonor mission on the evaluation of the food supply, agriculture and livestock situation.* Rome: FAO Office for Special Relief Operations Report 01/86/E.

FAO & LUND (1989) *Woody biomass distribution. Kassala Province, The Democratic Republic of The Sudan. (1:1,000,000 map based on 1987 and 1988 LANDSAT TM scenes).* Khartoum: prepared by the LUND team for FAO, NFC and FAO-FDES, GCP/SUD/033/NET.

FARAH, G.M. (1990) *The financial allocation; the Eastern Region as a case study.* Khartoum: University of Khartoum/DSRC (diploma thesis).

FARMER, G. & T.M.L. WIGLEY (1985) *Climatic trends for tropical Africa. A research report for the ODA.* Norwich: Climatic Research Unit, School of Environmental Sciences, University of East Anglia.

FATTOVICH, R. (1990) *Archaeology and history in the Gash Delta (Kassala Province, Sudan). Instituto Universitario Orientale Napoli.* Paper delivered at the 7th international conference, International Society for Nubian Studies, Geneva 3-8 September 1990.

FLEMING, G.J. (1922) "Kassala". In: *Sudan Notes and Records*, Vol.5, No.2, pp.65-77.

FLOOR, W. & J. GORSE (1988) "Household energy issues in West Africa". In: F. FALLOUX & A. MUKENDI, eds., *Desertification control and renewable resource management in the Sahelian and Sudanian Zones of West Africa.* World Bank Technical Paper No.70. Washington: World Bank.

FRE, Z. (1991) *Wetlands in drylands: the agroecology of savanna systems in Africa. Part 3d: Khor Baraka,a key resource in eastern Sudan and Eritrea.* IIED Drylands Programme. London: International Institute for Environment and Development.

GAMACHU, D. (1977) *Aspects of climate and water budget in Ethiopia.* Addis Ababa: Addis Ababa University Press.

GDAC (Gash Delta Agricultural Corporation) *Annual reports 1980-1990.* Aroma: GDAC.

GILL, G.J. (1993) *O.K. The data's lousy, but its all we've got (being a critique of conventional methods).* IIED Gatekeeper series 38. London: International Institute for Environment and Development.

GLEICHEN, S. (1898) *Report on the Nile and country between Dongola, Suakin, Kassala and Omdurman describing the various routes bearing on this country.* London: War Office.

GOOSSENS, E. (1992) "Étude de la dégradation environnementale dans le Parc Regional W au Nord Bénin. Rapport final, Faculté des Sciences Agronomiques, Université de Gand". In: *Rapports entre agriculteurs et éleveurs au Nord Bénin: Écologie et Interdependance Transformeé.* Université d'Amsterdam et Université National de Bénin, Rapport Final Tome II Annexes, pp.2-14. Amsterdam: Université d'Amsterdam.

GOVERNMENT OF THE SUDAN (GOS) (1982) *Upper Atbara Feasibility Study. Final feasibility report. Appendices. Volume A: previous studies and survey results. Volume B: main features of projected schemes.* Khartoum: Ministry of Irrigation & SOGREAH Consulting Engineers.

GOS (1986) *Strategy for development of rainfed agriculture, Annex 1. Technology Task Force Main Report.* Khartoum: Ministry of Agriculture.

GOS/GON (1988) *Joint Sudanese-Netherlands evaluation mission Kassala Area Development Activities KADA Sudan. Main report October 1988.* Khartoum/The Hague: Government of The Sudan and Government of The Netherlands.

GOS/GON (1990a) *KADA (Kassala Area Development Activities project). Joint evaluation November 1990.* Khartoum/Utrecht: Government of The Sudan and Government of The Netherlands. (including seperate "Annex to logical Framework Project Design Summary").

GOS/GON (1990b) *Gash Delta Corporation Project Sudan. Joint evaluation October 1990.* Khartoum/Utrecht: Government of The Sudan and Government of The Netherlands.

GOS/GON (1992) *Formulation Mission Gash Delta Project, Sudanese-Netherlands formulation mission October 1992.* Khartoum/The Hague: Government of The Sudan and Government of The Netherlands.

GREEN MISSION (1985) *Green Mission to Kassala, March 19-April 2, 1985.* The Hague: Ministry of International Co-operation.

GUGLER, J. & W.G. FLANAGAN (1978) "Urban-rural ties in West Africa: extent, interpretation, prospects and implications". In: W.M.J. VAN BINSBERGEN & H.A. MEILINK, eds., *Migration and the transformation of modern African society*. African Perspectives 1978/1, pp.67-78. Leiden: Afrika Studie Centrum.

GULHATI, R. (1990) "Who makes economic policy in Africa and how ?" In: *World Development*, Vol.18, No.8, pp.1147-1161.

HAAGSMA, B. & C. REIJ (1993) "Frentes de trabalho: potentials and limitations of large scale labour employment for soil and water conservation in Cape Verde". In: *Land Degradation and Rehabilitation* 4 (2), pp.73-85.

HALE, G.A. (1966) *Cultivation terraces in Western Darfur, Sudan: a study in an agricultural form and practice*. Los Angeles: University of California (Ph.D. thesis).

HALLAM, G.M., W. VAN CAMPEN & E. ROOSE (1985) *Reacting to farmers' complaints of agriculturally intensive farms in southern Mali: from fixed answers to flexible responses. Paper to the IVth Int. Conf. on Soil Conservation, Maracay Venezuela, November 1985.*

HALL, A.E., G.H. CANNELL & H.W. LAWTON, eds., (1979) *Agriculture in semi-arid environments*. Ecological Studies 34. Berlin: Springer Verlag.

HAMMOND, R. & P. McCULLAGH (1982) *Quantitative techniques in geography*. Oxford: Clarendon Press.

HARMOND ASSOCIATES (1991) *Kordofan Area Development Scheme, water resources consultancy. First impact report*. Khartoum: Harmond Associates SUD 90/002/99.

HARRISON, P. (1987) *The greening of Africa. Breaking through in the battle for land and food*. London: Paladin Grafton Books with IIED.

HARRISON, M.N. & J.K. JACKSON (1958) *Ecological classification of the vegetation of the Sudan*. Forests Bulletin No.2. Khartoum: Ministry of Agriculture, Forests Department.

HARRISS, B. & J. HARRISS (1984) "Generative or parasitic urbanism ? Some observations from the recent history of a South Indian market town". In: *Journal of Development Studies* 20(3), pp.82-101.

HARVEY, D. (1973) *Social justice and the city*. London: Edward Arnold.

HASHIM, M.H. (1987) *Hedadeib water spreading innovation versus traditional practices*. Kassala: Ministry of Agriculture and Natural Resources, Department of Soil Conservation, Land Use and Water Programming.

HEATHCOTE, R.L. (1983) *The arid lands: their use and abuse*. London: Longman Group Ltd.

HOLMBERG, J. & R. SANDBROOK (1992) "Sustainable development: what is to be done ?" In: J. HOLMBERG, ed., *Policies for a small planet*, pp.19-38. London: International Institute for Environment and Development & Earthscan.

HOLT, P.M. & M.W. DALY [1961](1988) *A history of The Sudan from the coming of Islam to the present day*. London: Longman.

HOROWITZ, M.M. & M. SALEM-MURDOCK (1987) "The political economy of desertification in White Nile Province Sudan". In: P.D. LITTLE, M.M. HOROWITZ, & A.E. NYERGES, eds., *Lands at risk in the Third World: local-level perspectives*, pp.95-113. London: Westview Press.

HOSELITZ, B.F. (1960) *Sociological aspects of economic growth*. New York: Academic Press.

HUBACH, A.B. & S.A. MAROUF (1985) *Kassala urban and adjacent area development project. Identification off-farm activities First phase 1984*. Kassala: KADA/DHV Consulting Engineers BV.

HUDSON, N.W. (1971). *Soil conservation*. London: Batsford Ltd.

HUDSON, N.W. (1987). *Soil and water conservation in semi-arid areas*. Soils Bulletin No.57. Rome: FAO.

HUIJSMAN, A. (1986) *Choice and uncertainty in a semi-subsistence economy. A study of decision making in a Philippine village*. Amsterdam: Royal Tropical Institute (Ph.D. thesis).

HULME, M. (1989) "The changing rainfall resources of Sudan". In: *Trans. Inst. Br. Geogr., New Series*, Vol.15, pp.21-34.

HULME, M. (1991) "Rainfall monitoring and forecasting in Sudan. Paper presented at the 2nd International Conference on Sudan, University College, Durham 8-11 April 1991". In: *Second International Sudan Studies Conference Papers Sudan: Environment and People held at University of Durham 8-11 April 1991*, Vol.2, pp.70-75. Durham: University of Durham.

HULME, M., Y. BIOT, J. BORTON, M. BUCHANAN-SMITH, S. DAVIES, C. FOLLAND, N. NICHOLDS, D. SEDDON, & N. WARD (1992) "Seasonal rainfall forecasting for Africa. Part 1: Current status and future developments". In: *Intern. J. Environmental Studies*, Vol.39, pp.245-256.

HURNI, H. (1986) *Guidelines for development agents on soil conservation in Ethiopia. Community Forests and Soil Conservation Development Department*. Adis Abbeba: Ministry of Agriculture.

HUXLEY, P.A., ed., (1983) *Plant research and agroforestry. Proceedings of a consultative meeting held in Nairobi 8-15 April 1981*. Nairobi: ICRAF.

HVA (1990) *Gash Delta Agricultural Corporation Agricultural. Report cropping season 1989/1990*. Diemen: HVA International BV.

HVA (1992) *Gash Delta Agricultural Corporation. Irrigation report*. Diemen: HVA International BV.

HVA (1993) *Gash Delta Project. Vol.1 Eco-Socio-Economic Survey. Vol.2 Codebooks and SPSS data*. Diemen: HVA International BV.

IBN HAWQAL [988 AD] (1975) Untitled. In: G. VANTINI, ed., *Oriental sources concerning Nubia. Published as field manual for excavators at the request of the Society for Nubian Studies*, pp.149-170. Warshaw/Heidelberg: Polish Academy of Sciences & Heidelberger Akademie (translation from Arabic).

IBRAHIM, A.A. (1980) *River Gash behaviour and training*. Khartoum: University of Khartoum, Department of Civil Engineering (M.Sc. thesis).

IBRAHIM, F.N. (1988) "Water harvesting and water spreading für Flutanbau in der Sahelzone des Sudan". In: *Die Erde*, Vol.119, pp.179-184.

IBRD (1992) *World Development Report 1992. Development and environment, world development indicators*. Washington: Oxford University Press.

IES (Institute of Environmental Studies) (1991) *Kassala Province Environmental Profile*. Khartoum: University of Khartoum, IES.

IFAD (1986) *Soil and water conservation in Sub-Saharan Africa. Issues and options*. Amsterdam: Free University Amsterdam/CDCS with IFAD.

IFAD (1988) *Guinee. Réhabilitation agricole et conservation des sols dans le Fouta Djallon. Rapport de la mission de préparation. Rapport No. 0085-Gu*. Rome: IFAD.

IFAD (1990) *Small-scale water control project*. Appraisal report, working papers. Rome: IFAD.

IFAD (1992) *Soil and water conservation in Sub-Saharan Africa. Towards sustainable production by the rural poor*. A report prepared for the IFAD by the CDCS Free University Amsterdam. Amsterdam: Free University/CDCS.

ILACO (1981) *Agricultural compendium for rural development in the tropics and subtropics*. Amsterdam: Elsevier Scientific Publishing Company.

ILO (International Labour Office) (1986) *Proposals for a labour-intensive public works programme in Aroma and Kassala Districts. Eastern region, Sudan*. Geneva: ILO SUD/85/006.

INGEBRITSEN, S.E. & R.L.P. LYON (1985) "Principle component analysis for multitemporal image pairs". In: *International Journal of Remote Sensing*, Vol.6, No.5, pp.687-696.

IOV (Inspectie Ontwikkelingssamenwerking te Velde) (1991) *Sectorprogramma Plattelandsontwikkeling. Een programma-evaluatie met projectonderzoek in Indonesië, Soedan, Rwanda, Tanzania en Nicaragua*. 's Gravenhage: DGIS.

IOV (1992) *The sector programme for rural development. Programme evaluation with special reference to Indonesia, Sudan, Rwanda, Tanzania and Nicaragua. Summary evaluation report*. The Hague: DGIS.

IOV (1993) *Evaluatie en monitoring. De rol van project evaluaties en monitoring in de bilaterale hulp*. 's Gravenhage: DGIS.

IUCN (International Union for Conservation of Nature and Natural Resources) (1980) *World conservation strategy. Living resource conservation for sustainable development*. Gland: IUCN with UNEP & WWF in collaboration with FAO & Unesco.

JAMES [1883](1969) *The wild tribes of the Soudan. An account of travel and sport chiefly in the Basé country. Being personal experiences and adventures during three winters in the Soudan*. New York: Negro Universities Press.

JEFFERSON, J.H.K. (1952) *Soil conservation in the Sudan, development and projects*. Khartoum: Ministry of Agriculture.

JONES, W.I. & R. EGLI (1984) *Systèmes de culture en Afrique. Les hautes terres de Zaire, Rwanda et Burundi*. Document technique de la Banque Mondial, No.27F. Washington: La Banque Mondial.

KADA (Kassala Area Development Activities) (1985) *Report on the KADA Quarterly Seminar no.1 held on 14-15 December 1985 in Kassala Sudan*. Kassala: KADA/DHV Consulting Engineers BV.

KADA (1986) *Report KADA Health mission*. Kassala: KADA/DHV Consulting Engineers BV.

KADA (1987a) *Workplan 1987 sector agriculture*. Kassala: KADA/DHV Consulting Engineers BV.

KADA (1987b) *Workplan 1988*. Kassala: KADA/DHV Consulting Engineers BV.

KADA (1988) *Workplan 1989-1990*. Kassala: KADA/DHV Consulting Engineers BV.

KADA (1989a) *Progress report 23*. Kassala: KADA/DHV Consulting Engineers BV.

KADA (1989b) *Progress report 25*. Kassala: KADA/DHV Consulting Engineers BV.

KATES, R.W. & V. HAARMANN (1992) "Where the poor live. Are the assumptions correct ?" In: *Environment*, Vol.34, No.4, pp.4-28.

KEJELA, K. (1992) "The costs of soil erosion in Anjeni, Ethiopia". In: K. TATO & H. HURNI,eds., *Soil conservation for survival. Selected papers presented at the 6th Int. Soil Conservation Conference held in Ethiopia and Kenya 6-18 November 1989*, pp.219-230. Ankeney: SWCS with ISCO & WASWC.

KERR, J.M. & N.K. SANGHI (1991) "Economic determinants of soil and water conservation investments". In: *Farmers' Practices and Soil and Water Conservation Programs. Summary proceedings of a workshop, 19-21 June 1991*, pp.15-17. Andhra Pradesh: ICRISAT Center India, ICRISAT/ Winrock International Institute for Agricultural Development.

KHALAF, N. (1965) *British policy regarding the administration of the Northern Sudan 1899-1951*.Ann Arbor: UMI microfilm (Ph.D. thesis).

KHOGALI, M.M. (1980) *Sedentarization of nomadic tribes in the North and Central Sudan*. Khartoum: University of Khartoum (Ph.D. thesis).

KHOGALI, M.M. (1981) "Sedentarization of the nomads: Sudan". In: J.G. Galaty, D. Aronson, P.C. Salzman & A. Chovinard, eds., *The future of pastoral peoples. Proceedings of a Conference held in Nairobi, Kenya, 4-8 August 1980*, pp.302-317. Ottawa: IDRC.

KIEFT, H. (1993) "The potential of Low External Input Agriculture in Sub-Saharan Africa". In: H. van Reuler & W.H. Prins, eds., *The role of plant nutrients for sustainable food crop production in Sub-Saharan Africa*, pp.127-140. Leidschendam: Dutch Association of Fertilizer Producers VKP.

KLEMM, W & D. PRINZ (1989) "Utilization of the factor soil through run-off farming: a case study from Mali". In: *Natural Resources and Development*, Vol.30, pp.76-87.

KRAAYENHAGEN, J. (1987) *Foothill rehabilitation through waterspreading*. Kassala: KADA/DHV Consulting Engineers BV.

KUHLMAN, T., SALAH EL-DIN EL-SHAZALI IBRAHIM, & W. KOK (1987) *Refugees and regional development. Final report of the research project "Eritreans in Kassala"*. Amsterdam: Free University of Amsterdam/CDCS and University of Khartoum/DSRC.

KUHLMAN, T. (1990) *Burden or Boon ? A study of Eritrean refugees in The Sudan*. Anthropological Studies VU. Amsterdam: Free University Press.

KUHLMAN, T. (1994) *Asylum or Aid ? The economic integration of Ethiopian and Eritrean refugees in the Sudan*. Amsterdam: Free University (Ph.D. thesis).

KUTSCH, H. (1983) "Currently used techniques in rain-fed water-concentrating culture". In: *Applied Geography and Development*, Vol.21, pp.108-117.

LANDECK, J. (1992) *Technical support mission for the Fouta Djallon agricultural rehabilitation project Guinee-Conakry*. Amsterdam: Free University/CDCS.

LEBON, J.H.G. (1965) *Land Use in Sudan*. The World Land Use Survey Monograph 4. London: Geographical Publications Ltd.

LITTLE, P.D. (1992) *The elusive granery. Herder, farmer, and state in northern Kenya*. African Studies Series 73. Cambridge: Cambridge University Press.

LIPTON, M. (1977) *Why poor people stay poor. A study of urban bias in world development*. London: Temple Smith.

LIVERENZ, I. (1989) *"Without injections I can't find any comfort". On preconditions for a Swiss Red Cross supported MCH and health education programme in Awad*. Kassala: Swiss Red Cross.

LRDC (Land Resources Development Centre) (1987) *Sudan, profile of agricultural potential*. Surrey: Overseas Development Administration.

LUDWIG, H.D. (1968) "Permanent farming on Ukara. The impact of land shortage on husbandry practices". In: H. RUTHENBERG, ed., *Smallholder farming and smallholder development in Tanzania*, pp.88-135. Afrika Studien 24. München: Weltforum Verlag.

MAFNR (Ministry of Agriculture Food and Natural Resources) (1976a) *Sudan National Livestock Census and Resource Inventory, Vol.28: the results of the household questionnaire survey 1975-1976*. Khartoum: MAFNR.

MAFNR (1976b) *Sudan National Livestock Census and Resource Inventory, Vol.31: the results of an aerial census of resources in Sudan from August 1975 to January 1976*, Khartoum: MAFNR.

MANR (Ministry of Agriculture and Natural Resources) (1985) *Agricultural prices in Sudan. A historical review and analysis 1970-1984*. Khartoum: MANR/Department of Agricultural economics, Planning and Agricultural Economics Administration,

MANR (1988) *Border Area Pilot Project, Animal Resources Survey Part 1: Dry Season Survey Februari-April 1988*. Kassala: MANR/Department of Range and Pasture Management.

MANR (1991) *Agricultural situation and outlook. Annual Report 1989/1990. Ministry of Agriculture and Natural Resources*. Khartoum: MANR/Department of Agricultural Economics, Planning and Agricultural Economics Administration.

MARSHALL, A.H. (1949) *Report on local government in The Sudan*. Khartoum: Sudan Government.

MARTIN, A. (1985) *Farming systems in the Kebkabiya area, North Darfur Province, Sudan. The role and potential of small scale irrigated gardening*. Khartoum: OXFAM.

MARTIN, P. (1990) *Mission d'appui au projet de réhabilitation agricole dans le Fouta Djallon (Guinee)*. Amsterdam: Free University Amsteradm /CDCS.

MATTSSON, J.O. & A. RAPP (1991) "The recent droughts in Western Ethiopia and Sudan in a climatic context". In: *Ambio*, Vol.20, No.5 pp.172-175.

McCARTHY, J.W., C. CLAPP-WINCEK & S. LONDNER (1985) *A soil and water conservation project in two sites in Somalia: 17 years later*. AID Project Impact Evaluation Report No.62 for USAID. Mogadishu: USAID.

MEEUWS, R. (1990) *Summary of a report on planning in the Eastern Region*. Report by the DGIS Planning Advisor. Kassala: MFEP (mimeo).

MFEP (Ministry of Finance and Economic Planning) (1982) *Project proposal co-operative mixed farming Salem Aleikum*. Kassala: MFEP.

MFEP (1985) *Development programme estimates for the fiscal year 1986/1987*. Khartoum: MFEP.

MFEP (1989a) *A report on the visits by the Mininistry of Finance mission to KADA projects in the Eastern Region*. Kassala: MFEP/Directorate of Development and Economic Planning.

MFEP (1989b) *The Four Year Salvation, Reform and Development Plan, Development Programme estimates for the fiscal year 1988-1989*. Khartoum: MFEP.

MFEP (1990) *Cost of living index, Khartoum low and high income groups, 1985-1988*. Khartoum: MFEP/Department of Statistics, Division of Internal Trade.

MFEP & UNDP (1990) *Baseline survey of the Eastern Region* Vol. 1-5 (draft report). Khartoum: University of Khartoum/DSRC, Department of Geography.

MFEP & UNDP (1991) *A strategy for the rehabilitation and development of the Eastern State*. Khartoum: MFEP, UNDP/DTCD Sud/86/078.

MFNE (Ministry of Finance and National Economy) (1987) *Population and housing census 1983*. Khartoum: MFEP/Department of Statistics, Population Census Office.

MFPNE (Ministry of Finance, Planning and National Economy) (1976) *Second population census 1973, Kassala Province*. Khartoum: Department of Statistics, Population Census Office.

MIEHE, S. (1988) *Vegetation ecology of the Jebel Marra Massif in the semiarid Sudan*. Dissertationes Botanicæ, Band 113. Berlin: Cramer.

MILLINGTON, A. (1987a) "Soil conservation and shifting cultivators part 1". In: *ILEIA* Vol.3, No.1, pp.15-16.

MILLINGTON, A. (1987b) "Soil conservation and shifting cultivatorsn part 2". In: *ILEIA* Vol.3, No.2, pp.18-19.

Ministry for Social Affairs (1958) *First population census of Sudan 1955/56. Methods Report Vol.II*. Khartoum: Population Census Office.

MITCHELL, C.W. (1991) "Physiography, geology and soils". In: G.M. CRAIG, ed., *The agriculture of The Sudan*, pp.1-18. Oxford: Oxford University Press.

MoA (Ministry of Agriculture) (1963) *Roseires soil survey report No.2, Blue Nile east bank. Guneid to Khartoum. Soils and engineering reconnaissance*. London: Hunting Technical Services & Sir M. MacDonald & Partners.

MoA (1965) *Roseires soil survey report No.14, White Nile east bank, Rabak to Khartoum. Semi-detailed soil survey and land classification Vol.1: the report.* London: Hunting Technical Services & Sir M. MacDonald & Partners.

MOHAMED, A.E. (1983) *Some impacts of desertification on the Gash delta region.* Khartoum: University of Khartoum/IES (M.Sc. thesis).

MOHAMED, S.A. (1992) *Evaluation of the community subsistence cultivation programme in Red Sea Hills.* Report prepared for UNICEF. Khartoum.

MOKHTAR, G. (1990), ed., *Ancient civilizations of Africa.* UNESCO general history of Africa series volume II. Dar es Salaam: Tanzania Publishing House.

MORGAN, W.T.W. (1974) "The South Turkana Expedition. Scientific Papers X. Sorghum gardens in south Turkana: cultivation among a nomadic pastoral people". In: *The Geographical Journal,* Vol.140, Part 1, pp.80-94.

MORIS, J. & R. NORMAN (1984) *Prospects for small-scale irrigation development in the Sahel.* USAID Water Management Synthesis II Project report 26. Utah: WMR.

MORIS, J. (1987) "Irrigation as a priviliged solution in African development". In: *Development Policy Review,* Vol.5, pp.99-123.

MORTON, J.F.M. (1989) *Descent, reciprocity and inequality among the Northern Beja.* Hull: University of Hull (Ph.D. thesis).

MORTON, J.F.M. (1993) "Pastoral decline and famine: the Beja case". In: J. MARKAKIS, ed., *Conflict and the decline of Pastoralism in the Horn of Africa,* pp.31-44. London: The Macmillan Press Ltd and ISS The Hague.

MORTON, J. & Z. FRE (1986) *Red Sea Province and the Beja. A preliminary report for Oxfam.* Port Sudan: OXFAM.

MULDER, P.J.M. (1990) *Water spreading and related water management activities in the Border Area of Kassala Province, Sudan. Recommendations for improvement.* MFEP/KADA/DHV Consulting Engineers, Amersfoort. Kassala: MFEP/KADA/DHV Consulting Engineers BV.

MUSA A.M., M.H. AHMED, & O.K. MOHAMED (1989) *A review of Hadadeib waterspreading experiment.* Kassala: Department of Soil Conservation, Land Use and Water Programming.

NADEL, S.F. (1945) "Notes on Beni Amer society". In: *Sudan Notes and Records,* Vol.25, No.1. pp.51-94.

NADEL, S.F (1949) *The Nuba. An anthropological study of the hill tribes in Kordofan.* Oxford: Oxford University Press.

NEI (Netherlands Economic Institute) (1991) *Gash Delta Rehabilitation Project (Kassala block). Pre-Feasibility Study.* Rotterdam: NEI.

NEWBOLD, D. (1935) "The Beja tribes of the Red Sea hinterland". In: J.A. de C. Hamilton, ed., *The Anglo-Egyptian Sudan from within,* pp.140-164. London: Faber & Faber Ltd.

NIEMEIJER, D. (1993) *Indigenous runoff farming in a changing environment. The case of Kassala's Border Area, Sudan.* Amsterdam: University of Amsterdam, Wageningen Agricultural University (M.Sc. thesis).

NORÉN, S. & L.G. NORDÉN (1983) *Soil conservation and tree planting in Zambia: proposal of activities within Agricultural Sector Support Programme.* Working Paper 10. Uppsala: Swedish University of Agricultural Sciences, International Rural Development Centre.

NSIBANDZE, B.M. (1987) *The history of soil conservation in Swaziland.* A paper presented to a Southern African Development Co-ordinating Conference (SADCC) Seminar held in Maputo March 1987.

NORUSIS, M.J. (1988) SPSS/PC+ V2.0 base manual. Chicago: SPSS Inc.

OAKLEY, P. (1987) *People's participation in soil and water conservation.* SADCC Seminar March 2-6, 1987, Report No.10. Maputo: SADCC.

OECD (Organization for Economic Cooperation and Development) (1992) *Development co-operation: efforts and policies of the development assistance committee.* Paris: OECD.

OLDEMAN, L.R., R.T.A. HAKKELING & W.G. SOMBROEK (1990) *World map of the status of human-induced soil degradation. An explanatory note.* Wageningen: GLASOD, ISRIC/UNEP.

OMER, M.A. & G. MADIBO (1990) *Development and environmental control measures and cropping practices for the gardud soils at Megabeila and El Shaigeia villages (N. Kordofan).* Final report of the Western Sudan Agricultural Research Project. El Obeid: Ministry of Agriculture, UNICEF.

OMER, M.A. & E.T.M. EL AMIN (1991) *Evaluation of water harvesting techniques for improving crop production in North Kordofan*. Final report to the National Council for Research and the Ford Foundation. El Obeid: Western Sudan Agricultural Research Project, Agricultural Research Corporation El Obeid.

OSMAN, M.E.M. (1989) *BAPP agriculture experimental programme*. Kassala: Kassala Agricultural Research Station.

OSMAN, K. & S. DABALOB (1990) "Water management in the Red Sea area". In: *Proceedings from the second annual workshop in Khartoum, The Red Sea Area Programme RESAP*, 17 pages. Khartoum: University of Khartoum, University of Bergen.

OWEN, T.R.H. (1937) "The Hadendowa Part II". In: *Sudan Notes and Records*, Vol.20, p.183-208.

OXFAM (1988) *Tour Report 10 November 1988*. Port Sudan: OXFAM.

PACEY, A. & A. CULLIS [1986] (1989) *Rainwater harvesting. The collection of rainfall and run-off in rural areas*. London: Intermediate Technology Publications.

PAUL, A. (1950) "Notes on the Beni Amer". In: *Sudan Notes and Records*, Vol.31, pp.223-245.

PAUL, A. [1954] (1971) *History of the Beja tribes of The Sudan*. Cambridge: Cambridge University Press.

PEARCE, D., E.B. BARBIER & A. MARKANDYA (1990) *Sustainable development. Economics and environment in the Third World*. London: Earthscan Publications Ltd.

PENNING DE VRIES, F.W.T. & M.A. DJITEYE (1991), eds., *La productivité des pâturages sahéliens. Une étude des sols, des végétations et de l'exploitation de cette ressource naturelle*. Wageningen: PUDOC.

PFLAUMBAUM, H. (1991) "Dürrebedingte Veränderungen des Weidewirtschaftlichen Nutzungspotentials in der Butana". In: H. PFLAUMBAUM & K.H. PÖRTGE, eds., *Göttinger Forschungsprojekte im Sudan*, pp.93-112. Göttingen: Geographisches Institut der Universiät Göttingen.

PIERI, C. (1989) *Fertilité des terres de savanes. Bilan de trente ans de recherche et de développement agricoles au sud du Sahara*. Paris: Ministère de la Coopération et Du Développement CIRAD/IRAT.

PIQUEMAL, D. (1991) "Inventaire et bilan des retenues d'eau au Burkina Faso". In: *Cahier d'Outre-Mer*, No.44 (175), pp.259-280.

POST, J. (1987) T*he future development of Kassala's town centre. Final report: the outlines of a plan for the central area*. Amsterdam: University of Amsterdam, Department of Planning and Demography.

PRETTY, J.N. & R. CHAMBERS (1993) "Towards a learning paradigm: new professionalism and institutions for agriculture". In: *Rural people's knowledge, agricultural research and extension practice. Overview papers*, pp.48-84. IIED Research Series Vol.1, No.1. London: International Institute for Environment and Development.

PRICE GITTINGER, J. (1984) *Economic analysis of agricultural projects*. EDI Series in Economic Development. Baltimore: The John Hopkins University Press.

PRINCE, S.D., C.O. JUSTICE & S.O. LOS (1990) *Remote sensing of the Sahelian environment. A review of the current status and future prospects*. Brussels: CEC Directorate General VIII, Joint Research Centre Ispra Establishment/CTA.

RANDELL, J.R. (1963) "Land Use on the arid margin of the clays in Blue Nile and Kassala Provinces". In: J.H.G. Lebon, ed., *Proceedings of the Ninth Annual Conference on the Theme of Surveying for Development in Sudan held in Khartoum 12-13 January 1961*, pp.67-81. Khartoum: Sudan Survey Department.

RANDS, B.C. (1992) "Experiences in soil and water conservation work amongst pastoral peoples of northeastern Mali". In: K. TATO & H. HURNI, eds., *Soil conservation for survival. Selected papers presented at the 6th Int. Soil Conservation Conference held in Ethiopia and Kenya 6-18 November 1989*, pp.345-355. Ankeney: SWCS, ISCO and WASWC.

RASMUSSON, E.M. (1987) "Global climate change and variability: effects on drought and desertification in Africa". In: M.H. GLANTZ, ed., *Drought and hunger in Africa: denying famine a future*, pp.3-22. Cambridge: Cambridge University Press.

REIJ, C. (1984) *Back to the grass strips: an analysis of soil conservation policies in Swaziland*. RDRP Research Paper 12. Kwaluseni: SSRU.

REIJ, C., P. MULDER, & L. BEGEMANN (1988). *Water harvesting for plant production*. World Bank Technical Paper 91. Washington: World Bank.

REIJ, C. (1990) *Indigenous soil and water conservation in Africa: an assessment of current knowledge.* Paper presented to the Workshop on conservation in Africa: indigenous knowledge and conservation strategies, Harare, Zimbabwe, December 2-7 1990. Social Science Research Council/American Council of Learned Societies, The Joint Committee on African Studies. Amsterdam: Free University Amsteradm/CDCS.

REIJ, C. (1992) *Indigenous soil and water conservation in Africa.* IIED Gatekeeper series 27. London: International Institute for Environment and Development.

REIJNTJES, C. (1986) "Dryland management, take a fresh look at local traditions". In: *ILEIA*, Vol.5, pp.4-13.

REIJNTJES, C., B. HAVERKORT & A. WATERS-BAYER (1992) *Farming for the future. An introduction to Low-External-Input and Suistainable Agriculture.* London: The MacMillan Press Ltd. and ILEIA.

REITSMA, H.A., A.J. DIETZ, & L. DE HAAN, eds., (1992) *Coping with semiaridity. How the rural poor survive in dry-season environments.* Netherlands Geographical Studies 146. Amsterdam: Department of Human Geography.

REMONDE, R., L. VILLAMORA, & E.J. SIMONIDES (1992) "Labour demand for organic contour farming". In: *ILEIA* Vol.8, Nr.4, pp.6-8.

RICHARDS, G.H. (1951) *The Gash Delta.* Bulletin no.3. Khartoum: Ministry of Agriculture.

RICHARDS, J.A. (1986) *Remote sensing, digital image analysis. An introduction.* Berlin: Springer-Verlag.

RICHARDS, P. (1985) *Indigenous agricultural revolution. Ecology and food production in West Africa.* London: Hutchinson & Co.

ROCHETTE, R.M., ed., [1989] (1990) *Le Sahel en lutte contre la désertification. Leçons d'expériences.* Eschborn: CILSS-PAC, GTZ.

ROSS, J.C. (1947) *Land utilization and soil conservation in the union of South Africa. A review of progress to date and developments contemplated.* Pretoria: State Information Office.

RUTHENBERG, H. [1971] (1983) *Farming systems in the tropics.* Oxford: Oxford University Press and Clarendon Press.

RUTTEN, M.M.E.M. (1992) *Selling wealth to buy poverty. The process of the individualization of landownership among the Masaai pastoralists of Kajiado District, Kenya, 1890-1990.* Saarbrücken: NICCOS/Verlag Breitenbach Publishers.

SAEED, E.M. (1969) *Ground water appraisal of the Gash river basin at Kassala, Kassala Province.* Bulletin No.17. Khartoum: Geological and Mineral Resources Department.

SAEED, E.M. (1972) *Hydrogeology of Kassala District, Kassala Province.* Khartoum: Ministry of Industry and Mining, Geological and Mineral Resources Department.

SALIH, H.M. (1971) *Some aspects of Hadendowa social organization.* Khartoum: University of Khartoum (M.Sc. thesis).

SALIH, H.M. (1980) "Hadanduwa traditional territorial rights and inter-population relations within the context of the Native Administration System (1927-1970)". In: *Sudan Notes and Records*, Vol.61, pp.118-133.

SALIH, M.M. (1982) *Labour migration: a case study of Kassala township.* Khartoum: University of Khartoum/DSRC (diploma thesis).

SALIH, M.A.M. (1990) "Symbolic identification among the Hadendowa of Eastern Sudan". In: G. Pálsson, ed., *From water to world-making. African models and arid lands*, pp.109-120. Uppsala: The Scandinavian Institute of African Studies.

SALIH, A.M.A & M.A.A. KHADAM (1982) *Water resources in Sudan.* Khartoum: National Council for Research, Council for Scientific and Technological Research.

SANGHI, N.K. & J.M. KERR (1991) "The logic of recommended and indigenous soil and water conservation practices". In: *Farmers' Practices and Soil and Water Conservation Programs. Summary proceedings of a workshop, 19-21 June 1991, ICRISAT Center India*, pp.13-14. Andhra Pradesh: ICRISAT/Winrock International Institute for Agricultural Development.

SCHWAB, G.O., R.K. FREVERT, T.W. EDMINSTER, & K.K. BARNES (1981). *Soil and water conservation engineering.* New York: John Wiley & Sons.

SCLUWP (Department of Soil Conservation, Land Use and Water Programming) *Annual Reports 1982-1993.* Kassala: Ministry of Agriculture and Natural Resources, Department of Soil Conservation, Land Use and Water Programming.

SCLUWP (1987) *Grain and fodder production by water spreading*. Kassala: Department of Soil Conservation, Land Use and Water Programming.

SCOONES, I. & J. THOMPSON (1993) "Beyond farmer first. Rural peoples's knowledge, agricultural research and extension practice: towards a theoretical framework". In: *Rural people's knowledge, agricultural research and extensiion practice. Overview papers*, pp.1-20. IIED Reserach Series Vol.1, No.1. London: International Institute for Environment and Development.

SELIGMAN, C.G. & B.Z. SELIGMAN (1930) "Note on the history and present condition of the Beni Amer (Southern Beja)". In: *Sudan Notes and Records*, Vol.13, pp.83-97.

SHAHIN, M. (1985) *Hydrology of the Nile basin*. Development in water sciences 21. Amsterdam: Elsevier Publishers.

SHEPERD, A. (1983) "Capitalist Agriculture in the Sudan's dura prairies". In: *Development and Change*, Vol.14, pp.297-321.

SHEPERD, A., M. NORRIS & J. WATSON (1987) *Water planning in arid Sudan*. Middle East Science Policy Studies No.10. Development Administration Group, Institute of Local Government Studies, University of Birmingham. London: Ithaca Press and Atlantic Highlands.

SIM (Stuurgroep Impactstudie Medefinancieringsprogramma) (1990) *Terms of Reference. Field research impact study* (mimeo).

SIM (1991) *Betekenis van het Medefinancieringsprogramma: een verkenning, eindrapport*. Utrecht: Libertas.

SIMONS, W. (1994) *The assessment of indigenous water harvesting systems by multi-temporal analysis of SPOT and LANDSAT imagery. The case of the Border Area in Eastern Sudan*. Amsterdam: University of Amsterdam/Department of Physical Geography and Soil Sciences.

SMALING, E. (1993) *An agro-ecological framework for integrated nutrient management with special reference to Kenya*. Wageningen: Agricultural University Wageningen/Staring Centrum (Ph.D. thesis).

SOGREAH (1982) *Upper Atbara feasibility study. Final feasibility report, appendices. Vol.B: main features of projected scheme*. Khartoum: Ministry of Irrigation.

SOMMERHALTER, T. (1987) *Boden- und Wasserkonservierende Maβnahmen im Ouaddai Geographique des Tschad*. Ouahigouya: Projet Agro-Ecologie.

SORBO, G.M. (1985). *Tenants and nomads in Eastern Sudan. A study of economic adaptations in the New Halfa Scheme*. Uppsala: Scandinavian Institute of African Studies.

SORBO, G.M. (1991) "Systems of pastoral and agricultural production in eastern Sudan". In: G.M. CRAIG, ed., *The agriculture of The Sudan*, pp.214-229. Oxford: Oxford University Press.

SPENCER TRIMINGHAM, J. [1948] (1984) *Islam in the Sudan*. London: Oxford University Press.

SRIVASTAVA, J.P., P.M. TAMBOLI, J.C. ENGLISH, R. LAL & B.A. STEWART (1993) *Conserving soil moisture and fertility in the warm seasonally dry tropics*. World Bank Technical Paper Number 221. Washington: The World Bank.

STERKENBURG, J.J. (1987) *Rural development and rural development policies: cases from Africa and Asia*. Netherlands Geographical Studies 46. Utrecht: Geografisch Instituut Rijksuniversiteit Utrecht.

STERN, P. (1985) *Small scale irrigation*. London: Intermediate Technology Publications.

STOCKING, M. & N. ABEL (1992) "Labour costs: a critical element in soil conservation". In: K. TATO & H. HURNI, eds., *Soil conservation for survival. Selected papers presented at the 6th Int. Soil Conservation Conference held in Ethiopia and Kenya 6-18 November 1989*, pp.206-218. Ankeney: SWCS with ISCO and WASWC.

STONE, J. (1955) *Sudan economic development 1899-1913*. Khartoum: Sudan Economic Institute.

STRAHLER, A.N. (1975) *Physical geography*. New York: John Wiley & Sons.

STREN, R.E & R.R. WHITE, eds., (1989) *African cities in crisis: managing rapid urban growth*. Boulder: Westview Press.

STROOSNIJDER, L. (1992) "Nutriëntenstromen en duurzaam landgebruik in de tropen". In: *Verslag van de 64ste Tropische Landbouwdag. Nutriëntenstromen en Duurzaam Landgebruik in de Tropen, 11 november 1992*, pp.i-ii. Wageningen: Nili-KGvL & KIT.

STROOSNIJDER, L. (1994) "Improving land care in West Africa". In: *ILEIA*, Vol.10, No.2, pp.12-13.

SUDAN GOVERNMENT (1926) *The Sub-Māmūr's handbook, lectures on the penal, criminal procedure and civil codes and notes on accounts, agriculture, forestry, sanitation*. Khartoum: Sudan Government.

SUDAN GOVERNMENT (1944) *Report of the Soil Conservation Committee Sudan Government*. Khartoum: Sudan Government.

TAUER, W. & G. HUMBORG (1992) *Runoff irrigation in the Sahel Zone. Remote sensing and geographical information systems for determining potential sites*. Weikersheim: Verlag Josef Margraf.

TIFFIN, M., M. MORTIMER & F. GICHUKI (1994) *More people, less erosion. Environmental recovery in Kenya*. Chichester: John Whiley & Sons.

TILLEY, P. & M. McEWAN (1988) *Agricultural practices and rainwater harvesting potential among the Bisharine Beja of Khor Addib, Halaib*. ACORD Red Sea Hills Programme. Khartoum: ACORD.

TIMBERLAKE, L. (1991) *Africa in crisis. The causes, the cures of environmental bankruptcy*. London: Earthscan Publications Ltd.

TNO (Institute of Applied Geoscience) (1982) *Water resources assessment Kassala Gash Basin. Final technical report*. The Hague: Ministry of International Co-operation/TNO, Sudanese Ministry of Energy and Mining/ National Administration for Water.

TNO & NCDRWR (1990) *Water resources assessment and development project in the Sudan (WADS) Technical Committee Kassala (appendix D), Modelstudy Gash basin (appendix E)*. Khartoum: NCDRWR & TNO.

TOTHILL, J.D., ed., [1948] (1952) *Agriculture in the Sudan. A handbook of agriculture as practised in the Anglo-Egyptian Sudan*. London: Oxford University Press.

TOULMIN, C. (1983) *Economic behaviour among livestock-keeping peoples: a review of the literature on the economics of pastoral production in the semi-arid zones in Africa*. Development Series Occasional Paper 25. Norwich: University of East Anglia.

TOULMIN, C., I. SCOONES, & J. BISHOP (1992) "The future of Africa's drylands: is local resource management the answer?". In: J. HOLMBERG, ed., *Policies for a small planet*, pp.225-257. London: International Institute for Environment and Development & Earthscan.

TRILSBACH, A. (1991) "Historical development of agriculture". In: G.M. CRAIG, ed., *The agriculture of The Sudan*, pp.177-192. Oxford: Oxford University Press.

TUBIANA, M. & J. TUBIANA (1977) *The Zaghawa from an ecological perspective. Food gathering, the pastoral system, tradition and development of the Zaghawa of the Sudan and the Chad*. Rotterdam: Balkema.

UN-NGLS (Non-Governmental Liason Service) (1990) *Voices from Africa*. Issue Number 2: NGOs and grassroots development. Geneva: UN-NGLS.

UNDP & FAO (1968) *Land and water resources survey in the Jebel Marra area, The Sudan. Final Report*. Rome: FAO SF:48/SUD-17.

UNDP (1990) *Project document Area Development Scheme Central Butana Eastern State (Sudan)*. Khartoum: UNDP SUD/90/006/A/01/99.

UNEP (1979) "Autralia combats desertification". In: *Desertification Control Bulletin*, Vol.2, No.1, pp.13-16.

UNEP (1987) "Rehabilitating Kenya's North. The Baringo pilot semi-arid area project". In: *Desertification Control Bulletin*, No.14, pp.55-58.

UNGER, P.W. (1984) *Tillage systems for soil and water conservation*. FAO Soils Bulletin 54. Rome: FAO.

UNSO (1992) *Assessment of desertification and drought in the Sudano-Sahelian Region 1985-1991*. New York: UNDP/UNSO.

USACERL (US Army Construction Engineering Research Laboratories) (1993) *GRASS 4.1 User's reference manual. Part 1 and Part 2*. Champaign: USACERL.

VAGENES, V. (1990) "Change in gender roles. With special reference to the Hadendowa pastoral society". In: *Red Sea Area Programme (RESAP), Proceedings from the Second Annual Workshop in Khartoum of the University of Khartoum and the University of Bergen, March 1990*, 26 pages.

VAN DAM, A.J. & J.A. HOUTKAMP (1992) *Livelihood and runoff farming. The case of the Border Area village of Hafarat Eastern Region Sudan*. Amsterdam: University of Amsterdam (M.A. thesis).

VAN DER KEVIE, W. & I. BURAYMAH (1976) *Exploratory soil survey of the Kassala province. A study in physiography, soils and agricultural potential*. Wad Medani: Soil Survey Administration.

VAN DER POL, F. (1992) *Soil mining. An unseen contributor to farm income in southern Mali*. Bulletin 325. Amsterdam: Royal Tropical Institute.

VAN DER WAL, A. & F. ZAAL (1990) *Bibliography on indigenous soil and water conservation with special reference to Africa*. Amsterdam: Free University Amsterdam/CDCS.

VAN DIEPEN, C.A., H. VAN KEULEN, J. WOLF, & J.A.A. BERKHOUT (1991) "Land evaluation: from intuition to quantification". In: *Advances in Soil Science*, Vol.15, pp.139-204.

VAN DIJK, J.A. (1990) *Position paper Hedadeib water spreading pilot scheme.* WARK occasional paper. Kassala: WARK.

VAN DIJK, J.A. (1991) *Water spreading by broad-based earth embankments. WARK final report: technical performances and socio-economic aspects in the Border Area, Eastern Region, Sudan.* Report to: National Council for Research, The Ford Foundation. Amsterdam.

VAN DIJK, J.A. (1992) "Rainwater harvesting in the Sudan". In: *Baobab* 8, Arid Lands Information Network RITA/ALIN/OXFAM, pp.14-15.

VAN DIJK, J.A. & M.H. AHMED (1993a) *Opportunities for expanding water harvesting in Sub-Saharan Africa. The case of the teras of Kassala.* IIED Gatekeeper series No.40. London: International Institute for Environment and Development.

VAN DIJK, J.A. (1993b) *Soil and water conservation among agro-pastoralists in Eastern Sudan; water and nutrients, labour and institutions.* Paper presented at the Regional Development Research Seminar, Institute of Social Studies, The Hague, 18 May 1993.

VAN DIJK, J.A. (1994) *Soil and water conservation in the multi-resources economy of Beja agro-pastoralists in eastern Sudan.* Paper presented at the 2nd Maa Conference on Culture and Development: Environment and Sustainable Pastoral Development, 30 May-3 June 1994, Arusha, Tanzania.

VAN DIJK, J.A. & C. REIJ (1995) "Indigenous water harvesting techniques in Sub-Saharan Africa: examples from Sudan and the West African Sahel". In: *Proceedings of the FAO Expert Consultation on Water Harvesting for Improved Agricultural Production, Cairo, 21-25 November 1993.* Rome: FAO. (forthcoming)

VAN DRIEL, W.F. & J.C.J. VLAAR (1991) "Impact des digues filtrantes sur le bilan hydrique et sur les rendements agricoles dans la région de Rissiam, Burkina Faso". In: M.V.K. SIVAKUMAR, ed., *Soil water balance in the Sudano-Sahelian zone. Proceedings of an International Workshop held at Niamey, Niger 18-23 February 1991*, pp.299-309.

VAN KEULEN, H. & J. WOLF (1986) *Modelling of agricultural production: weather, soils and crops.* Simulation Monographs. Wageningen: Pudoc.

VAN REULER, H. & W.H. PRINS, eds., (1993) *The role of plant nutrients for sustainable food crop production in Sub-Saharan Africa.* Leidschendam: Dutch Association of Fertilizer Producers VKP.

VAN SCHAIK, A. & H. REITSMA (1992) "Agriculture in dry-season environments" In: H. REITSMA, A.J. DIETZ & L. DE HAAN, eds., *Coping with semiaridity. How the rural poor survive in dry-season environments*, pp.19-36. Amsterdam: Netherlands Geographical Studies 146.

VINK, A.T. (1991) *Management of natural forests: the Rawashda experience.* FAO Technical Report, FDES/Forests Administration Technical Report. Khartoum: FAO GCP/SUD/033/NET.

VON THUNEN, J.H. (1826) *Der isolierte Staat in Beziehung auf Landwirtschaft und Nationalökonomie.* Rostock.

WALSH, R.P.D. (1991) "Climate, hydrology, and water resources". In: G.M. CRAIG, ed., *The agriculture of The Sudan*, pp.19-53. Oxford: Oxford University Press.

WALSH, R.P.D., M. HULME & M.D. CAMPBELL (1988). "Recent rainfall changes and their impact on hydrology and water supply in the semi-arid zone of The Sudan". In: *The Geographical Journal* Vol.154, No.2, pp.181-198.

WARK (Water Spreading Research Kassala) (1989) *Progress Report 1 (May-November 1989).* Kassala: WARK.

WARK (1990a) *Progress Report 2 (December 1989 - May 1990).* Kassala: WARK.

WARK (1990b) *Progress Report 3 (June-November 1990).* Kassala: WARK.

WARK (1990c) *Field visit report of Kassala and Port Sudan SCLUWP department staff to the Red Sea and Kassala provinces, February 12-16 1990.* Kassala: WARK.

WCED (World Commission on Environment and Development) (1987) *Our common future.* Oxford: Oxford University Press.

WESTHOFF, L. (1985). *Socio-economic development in Eastern Sudan. A case study of the Kassala Rural District.* Utrecht: University of Utrecht (M.A. thesis).

WHITEMAN, A.J. (1971) *The geology of the Sudan Republic.* London: Clarendon Press.

WIDANAPATHIRANA, A.S. (1986) *Study on present status and proposed development strategy of wadi agriculture in Northern Darfur, Sudan.* El Fasher: Ministry of Agriculture and Natural Resources.

WILLCOCKS, T.J. (1979) *What form of tillage is necessary in the semi-arid soils of Botswana ?* Paper presented at the FAO Symposium on Rainfed Agriculture, Amman, May 1979.

WILSON, R.T. (1986) *Livestock production in central Mali: long term studies on cattle and small ruminants in the agropastoral system.* ILCA Research Report No.14. Addis Ababa: ILCA.

WITSENBURG, K. (1994) *On the slopes of the Mandara mountains. A case-study on indigenous land uses among the Mada of Tazang.* Amsterdam: University of Amsterdam (M.A. thesis).

WOLF, J., H. BREMAN, & H. VAN KEULEN (1991) *Bio-economic capability of West-African drylands.* CABO-DLO Report 147, Wageningen: CABO-DLO.

YASEIN, M.O. (1967) "Social, economic and political role of urban agglomerations in the developing states. The Sudan's experience". In: *Les agglomérations urbaines dans les Pays du Tiers Monde. Leur rôle politique, social et économique,* pp.335-355. Bruxelles: Institut International des Civilisations Differentes INCIDI, Editions de L'Institut de Sociologie, Université Libre Bruxelles.

ZAAL, A.F.M. (1994) *Growing market dependence of food security arrangements for pastoralists in S. Kenya and N. Burkina Faso.* Amsterdam: University of Amsterdam.

Summary

The Border Area is a 150-400 mm arid to semi-arid region in eastern Sudan. The nomadic Beja have been engaged in crop production in this area using seasonal floodwater since at least the 19th century. They were forced to take up more sedentary lifestyles in the course of the 20th century. This was mainly a result of external societal processes. These include (i) infrastructural development and market integration; (ii) the construction of irrigation schemes on their wet season grazing lands in the Gash and Tokar deltas; (iii) Colonial pacification of the tribal domain; (iv) government sedentarization policies; (v) the development of large-scale mechanized-farming schemes on their dry season grazing lands in the Gedaref region; and (vi) repeated occurrences of drought. Religious organization is a notable internal societal process which induced their sedentarization in the Border Area. At the time of transition to a more settled life, the Beja have also adopted soil and water conservation (SWC) techniques for crop production. These have a dominant water-harvesting component. The *teras* is the most elaborate technique. It has been in use locally since at least the 1950s. Its origins probably lie in the Nile region. The adoption of *teras* in the sparsely populated Border Area is not significantly forced by pressure of population on scarce resources. Adoption, on the contrary, is likely to have been facilitated by two factors. These are (i) demonstration effects of early SWC applications by West-African migrants and early government SWC interventions in the area; and (ii) the climatologically relatively wet period of the 1950s. The Ministry of Agriculture introduced several new SWC techniques in the Border Area from the 1970s onwards. The level of these interventions further increased after the catastophic 1984-1985 drought. From then onwards, support was also received in this field from international organizations. The experiences in the Border Area with SWC techniques which were introduced had been disappointing. A research programme was therefore set up to evaluate (i) the achievements of interventions in more detail; and (ii) to assess the potential of indigenous SWC as a possible alternative. The scope of the latter question was narrowed down to an assessment of the importance of indigenous SWC and crop production in Beja household livelihoods. The primary sources for research consist of (i) socio-economic data collected by means of questionnaire surveys. Some of these data were partly modelled; (ii) geographic data available from aerial photographs and satellite imageries. The latter data was processed in a GIS. The secondary sources for research consist of (i) archives of the Anglo-Egyptian Administration; (ii) reports of the government and international development projects; and (iii) reports of international evaluation missions to the Border Area. The first part of the study on the achievements of government SWC interventions follows the DAC monitoring and evaluation framework. It mainly uses evaluation mission reports and socio-economic field data for this purpose. It was found in two of the SWC projects studied, introducing earth dams and contour earth embankments respectively, that the objectives of the projects tally with the general policy directives of the government. However, the use of both strategies of "Transfer-of-Technology" and "Participatory Approaches" was contradictory. The effectiveness of interventions and their impact on the economic positions of the target group were poor. There were also negative impacts on the environment. The intervention efficiency could not unequivocally be judged. The SWC projects implemented in the Border Area, finally, proved not sustainable and replicable. The second part on the study of the relative importance of indigenous SWC and crop production mainly uses geographic and socio-economic field data. The latter were processed to evaluate the role of indigenous SWC and crop production in terms of their (i) contributions to the household income in monetary terms; (ii) amount of labour time (potentially) allocated; (iii) share of cultivated lands allocated; and (iv) priority rankings over a range of different livelihood engagements. It was found for 244 households in four research villages in the Border Area that more than half uses indigenous SWC. However, the variations between the villages and main research years 1983 and 1988 are great. These range from 9 % to 82 %. In the group of indigenous SWC-using households, these techniques contributed on average to about 75 % of the total household crop production income. This crop production income, in turn, contributed about 50 % to the grand total household income. The remainder was earned in livestock husbandry, labour migration, local off-farm employment, informal transfers or "networking", and was gained from emergency relief aid. The contribution of these indigenous SWC techniques, alternatively expressed as a percentage of the grand total household income, was in this same group about 25 %. The socio-economic data indicate that these households allocated about 90 % of their labour time for selected land preparation and cultivation activities to indigenous SWC. The remainder was allocated to various non-SWC techniques. Of the cultivated lands, some 75 % was dedicated to indigenous SWC by these households. The perceived importance of these techniques and of crop production, as expressed by land users by means of rank scores, was high. Given this important role of indigenous SWC and its low level of crop production with sorghum yields of less than 400 kg/ha, the significance of these techniques is most likely to be understood from their contribution to livelihood diversification and the spreading of subsistence risk. Because the application of SWC demands the allocation of additional labour which generates only low and uncertain returns, labour-opportunity costs were expected to exercise a decisive influence on the spatial pattern of the significance of indigenous SWC in the Border Area. This could not unequivocally be assessed. In the two different patterns studied in this research, opposite outcomes have been found. However, the pattern in which the importance of indigenous SWC increases with declining levels of labour-opportunity costs, and increasing distance to the regional urban centre, is more likely to typify the situation in the Border Area. Indigenous SWC, and particularly the *teras*, provides a good alternative for the SWC techniques introduced by the government. The advantages of the *teras* derive from being a private technique, based on the use of rainwater run-off, which can be entirely farmer-managed. Common environmental and socio-economic constraints to SWC interventions are also less pronounced when programmes rely on this particular technique. Several improvements can still be made to enhance the performance of the *teras*. Indigenous SWC, accordingly, may support the Beja livelihoods during the ongoing process of their transformation. In the long term, these livelihoods can only be based on substantial contributions of other economic sectors, and not on crop production in the Border Area alone.

Résumé

La Zone frontalière du Soudan oriental est une région aride à semi-aride dont la pluviométrie se situe entre 150 et 400 mm. Depuis au moins le XIXe siècle la tribu nomade des Beja se sert d'inondations saisonnières pour la production agricole. Dans le courant du XXe siècle, des processus de société internes ont obligés les Beja à adopter une vie plus sédentaire. Ces processus comprennent (i) le developpement d'infrastructures et l'intégration des marchés; (ii) la construction des périmètres d'irrigation dans les deltas du Gash et de Tokar, sur les pâturages de saison des pluies; (iii) la pacification coloniale des terres tribales; (iv) la politique gouvernementale de sédentarisation; (v) le développement de projets de grandes fermes mécanisées sur leurs pâturages de saison sèche dans la région de Gedaref; et (iv) les sécheresses répétées. L'organisation religieuse est un processus interne de societé qui a incité leur sédentarisation dans la Zone frontalière. Au moment de la transition vers un mode de vie plus sédentaire les Beja ont également adopté des techniques de Conservation des Eaux et des Sols (CES) avec comme composante dominante la collecte d'eau pour la production agricole. La *teras* est l'unique technique élaborée en usage localement depuis au moins les années cinquante. Elle est vraisemblablement originaire de la région du Nil. L'apparition de la *teras* dans la Zone frontalière peu peuplée n'a pas été imposée par une pression démographique sur des ressources rares. Au contraire, son adoption par les Beja aurait plutôt été facilitée par (i) des effets de démonstration de l'utilisation des CES par des immigrants de l'Afrique occidentale et par des interventions gouvernementales locales dans la zone; et (ii) la pluviométrie élevée des années cinquante. Depuis les années soixante-dix le Ministère de l'Agriculture a introduit plusieurs nouvelles techniques dans la Zone frontalière. L'intensité d'intervention en CES a encore augmenté après la sécheresse catastrophique de 1984-1985. A partir de cette période des organisations internationales ont également développé des activités dans ce domaine. Les essais d'introduction de techniques de CES dans la Zone frontalière ont été décevants. Pour cette raison, un programme de recherche a été mis au point a fin d'évaluer (i) en profondeur les résultats; (ii) le potentiel de techniques de CES endogènes en tant qu'alternative. La portée de cette dernière question a été limitée à l'évaluation de l'importance de l'utilisation de techniques et de production agricole endogènes par rapport à l'ensemble des moyens de subsistance des familles Beja. Les principales sources de recherche se composent de (i) données socio-économiques recueillies à l'aide d'enquêtes et partiellement mises en modèle; (ii) données géographiques obtenues à partir de photographies aériennes et d'images satellites traîtées par un programme de SIG. Les sources secondaires sont constituées par (i) des archives de l'Administration anglo-égyptienne; (ii) des rapports de projets de développement gouvernementaux et internationaux; et (iii) des rapports de missions internationales d'évaluation effectuées dans la Zone frontalière. L'étude des résultats des interventions gouvernementales en CES suivent la structure de monitoring et d'évaluation DAC et se base surtout sur les rapports de missions d'évaluation et sur des données socio-économiques du terrain. Dans deux projets de CES introduisant des barrages et des banquettes de contours en terre, les objectifs correspondent aux directives de politique générale. Cependant, l'usage des deux stratégies de "Transfert de Technologie" et d'"Approche avec Participation" était contradictoire. L'efficacité des interventions et leur impact sur la situation économique du groupe cible étaient faibles. Des effets négatifs sur l'environnement étaient également observés. L'efficacité de l'intervention n'a pas pu être clairement établie. Les projets de CES réalisés dans la Zone frontalière se sont avérés sans continuité et sans possibilité de propagation, et leur fonctionnement a cessé dès que le gouvernement a mis un terme à sa participation. L'étude de l'importance relative de la CES et de la production agricole endogènes se base principalement sur les données géographiques et socio-économiques du terrain. Ces dernières ont été traîtées de manière à évaluer le rôle des techniques CES et de production agricole endogènes en termes de (i) contribution en espèces aux revenus familiaux; (ii) durée de temps de travail (potentiel) investi; (iii) proportion des parcelles cultivées affectées; et (iv) classification des priorités accordées aux différents moyens de subsistance. Plus de la moitié de 244 familles des quatre villages étudiés dans la Zone frontalière, pratiquent des techniques de CES endogènes. Les variations entre villages et entre les années principales de l'étude (1983 et 1988) sont considérables: de 9 à 82 %. Dans le groupe des familles utilisant les techniques de CES endogènes, celles-ci ont contribué en moyenne à près de 75 % des revenus totaux agricoles. Ces revenus agricoles représentent à leur tour près de 50 % du revenu familial total; le reste provenant de l'élevage, de la migration liée au travail, de l'emploi local hors-ferme, des transferts informels ou "réseaux" et de l'aide d'urgence pour le groupe consideré. La part des techniques de CES endogènes exprimée en pourcentage des revenus familiaux totaux est de l'ordre de 25 %. Les données socio-économiques indiquent en outre que près de 90 % du temps de travail pour une préparation et une culture sélectionnées des terres était consacré par ces familles aux techniques de CES endogènes; le reste étant pour l'application de différentes techniques non-CES. Ces familles pratiquaient sur 75 % des terres cultivées les techniques CES endogènes. L'importance de ces techniques et de la production agricole exprimée par les utilisateurs occupe un rang élevé. Au regard de ce rôle prépondérant des techniques de CES endogènes et des niveaux de production agricole restreints, avec des récoltes de sorgho de moins de 400 kg/ha, l'intérêt de ces techniques doit plutôt être interprété dans le sens qu'elles contribuent à diversifier les revenus et à repartir les risques liés aux moyens de subsistance des familles. En raison de ces engagements simultanés des familles dans différentes activités, et du travail additionnel dans la CES ne générant qu'un rendement faible et incertain, le côut d'opportunité pour le facteur travail était supposé exercer une influence décisive sur la distribution spatiale de l'importance de la CES indigène dans la Zone frontalière. Toutefois ceci n'a pas pu être démontré clairement, car dans les deux cas étudiés ici des résultats contradictoires ont été constatés. Cependant, le modèle selon lequel l'importance des techniques de CES s'intensifie avec la diminution du côut d'opportunité pour le facteur travail et l'accroissement de la distance par rapport au centre urbain régional, semble le mieux caractériser la situation rencontrée dans la Zone frontalière. Les techniques de CES endogènes, en particulier la *teras*, forment une bonne alternative aux techniques de CES introduites localement. Les avantages de la *teras* découlent du fait qu'il s'agit d'un travail personnel, où le fermier gère lui-même entièrement l'utilisation de l'eau de ruissellement. Les contraintes écologiques et socio-économiques dans les interventions de CES sont également moins prononcées quand les programmes s'appuient sur cette technique précise. Diverses améliorations permettraient une augmentation de l'efficacité de la *teras*. Par conséquent, les techniques de CES endogènes sont en mesure de soutenir les Beja dans leur subsistance durant les processus de transformation actuels. A long terme, leur mode de vie ne peut être basé uniquement sur la production agricole, d'autres activités économiques doivent fournir des contributions substantielles.

Samenvatting

Het Grensgebied is een semi-aride tot aride streek in oostelijk Soedan. De nomadische Beja bevolking heeft zich hier sinds de 19e eeuw met akkerbouw beziggehouden, daarbij voornamelijk gebruikmakend van rivierwater. In de loop van de 20e eeuw werden zij gedwongen om meer sedentaire levenswijzen te volgen. Dit was voornamelijk het gevolg van externe maatschappelijke ontwikkelingen, zoals (i) de aanleg van infrastructuur en opkomende marktintegratie; (ii) de aanleg van irrigatiesystemen in de weidegebieden van de Gash en Tokar die door de Beja in het natte seizoen worden bezocht; (iii) het Britse koloniale pacificatiebeleid in het Beja gebied; (iv) het vestigingsbeleid van de Soedanese overheid ten aanzien van nomaden; (v) de ontwikkeling van grootschalige gemechaniseerde landbouwprojecten in de weidegebieden rond de stad Gedaref die door de Beja in het droge seizoen worden bezocht; en (vi) regelmatig terugkerende droogten. Lokale organisatie op religieuse grondslag is een opmerkelijke interne maatschappelijke ontwikkeling die de Beja in het Grensgebied tot sedentarisatie heeft aangezet. Vanaf deze periode waarin de meer sedentaire levenswijzen werden ontwikkeld, hebben de Beja zich ook toegelegd op het gebruik van inheemse bodem- en waterconserveringstechnieken in de akkerbouw. Deze technieken zijn voornamelijk gebaseerd op het verzamelen en concentreren van regen- en rivierwater, ook wel *water harvesting* genoemd. De belangrijkste techniek is de *teras*, die in het Grensgebied vanaf de jaren vijftig wordt toegepast. De herkomst van deze techniek is waarschijnlijk het gebied van de Nijl. Het gebruik van *teras* in het Grensgebied werd niet afgedwongen door een bevolkingsdruk op schaarse natuurlijke hulpbronnen. Dit gebruik werd eerder mogelijk gemaakt, en wel door de volgende ontwikkelingen: (i) het voorbeeldgedrag van zowel migranten uit West-Afrika die lokaal al eerder bodem- en waterconservering toepasten, als van de koloniale overheid die in het Grensgebied interventies uitvoerde waarbij deze technieken werden gebruikt; en (ii) de klimatologisch relatief natte periode van de jaren vijftig. Het regionale Ministerie van Landbouw heeft sinds de jaren zeventig een aantal nieuwe technieken op het gebied van bodem- en waterconservering in het Grensgebied geïntroduceerd. De intensiteit van deze overheidsinterventies nam belangrijk toe na de katastrofale droogte van 1984-1985. Vanaf dat moment werd de Soedanese overheid voor dit doel ook gesteund door internationale hulpprogramma's. De resultaten die in deze projecten werden behaald zijn beneden verwachting gebleven. Een onderzoeksprogramma werd opgezet om, ten eerste, deze resultaten meer in detail te bestuderen. Ten tweede, om de mogelijkheden te verkennen die inheemse technieken bieden als alternatief voor de technieken die in deze projecten werden geïntroduceerd. Deze laatste onderzoeksvraag werd beperkt tot het vaststellen van de bijdrage van de inheemse technieken in het geheel van bestaansbronnen van de Beja in het Grensgebied. De primaire bronnen van informatie die voor dit onderzoek werden gebruikt zijn (i) sociaal-economische gegevens verzameld door middel van enquêtes. Een deel van deze gegevens werd in eenvoudige modellen verwerkt; (ii) geografische gegevens. Deze zijn beschikbaar in de vorm van luchtfoto's en satellietopnamen. De laatste gegevens werden bewerkt in een GIS. De secundaire bronnen zijn (i) archieven van het koloniale Brits-Egyptische Bestuur; (ii) rapporten van de Soedanese overheid en van internationale hulpprogramma's; en (iii) rapporten van de gezamenlijke Soedanees-Nederlandse evaluatiemissies die ten behoeve van deze programma's in het Grensgebied zijn uitgevoerd. In het eerste deel van het onderzoek naar de resultaten van de overheidsinterventies in bodem- en waterconservering werden de algemene richtlijnen voor evaluatie gevolgd zoals deze zijn opgesteld door het DAC. Hierbij werd voornamelijk gebruik gemaakt van de verschillende rapporten van de evaluatiemissies en van de sociaal-economische veldgegevens. In de twee bodem- en waterconserveringsprojecten van de overheid die in het Grensgebied werden bestudeerd, waarbij respectievelijk gebruik werd gemaakt van een systeem van dijken aangelegd op de hoogtelijnen en van een dam, kwamen de projectdoelstellingen overeen met het geformuleerde overheidsbeleid. Echter, het gebruik van zowel methoden van techniekoverdracht als van participatieve benaderingen, was tegenstrijdig. De doelmatigheid van deze interventies en het effect op de doelgroep bleken gering. Daarnaast traden, als gevolg van deze activiteiten, ook negatieve milieueffecten op. De doeltreffendheid van de uitgevoerde interventies kon met de beschikbare gegevens niet worden beoordeeld. De twee bestudeerde projecten bleken uiteindelijk niet duurzaam en repliceerbaar te zijn. Het tweede deel van het onderzoek naar het relatieve belang van de inheemse bodem- en waterconserveringstechnieken is voornamelijk gebaseerd op het gebruik van de geografische en sociaal-economische gegevens. De laatste werden bewerkt om dit belang van inheemse technieken en akkerbouw vast te stellen naar (i) hun bijdrage aan het huishoudinkomen uitgedrukt in geld; (ii) het aandeel van de (potentiële) arbeidstijd dat aan deze technieken wordt besteed; (iii) het aandeel van de akkerbouwgronden dat aan deze technieken wordt toegewezen; en (iv) het belang dat grondgebruikers zelf door middel van rangnummers toekennen aan deze technieken en aan akkerbouw binnen het geheel van hun bestaansbronnen. De resultaten van onderzoek onder 244 huishoudens in vier dorpen in het Grensgebied tonen aan dat meer dan de helft van deze huishoudens inheemse bodem- en waterconserveringstechnieken toepast. De verschillen tussen de dorpen en de belangrijkste onderzoeksjaren 1983 en 1988 zijn echter groot. Deze aandelen variëren namelijk tussen gemiddeld 9 % en 82 %. Binnen de deelgroep van huishoudens die inheemse technieken toepassen blijkt dat deze technieken gemiddeld 75 % bijdragen aan het huishoudinkomen uit akkerbouw. De huishoudinkomens verkregen uit akkerbouw dragen op hun beurt gemiddeld 50 % bij aan het totale huishoudinkomen gebaseerd op alle bestaansbronnen gezamenlijk. De andere inkomsten worden verkregen uit veehouderij, arbeidsmigratie, lokale niet-agrarische arbeid, informele inkomensoverdrachten die via sociale netwerken verlopen en internationale noodhulp. De bijdrage van inheemse bodem- en waterconserveringstechnieken, ditmaal uitgedrukt als percentage van het totale huishoudinkomen gebaseerd op alle bestaansbronnen gezamelijk, is in de deelgroep van huishoudens die deze technieken gebruiken gemiddeld ongeveer 25 %. De gegevens tonen verder aan dat binnen deze zelfde deelgroep, gemiddeld 90 % van de arbeidstijd voor een geselecteerd aantal activiteiten wordt besteed aan de inheemse bodem- en waterconserveringstechnieken. De resterende arbeidstijd wordt ingezet ten behoeve van andere akkerbouwtechnieken die niet gebaseerd zijn op bodem- en waterconservering. In dezelfde deelgroep wordt gemiddeld 75 % van de totaal beschikbare hoeveelheid akkerbouwgrond gereserveerd voor inheemse bodem- en waterconserverings technieken. Het belang van deze technieken en akkerbouw, zoals in deze deelgroep uitgedrukt door middel van rangnummers, is hoog. Gegeven deze belangrijke rol van inheemse bodem- en waterconserveringstechnieken, en gegeven ook hun betrekkelijk lage produktie met sorghum oogsten van minder dan 400 kg/ha, moet hun betekenis waarschijnlijk vooral begrepen worden uit hun bijdrage aan de diversificatie van bestaansbronnen en aan risicospreiding. Omdat de toepassing van bodem- en waterconserveringstechnieken de inzet van extra arbeid vereist die ook nog slechts lage en onzekere opbrengsten oplevert, werd verondersteld dat de kosten van het opgeofferde arbeidsalternatief een belangrijke rol zouden spelen in het ruimtelijk voorkomen van het belang van deze technieken in het Grensgebied. Dit kon echter niet ondubbelzinnig worden vastgesteld. In de twee voor dit doel bestudeerde ruimtelijke patronen werden tegengestelde uitkomsten gevonden. Het patroon waarbij het relatieve belang van inheemse bodem- en waterconserveringstechnieken toeneemt met afnemende kosten van het opgeofferde arbeidsalternatief, en met toenemende afstand tot het regionale urbane centrum, is waarschijnlijk het meest representatief voor de situatie in het Grensgebied. Inheemse bodem- en waterconservering, en met name de *teras*, biedt een goed alternatief voor de door de overheid geïntroduceerde technieken. De voordelen van de *teras* liggen in het feit dat het een techniek is voor individueel grondgebruik, voornamelijk gebaseerd op toepassing van neerslagafvoer in plaats van rivierwater, welke techniek geheel door huishoudens zelf kan worden beheerd. Beperkende factoren van fysisch en sociaal-economisch aard die vaak invloed hebben op bodem- en waterconserveringsinterventies spelen ook een minder grote rol wanneer deze interventies gebaseerd zijn op de *teras* techniek. Deze techniek kan op een aantal punten nog verder worden verbeterd. Inheemse bodem- en waterconservering kan aldus een bijdrage leveren in de ontwikkeling van de bestaansbronnen van de Beja in het Grensgebied. Echter, op de lange termijn kunnen deze bestaansbronnen alleen gebaseerd zijn op aanzienlijke bijdragen van andere economische sectoren, en niet op de lokale akkerbouw alleen.

تدعم انتاج الغذاء بالنسبه للبجه فى خلال عملية التحول التى تجرى لهم .

وسا ئل الحصول على الغذاء هذه يمكن ان تبنى فقط بمساهمه جاده من القطاعات الاقتصاديه الاخرى للاسر وليس فقط من انتاج المحاصيل فى مناطق الحدود .

٥ .

مساهمة الوسائل المحليه للحفاظ على التربه والمياه والتى يشار اليها تبادليا كنسبه من اجمالى دخل الاسره كانت فى نفس المجموعه حوالى 25٪ . البيانات الاجتماعيه الاقتصاديه تشير الى ان حوالى 90٪ من وقت العماله المخصص لتجهيز الارض وعمليات الزراعه الاخرى كان قد خصص للوسائل المحليه للحفاظ على التربه والمياه والوقت المتبقى خصص لبقية الاعمال .

وسائل الحفاظ على التربه والمياه خصصت لها 75٪ من الارض المزروعه بواسطة الاسر . الاهالى الذين يفلحون هذه الاراضى لديهم اعتقاد راسخ باهمية هذه الاساليب . عند اعتبار الدور المهم للوسائل المحليه للحفاظ على المياه والتربه وتدنى انتاجية المحصول بالنسبه للذره والتى بلغت اقل من 400 كجم \ هكتار فان اهمية هذه الاساليب يمكن فهمها من مساهمتها فيتعدد وسائل انتاج الغذاء وانتشار المخاطر المعيشيه .

نسبة لهذه الاعمال المتوازيه فى المناشط المختلفه والاحتياج لعماله اضافيه للحفاظ على التربه والمياه والتى ينتج عنها عائد غير مؤكد , فا ن تكلفة فرص العماله كان من المتوقع ان تحدث اثر حاسم بالنسبه لنمط الوسائل المحليه فى المناطق الحدوديه . هذه لايمكن تقييمها بكل الوضوح . بالنسبه للنمطين المختلفين اللزين اجريت لهما الدراسه فى البحث فان النتائج التى تم التوصل لها كانت مختلفه عن الاخرى .عموما فان النمط الذى تزداد فيه اهمية الوسائل المحليه مع انخفاض تكلفه فرص العماله وتزايد المسافه من مراكز الحضر الاقليميه فانه النمط الاكثر استخداما .

الوسائل المحليه للحفاظ على التربه والمياه وبالاخص اسلوب الترس توفر بديل جيد بالنسبه للاساليب التى ادخلت محليا . تبرز اهمية " الترس " فى كونه اسلوب يعتمد على انسياب مياه الامطار ويمكن السيطره على ذلك بواسطة المزارع .

العوامل البيئيه العاديه والضغوط الاقتصاديه الاجتماعيه بالنسبه للتدخلات للحفاظ على التربه والمياه فان تاثيراتهما تعتبر ضئيله عند استخدام اسلوب الترس .

هنالك عدة تحسينات يمكن ادخالها للارتقاء باداء الترس .الوسائل المحليه لحماية التربه والحفاظ على المياه على ضوء ذلك يمكن ان

وتجد ذلك فى مشروعين تمت دراستهما باستخدام سدود و عمل حواجز ترابيه على الارتفاعات المختلفه (كنتوريه)
عموما فان استخدام استراتيجيتى نقل التكنلوجيا واسلوب المشاركه لم تكونا بشكل متوافق .
فعالية التدخل او التصدى الخاصه با لموقف الاقتصادى بالنسبه لمجموعة الاهداف الخاصه بالمجموعه كانت بمستوى متدنى . كانت هنالك ايضا اثار سلبيه على البيئه . كفاءة التدخل او التصدى لم يكن من الممكن الحكم عليها بوضوح .
مشاريع الحفاظ على التربه والمياه التى تم تنفيزها على منطقه الحدود اتضح فى النهايه انه لايمكن استمرارها او تكرارها , وقد توقفت عندما انهت الحكومه اعمالها فى هذا الصدد .
دراسة الاهميه النسبيه الوسائل المحليه للحفاظ على التربه والمياه وانتاج المحاصيل تستخدم فيها بشكل رئيسى البيانات الجغرافيه والبيانات الميدانيه الاجتماعيه الاقتصاديه والاخيره تم تمحيصها لتقييم دور الوسائل المحليه
لحماية للحفاظ على التربه والمياه فى شكل :
أ . المساهمه فى دخل الاسره النقدى .
ب. حجم الوقت المستغرق فى العمل .
ج. نصيب الاراضى المزروعه التى تم تخصيصها .
د. ترتيب الاسبقيه بالنسبه لعدد من عمليات انتاج الغذاء .
اتضح انه بالنسبه لعدد 244 اسره فى اربعه قرى اختيرت للبحث فى منطقة الحدود فان اكثر من النصف يستخدم الوسائل المحليه .
الاختلافات التى وجدت فى القرى خلال سنوات البحث الرئيسيه 1983-1988 كانت كبيره وتختلف من9٪الى 82٪.
هذه الاساليب ساهمت فى مجموعة الوسائل المحليه لحماية التربه والحفاظ على المياه بحوالى 75٪ من اجمالى دخل الاسره من انتاج المحاصيل .
كما وان انتاج دخل المحاصيل بدوره ساهم بحوالى 50 ٪ من اجمالى دخل الاسره . البقيه تم الحصول عليها من الانتاج الحيوانى , هجرة العمال , العماله خارج المشروع , تحويلات غير رسميه او منتظمه والدخل من منظمات الاغاثه .

أ . التأثر بالاسلوب الذى ينتهجه سكان المنطقه ا لزين نزحوا من غرب افريقيا .
ب. سياسات الحكومه فى المنطقه .
ج. حقبة الخمسينات من هذا القرن والتى اتسم مناخها بوفرة المياه بنسبه معقوله .
وزارة الزراعه قامت بادخال عدة اساليب جديده للحفاظ على التربه و المياه اعتبارا من السبعينات وقد ازداد معدل تدخل الحكومه فى هذا المضمار بعد كارثة الجفاف فى عا مى 1984-1985 .
اعتبارا من هذا التاريخ فان المنظمات العالميه والطوعيه اصبحت تقدم المساعدات والدعم فى هذا المجال .
لقد كانت كانت تجربة الحفاظ على التربه والمياه (SWC) مخيبه للامال وعلى ضوء ذلك تقرر اجراء دراسه لتقييم :
أ. النجاح الذى تحقق من خلال التجارب السابقه بمزيد من التفصيل
ب. تقييم امكانية استخدام الوسائل المحليه الخاصه بالحفاظ على التربه والمياه (SWC
ج. تضييق نطاق السؤال فى الفقره (ب) لتقييم مدى اهمية الوسائل المحليه للحفاظ على التربه والمياه SWC وانتاج المحاصيل فى مناطق البجه التقليديه .
مصادر البحث الرئيسيه تتكون من من :
أ . البيانات الاجتماعيه الاقتصاديه والتى جمعت بواسطة الاستبيان .
ب.البيانات الجغرافيه المتوفره فى شكل صور جويه او لاقمار صناعيه واخيره تم اعداها بواسطة (GIS)

المصادر الثانويه للبحث تتكون من :
أ . الوثائق الخاصه بالحكم الانجليزى المصرى .
ب. التقارير الحكوميه والمشاريع التنمويه العالميه
ج . التقارير العلميه الخاصه ببعثات التقييم لمناطق الحدود .
د. تقارير المجهودات والانجازات الحكوميه الخاصه بمشاريع الحفاظ على التربه والمياه (SWC)
باستخدام اساليب (DAC) للبحث والتقييم والتى تستخدم تقارير بعثات التقييم والبيانات الاجتماعيه الاقتصاديه الميدانيه الخاصه بهذا الصدد .

خلاصـــــــــــــــــــــه

المناطق الجافه وشبه الجافه على الحدود بشرق السودان تبلغ مساحتها 150,400 الف م م سكان هذه هذه المنطقه من قبائل البجه الرعويه ظلوا يزرعون هذه المنطقه اعتمادا على الامطار الموسميه منذ القرن التاسع عشر

في خلال القرن العشرين اصبحوا مجبرين على تبنى نوع من الحياة اكثر استقرارا وقد حدث ذلك نتيجة لعوامل يمكن تلخيصها فى الاتى :

أ.تطور البنيه التحتيه وتكامل السوق .

ب. انشاء المشاريع المرويه حيث مناطق الرعى الموسميه عند الامطار .

ج. بسط الامن والسيطره الاستعماريه على الاراضى التى تقطنها هذه القبائل .

د. سياسات الحكومه الخاصه بتوطين الرحل .

هـ. انشاء مشاريع الزراعه الاليه فى مناطق الرعى فى موسم الجفاف بمنطقة القضارف وايضا بسبب :

و. تكرار حدوث الجفاف.

هـ. الرابطه الدينيه كان لها اثرها الاجتماعى الواضح والذى ادى الى للمزيد من من الاستقرار عى المناطق الحدوديه.

من اجل حياة اكثر استقرارا تبنى البجه اساليب المحافظه على التربه والمياه باستخدام الاسلوب الوحيد وهو "الترس " والذى كان مستخدمامنذ الخمسينات من هذا القرن و من المعتقد انه قد ظهر فى مناطق النيل . ان تبنى اسلوب الترس فى المناطق المأهوله بنسبة كثافه ضئيله على الحدود من الواضح انه لم يحدث بسبب الضغوط السكانيه او الامكانيات الشحيحه ولكن على العكس فلقد حدث ذلك بسبب الاتى :

Figure 5.5 **Unsupervised classification of the first 4 principle components set to 10 spectral classes, S. Border Area**

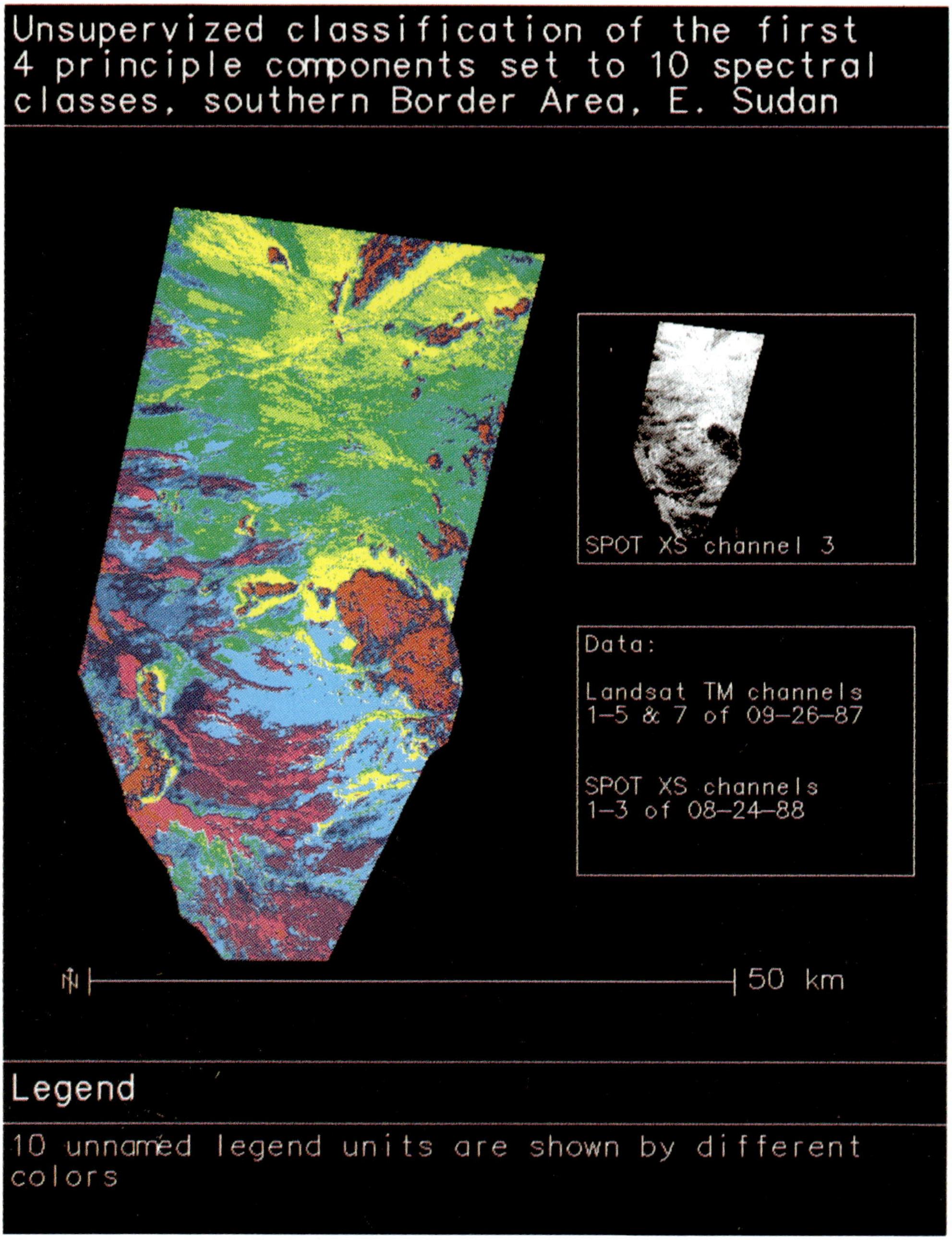

Figure 5.6 **Selected soil and water conservation land units, S. Border Area**

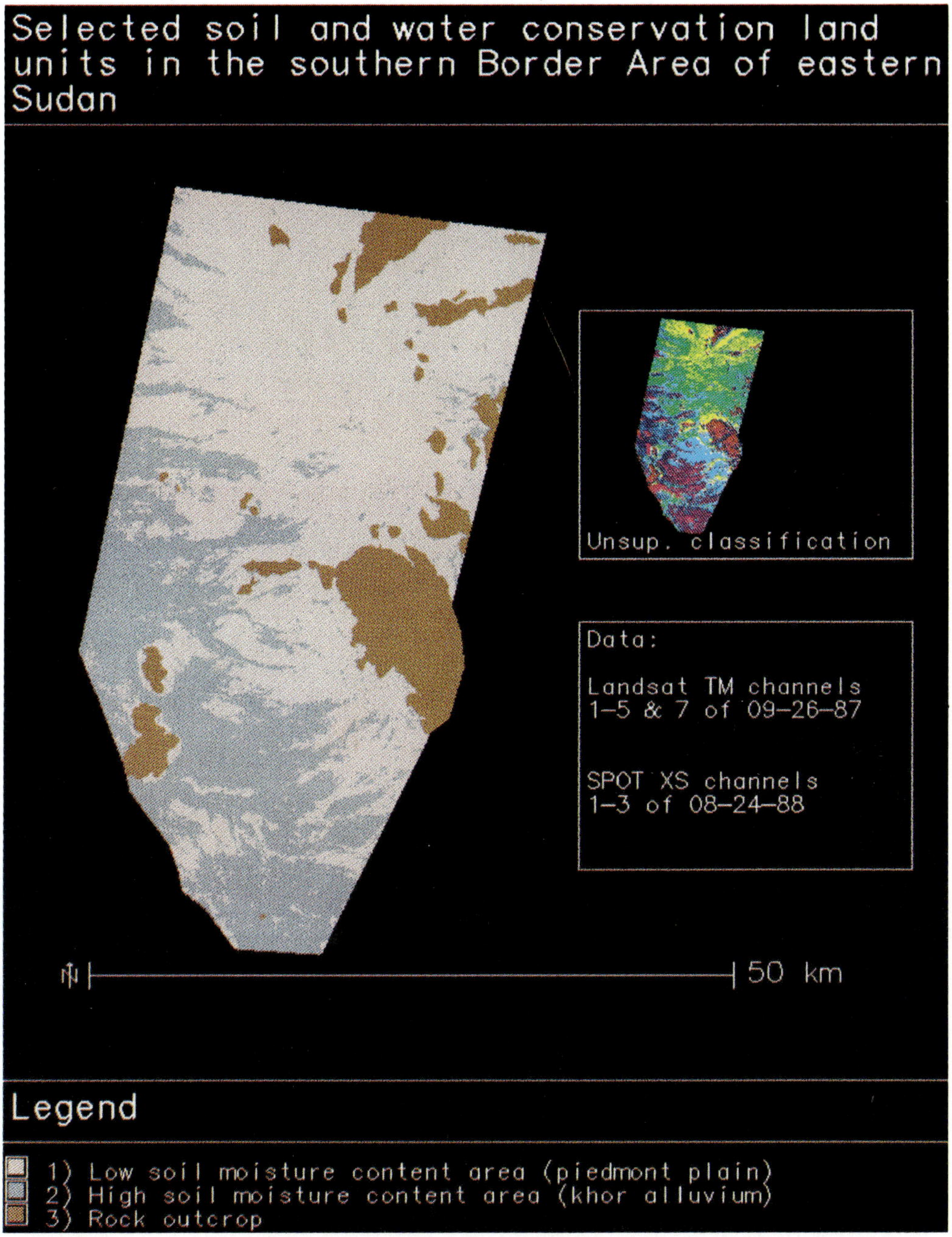

Figure 5.7 **Best-fit teras in the S. Border Area and biomass presence over combined years 1987 and 1988**

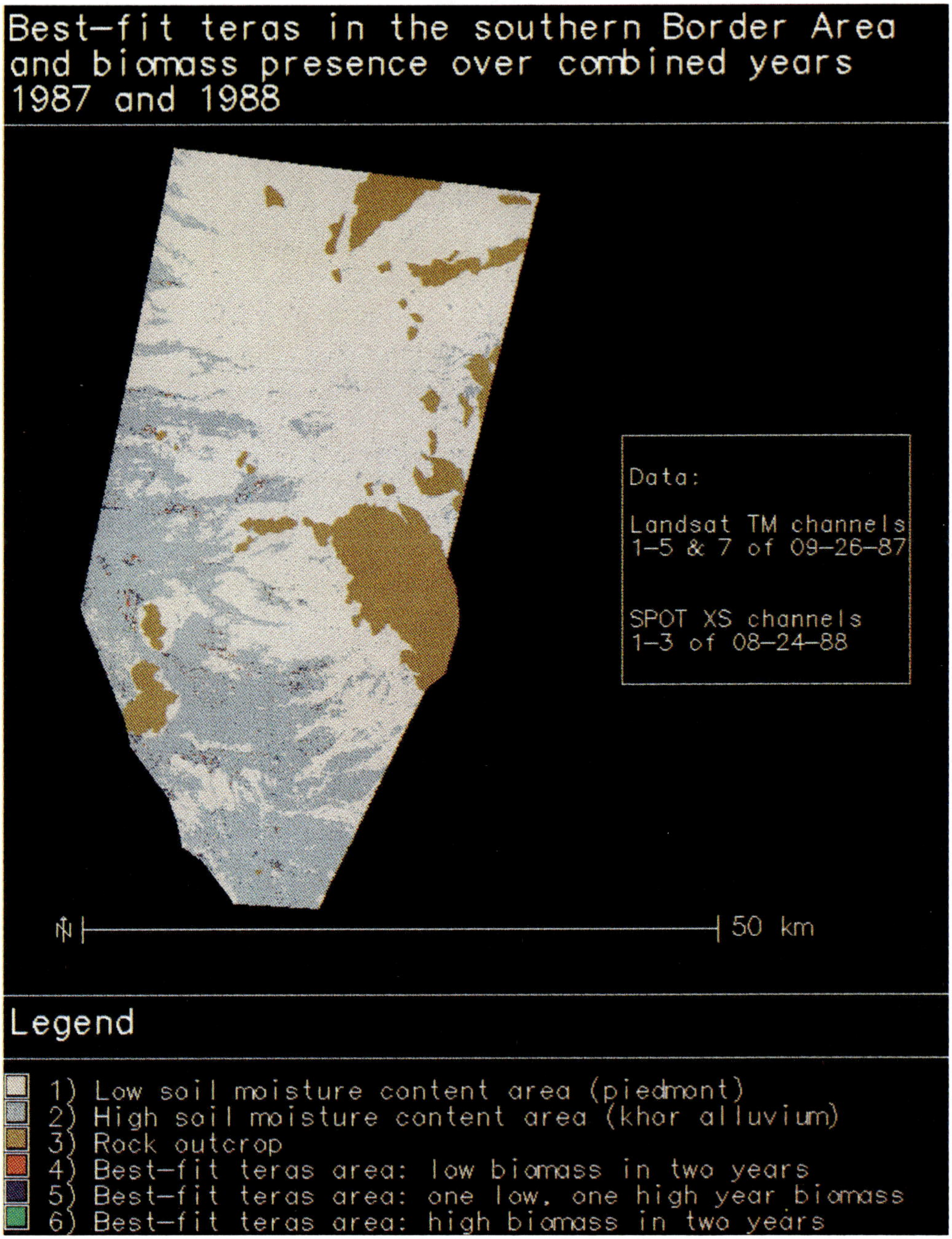

Figure 5.8 **Unsupervised classification of the first 4 principle components set to 10 spectral classes, N. Border Area**

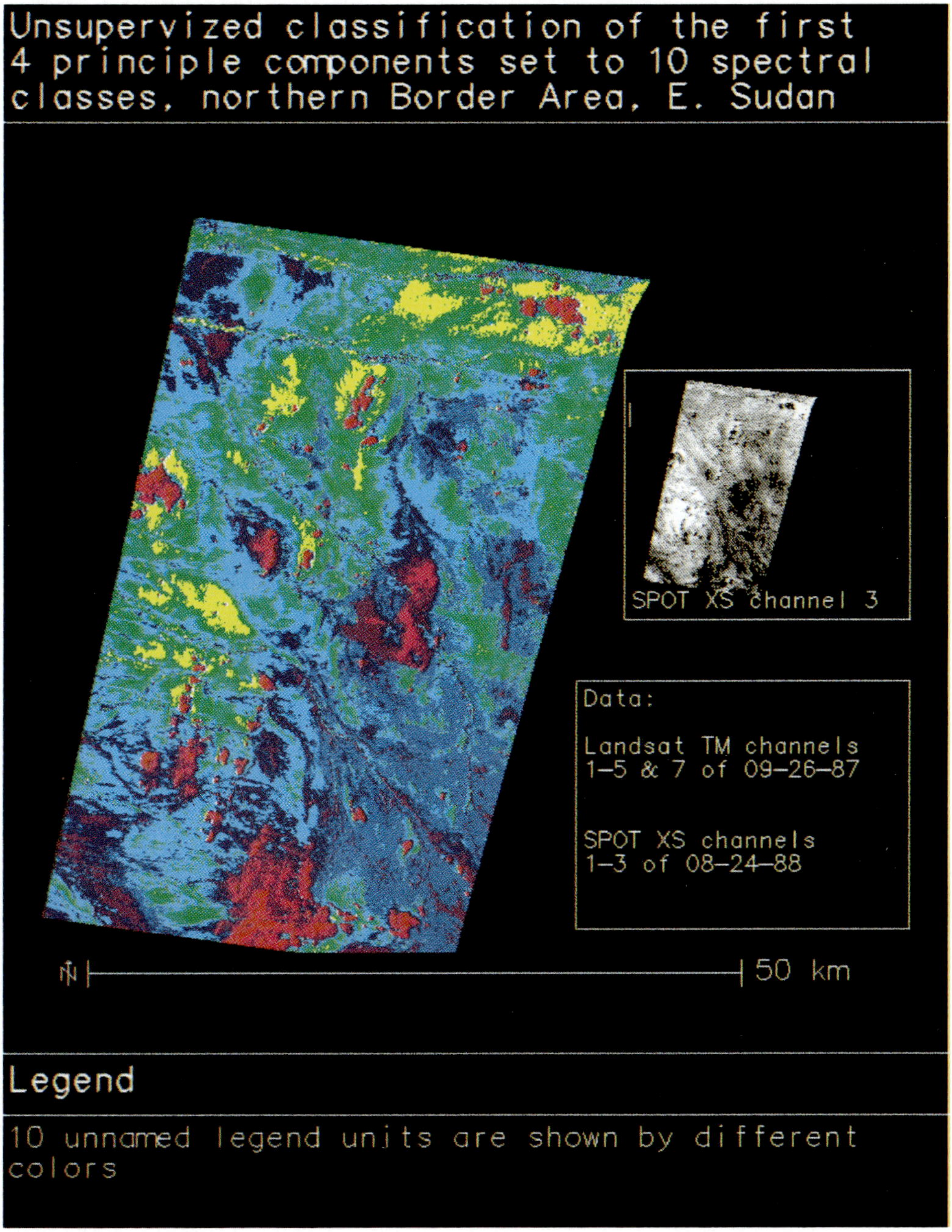

Figure 5.9 **Selected soil and water conservation land units, N. Border Area**

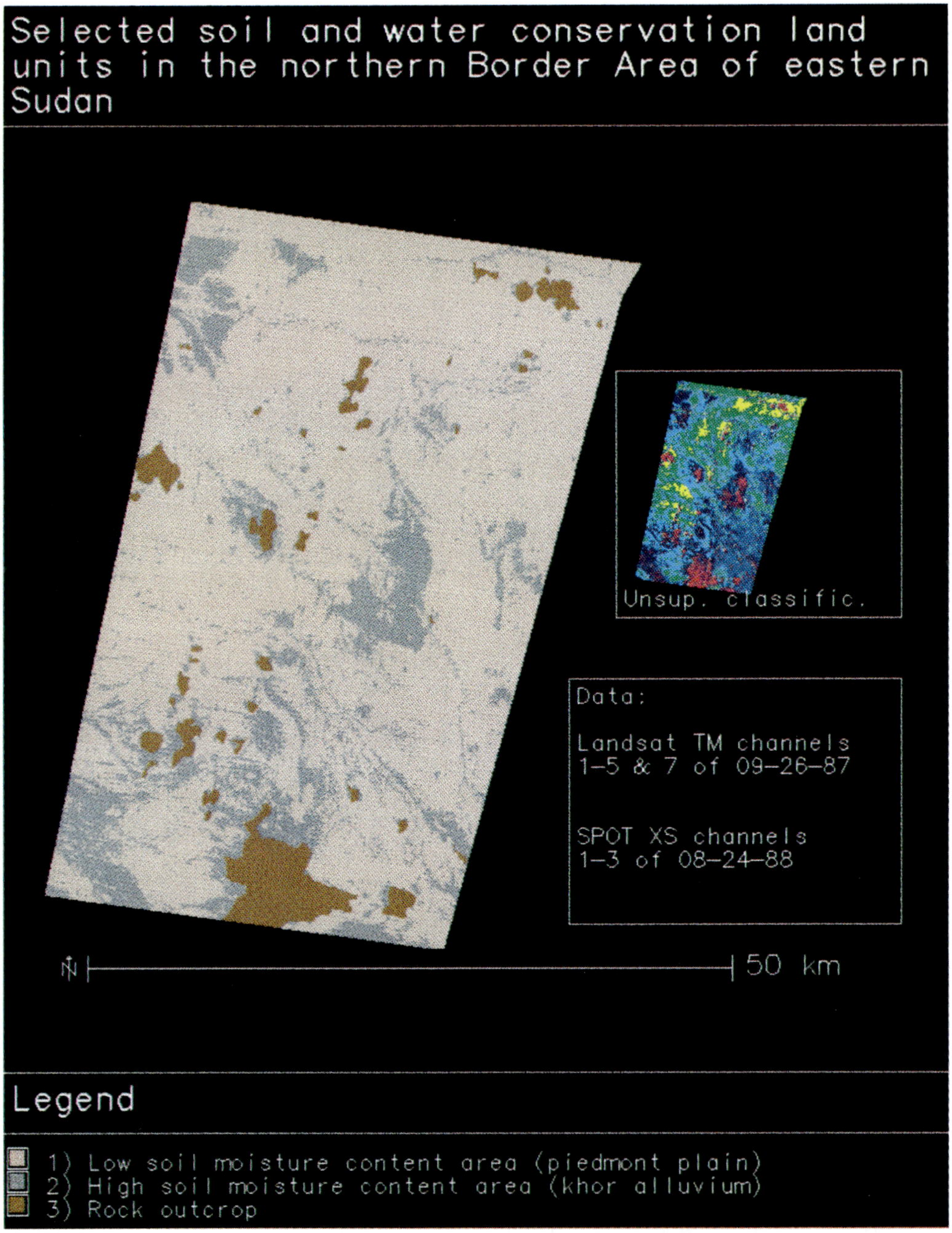

Figure 5.10 **Best-fit teras in the N. Border Area and biomass presence over combined years 1987 and 1988**

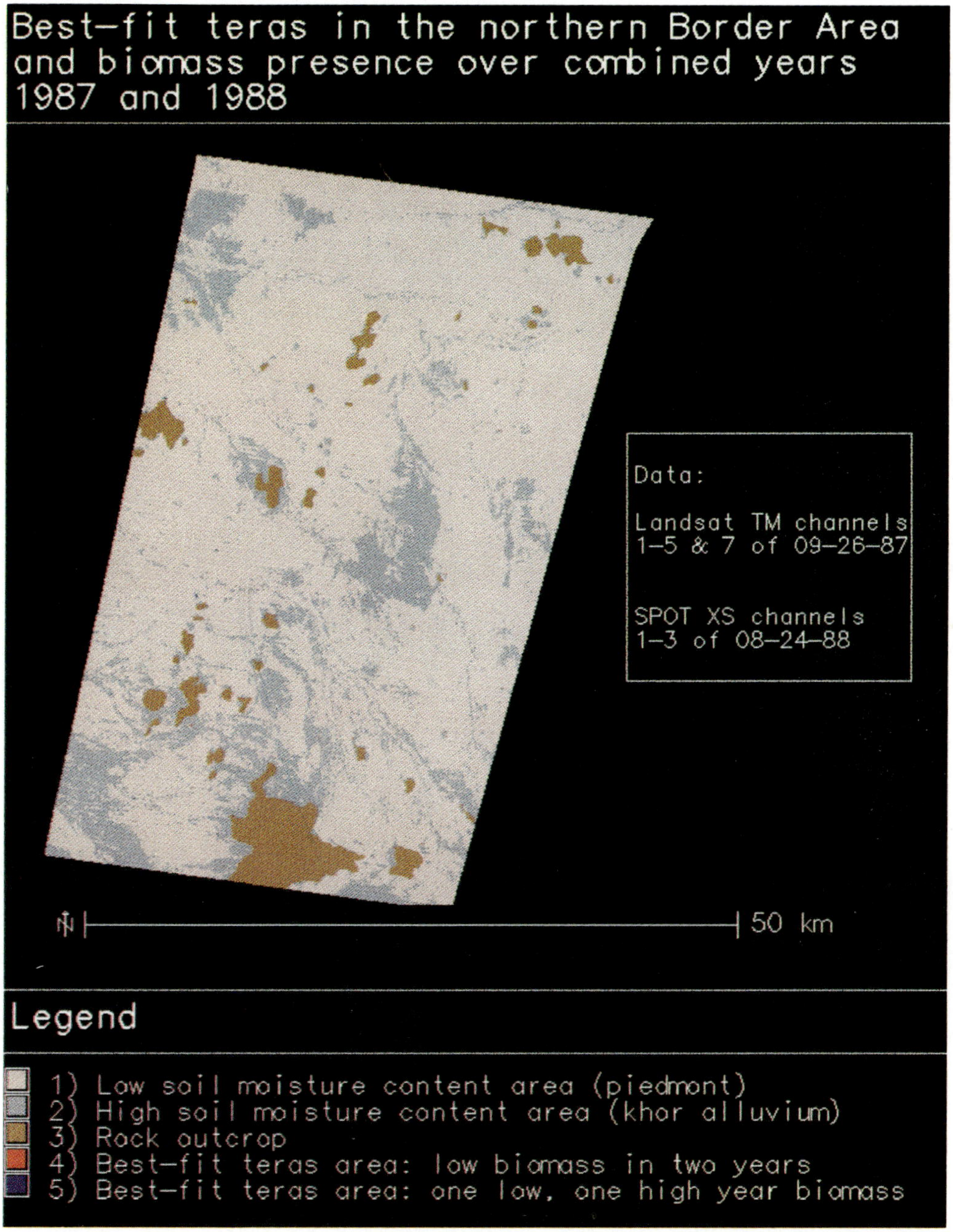

Figure 5.11 **Best-fit teras in the Hafarat area and biomass presence over combined years 1987 and 1988**

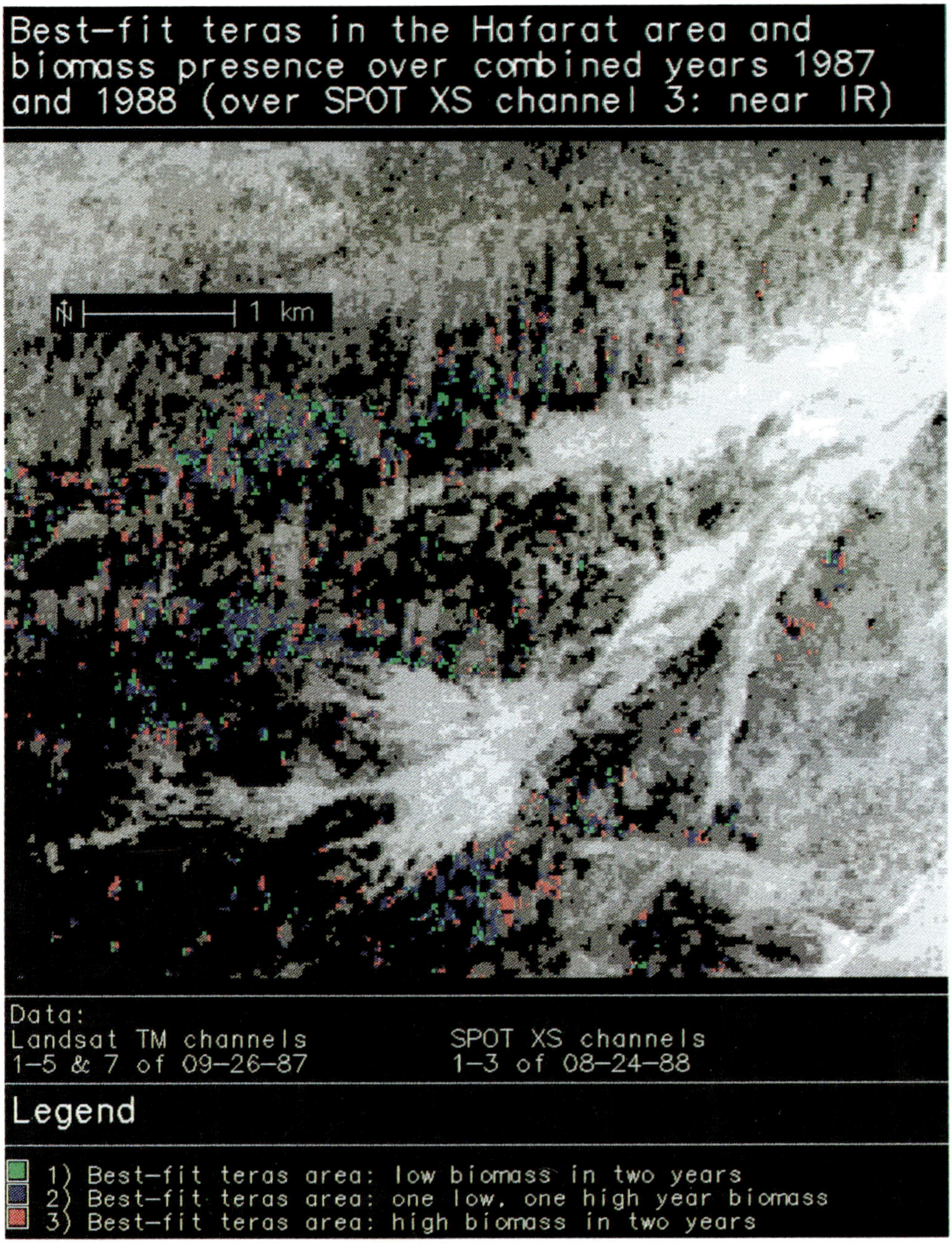